SPACE TO LIVE
The Search for an Alternative Home for Humanity

RODERICK J. HILL

Essex, Connecticut

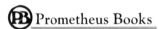
Prometheus Books
An imprint of The Globe Pequot Publishing Group, Inc.
64 South Main Street
Essex, CT 06426
www.globepequot.com

Distributed by NATIONAL BOOK NETWORK

Copyright © 2025 by Roderick J. Hill

Quote on pp. 246–47: This quotation is extracted with permission from P. Ayres, *Mawson. A Life* (Carlton: Melbourne University Press, 1999).

Quote on p. 305 (note 2): Used with the permission of Doubleday, an imprint of the Knopf Doubleday Publishing Group, a division of Penguin Random House LLC. All rights reserved.

All rights reserved. No part of this book may be reproduced in any form or by any electronic or mechanical means, including information storage and retrieval systems, without written permission from the publisher, except by a reviewer who may quote passages in a review.

British Library Cataloguing in Publication Information available

Library of Congress Cataloging-in-Publication Data available

ISBN 978-1-63388-950-7 (cloth : alk. paper)
ISBN 978-1-63388-951-4 (electronic)

∞™ The paper used in this publication meets the minimum requirements of American National Standard for Information Sciences—Permanence of Paper for Printed Library Materials, ANSI/NISO Z39.48-1992.

Contents

Prologue 1

Part I: Life on Earth: How Unusual It Is, Why Its Future Is Under Threat, and What Might Be Done to Ensure Its Survival . . . 4

Chapter 1: Are We Alone? The Search for Extraterrestrial Life 5

Chapter 2: Scenarios for the Demise of Humanity 67

Chapter 3: How Urgent Is the Imperative to Search for Other Locations for Humans to Live? 112

Conclusions to Part I 128

Part II: Potential Locations for Human Settlement 131

Chapter 4: The Record of Space Travel So Far 134

Chapter 5: Options for Establishing a Permanent Settlement in Earth's Immediate Environment 145

Chapter 6: Options for Settlement Beyond the Earth-Moon Setting but Within the Solar System 176

Chapter 7: Options for Settlement Outside Our Solar System . . .203

Conclusions to Part II 215

Part III: The Journey Off-Earth 217

Chapter 8: The Tyranny of Distance and Time 221

Chapter 9: Physical, Social, and Psychological Challenges of Space Travel 232

CONTENTS

CHAPTER 10: Spacecraft Options for Human Transportation. 257
CHAPTER 11: Options for Spacecraft Propulsion Technology. 271
CONCLUSIONS TO PART III. .288

Part IV: What Next?. 290
CHAPTER 12: A Summary of the Challenges for a
Human Settlement Off-Earth 291
CHAPTER 13: Humans or Machines? 300

Epilogue 318
Notes. 324
References 336
Additional Internet Sources .350
Acknowledgments 353
Index. 354
About the Author 369

Prologue

Life is too precious to be placed on a single planet, to be at the mercy of these planetary threats.
— Michio Kaku, in *The Future of Humanity* (2018)

Life on Earth and with it, humanity, is going to disappear one day. It could take as long as another 4 billion years or so, when the Sun swells in size and engulfs the Earth during its red giant phase of development. Or the end of life could be much sooner than that; indeed, at the risk of appearing to be somewhat alarmist, as soon as the next few seconds.

Among the reasons for life's shorter-term demise, we can include: impact of a large comet or asteroid; deadly radiation from the explosion of a nearby star; accidental or deliberate human-induced nuclear holocaust; global disease pandemic; runaway global warming (or cooling); explosion or mass flooding of basalt from a "super" volcano; resource depletion from population explosion and/or massive pollution.

Each of these scenarios has a different probability and each could occur as single or multiple events or as combinations of each other, from the immediate future to much longer time frames. Each has the individual or combined capacity to wipe out the whole of humanity. In one of his Reith Lectures on the BBC, Stephen Hawking mused that while the probability of a worldwide disaster in any given year is quite low, the probabilities add up to significant proportions over hundreds of years. He hoped that by that time, humanity will have moved off-Earth and will have avoided its demise (Shukman 2016).

These dangers are not idle or harebrained threats. The Earth has already experienced five mass extinctions at the level of 50 to 96 percent of all species in existence at the time. The worst of these occurred at the end of the Permian geological period, 248 million years ago, when 96 percent of all species died out because of the combined effects of an asteroid impact and massive volcanism. All life on Earth today is descended from the 4 percent of species that survived this event. Were this sort of event(s) to occur today, the impact and consequences might be even worse because of the massive human population and the interdependence of the fragile social and environmental systems that support us. Even if we were not wiped out, civilization could be set back thousands of years.

We could sit idly by and wait stoically for the end to come, or we could start thinking about how humanity might circumvent its inevitable demise. We might not have enough time to work out a solution if a largish asteroid or deadly radiation from a nearby exploding star is already headed in our direction, or current global warming really does pass a tipping point of no return and we are on our way to a Venus-like nightmare. Or we may have billions of years to get it sorted, by working out a method of preventing the disaster or, as described here, in finding an alternative location for humanity to settle and create a new home base.

In an encouraging (although indirect) sign of this effort, a great deal of energy and resources is being devoted to the exploration of the Solar System in search of life or evidence for its previous existence; to wit, the numerous space probes that have been to the Moon, Mars, and other planets and their moons since the first satellite was launched in 1957. We are also devoting significant resources to the search for evidence of planets attached to other stars in the Milky Way galaxy that might have the right conditions to harbor life—in the so-called Goldilocks zone—where water, the foundation stone for life as we know it, exists in liquid form.

But these are all endeavors motivated by the desire to increase our understanding of cosmology and the origins of life and/or to search for extraterrestrial life to find out if we are "alone" in the galaxy. Rarely is the concept seriously discussed in justification of these ventures that they might lead to the discovery of a platform to which humanity might migrate to escape its demise on Earth and thereby ensure its longer-term survival.

Not so this book. It examines the case for the absolute need—the imperative—to find another Goldilocks zone for humanity to survive through the settlement of other worlds within or outside our Solar System. It examines the origins and evolution of life on Earth and summarizes the efforts to find evidence of life elsewhere in the Solar System and beyond. It discusses potential targets for off-Earth occupation, along with the various means by which we might attempt to get there, and the magnitude of the technological, physical, psychological, social, and spiritual challenges that would need to be overcome in achieving this end, or should I say, beginning.

Part I

Life on Earth: How Unusual It Is, Why Its Future Is Under Threat, and What Might Be Done to Ensure Its Survival

We travel together, passengers on a little spaceship, dependent on its vulnerable reserves of air and soil; all committed, for our safety, to its security and peace; preserved from annihilation only by the care, the work and the love we give our fragile craft. We cannot maintain it half fortunate, half miserable, half confident, half despairing, half slave—to the ancient enemies of man—half free in a liberation of resources undreamed of until this day. No craft, no crew can travel safely with such vast contradictions. On their resolution depends the survival of us all. But, our days are numbered—even if we eventually learn to get on with each other.
—Adlai Stevenson, American politician and statesman, from a speech to the UN Economic and Social Council, Geneva, Switzerland, July 9, 1965

The three chapters in this section seek to understand the history of life on Earth, the fragility of its origins and existence, the significant threats to its continuation, and the likelihood, or otherwise, that life-forms exist elsewhere in the Solar System and beyond. They provide the basis for attempts to preserve human civilization in the face of those threats through the identification of an alternative platform for human presence beyond Earth.

CHAPTER ONE

Are We Alone?
The Search for Extraterrestrial Life

When I was a student almost nobody thought there was any life beyond Earth. Today it's fashionable to say that there is life all over the place, that the universe is teeming with it, but the scientific facts on the ground haven't really changed.
—PAUL DAVIES, THEORETICAL PHYSICIST, COSMOLOGIST, AND ASTROBIOLOGIST AND BEST-SELLING WRITER AND BROADCASTER

The question of whether we are alone has fascinated humanity since the 16th century, when the Earth-centered model of the universe, attributed to Claudius Ptolemy (AD 100–ca. 170), was overturned by the Sun-centered model proposed by Nicolaus Copernicus (1473–1543). Copernicus's model was later tested and refined by other astronomers, including Tycho Brahe (1546–1601), who carefully documented the movements of celestial objects, which were then used by the mathematician Johannes Kepler (1571–1630), who developed his laws of planetary motion.

The invention of the telescope by Hans Lippershey (ca. 1570–1619) in 1608 and the use of a variation of this design by Galileo Galilei (1564–1642) in 1611 for astronomical observations added further conclusive evidence in support of the Copernicus model. The prospect that other stars may also have planets revolving around them also fueled this fervor. In more recent times, the development of spaceflight and the ability to fly by, orbit, and land spacecraft on other bodies in the Solar System enabled the study of the surfaces of these bodies for telltale direct and indirect signs of the presence of life there in the past and/or present. Through these developments it has become clear that the Earth and humans are not the center of everything; that the heavens are not "perfect," moving in complete harmony; and that there may be other places in the universe that have harbored or do or could harbor life.

The potential for the existence of extraterrestrial life has precipitated a massive search for evidence of this life through direct examination by dozens of sophisticated probes that have been sent to the surfaces of the Moon (both manned and unmanned), Mars, Venus, and Titan, the largest moon of Saturn, and several asteroids and comets (see discussion below). All this effort has produced no evidence, direct or indirect, for the presence of life off-Earth, but it has shown that several environments exist with the conditions conducive to life as we know it. Also, despite more than 60 years of endeavor there has been no confirmed evidence of radio and other signals from putative intelligent beings outside the Solar System.

Furthermore, all but a few of the tens of thousands of reports of "unidentified flying objects" (UFOs) and/or "unexplained aerial phenomena" (UAPs) have been ascribed to natural events and none have been positively attributed to visits by alien civilizations. Close examination of all reported observations of large, engineered structures on other bodies (e.g., the "canals" on Mars) have also revealed them all to be false.

Thus, life on Earth, at any of its wonderful levels of complexity from single-cell amoebae and bacteria to *Homo sapiens*, appears to be unique in the universe, as far as we are currently aware.

Even so, with the billions of planets that surround stars in the Milky Way galaxy and the significant number of these that appear to bear a close resemblance to Earth, it is hard to believe that none of these planetary systems have harbored, or are harboring, simple life-forms that have already evolved, or might in the future evolve, into intelligent and technologically competent life at some point in the 13.8 billion years since the Big Bang started the ball rolling.

As Arthur C. Clarke (1917–2008) is attributed to saying, "Two possibilities exist: either we are alone in the Universe, or we are not. Both are equally terrifying."

What Do We Mean by "Life"?

Before proceeding any further with a discussion about the genesis and evolution of life on Earth and the search for possible environments in which it may have evolved elsewhere and/or where it might be possible for humans to live, it is important to define what "life" actually is.

There are more than 120 definitions across the literature, each restrained by our experience of life as we know it on Earth and the (reasonable) assumption of the universality of the laws of physics and chemistry. One of the more generic, simple, and practical definitions is from NASA (Brennan 2023), which describes life as "a self-sustaining chemical system capable of Darwinian evolution."

In the context of the search for extraterrestrial life and an answer to the question of whether we are alone in the galaxy and/or universe, the distinction should be made between the rudimentary, single-celled forms that evolved on Earth soon after the planet formed some 4.5 billion years ago; the more complex, multicelled organisms that emerged about 500 million years later; and the technologically advanced, intelligent life that emerged only a few thousand years ago, of which humans are the most evolved and recent example known to date.

A consensus has emerged that life (in all its foreseeable forms) requires three essential components (Cockell et al. 2016):

1. an energy source to drive metabolic reactions,
2. a liquid solvent to mediate these reactions, and
3. a suite of nutrients both to build biomass and to produce enzymes that catalyze the metabolic reactions.

On Earth, life functions through the specialized chemistry of carbon and water and builds largely upon four key families of chemicals: lipids (cell membranes), carbohydrates (sugars, cellulose), amino acids (protein metabolism), and nucleic acids (DNA and RNA). Any successful theory of the genesis of life (as we know it) must explain the origins and interactions of these classes of molecules.

Similarities between all present-day life-forms on Earth suggest that all known and extinct species have evolved from a "last universal common ancestor" (LUCA) of all life on Earth about 3.8 billion years ago. Universal common descent through an evolutionary process was first proposed by the British naturalist Charles Darwin in the concluding sentence of his 1859 book *On the Origin of Species*:

> *There is grandeur in this view of life, with its several powers, having been originally breathed into a few forms or into one; and that, whilst this planet has gone cycling on according to the fixed law of gravity, from so simple a beginning endless forms most beautiful and wonderful have been, and are being, evolved. (Darwin 2009)*

Darwin did not have knowledge of the molecular mechanism (DNA, RNA, genes, etc.) by which inheritance of traits and evolution occurs, so he could not know that all currently living organisms on Earth thus share a common genetic heritage. In the case of "complex" animals, more than 6,000 groups of genes common to all living animal species have been identified and these may have arisen from a single common ancestor that lived 650 million years ago in the Precambrian.

The origin of the LUCA remains a mystery, but several experiments have been undertaken to ascertain how the earliest molecules that led to life might have been formed on the primordial Earth. The classic 1952 Miller-Urey experiment used a mixture of water, methane, ammonia, and hydrogen as a proxy for the early Earth atmosphere (Miller and Urey 1959) and applied an energy source in the form of a continuous electric spark as a simulation of natural lightning and radiation. This experiment demonstrated that most amino acids can be synthesized from inorganic compounds under conditions "replicating" those of the early Earth.

Later experiments by Cleaves et al. (2008) demonstrated that the addition of iron and carbonate minerals could produce amino acids in the presence of an atmosphere containing carbon dioxide and nitrogen. Other approaches ("metabolism-first") focus on the presence of ordered, interlocked networks of chemical reactions that evolve in complexity over time to provide the precursor molecules necessary for self-replication. In any event, the emergence of polymers that could replicate, store genetic information, and exhibit properties subject to selection likely was a critical step in the emergence of prebiotic chemical evolution.

These simple, generic experiments demonstrate that the conditions for the emergence of (primitive) life are likely to be present on other planets and moons in the Solar System, and indeed on planets and moons in other solar systems.

So far, Earth is the only place that we know about that is host to life. Whether it has emerged elsewhere, and whether it has evolved into complex, intelligent forms that have developed advanced civilizations capable of spaceflight and/or interstellar communication, is a discovery that would change our concept of humanity and its place in the universe.

Does Life Necessarily Require a Biological Platform?

Alien life in popular culture, for example, in the films *Star Trek* and *Star Wars*, is depicted mostly as (1) having many features in common with humans, namely "biological" and upright, with a similar means of locomotion (often bipedal), etc., and (2) mostly being intent on the destruction of the Earth and/or its occupants. Examples of nonbiological aliens include the "monolith" in *2001: A Space Odyssey* and the title life-form in *Transformers*, and cyborgs (i.e., a hybrid of biological and inorganic) as depicted in the *Terminator* film franchise. Thus, it might be fruitful (and wise) to extend our search beyond life based solely on water and carbon chemistry.

This should not come as a surprise since we are witnessing an inexorable and accelerating trend toward the incorporation of inorganic transplants and all sorts of other "mechanical" substitutes for human biological tissue as advances in medicine and materials science address failures in the natural tissue of our bodies. These substitutes include eyeglasses, false teeth, heart valves, pacemakers, stents, dialysis, heart-lung machines, hearing aids (passive and active), joint and limb replacements, eye corneas and retinal light sensors, and even false eyelashes, fingernails, and hairpieces.

Rapid developments are also occurring in artificial intelligence (AI), machine learning, data storage and retrieval, interconnection, and robotics that are destined to change our work and leisure times beyond recognition.

The inevitable extrapolation and expansion of these changes suggest that human life is changing to become ever more "cyborg"-like, perhaps heralding the evolution of a new species of *Homo* that is much more capable (physically and intellectually) and resilient than *sapiens* (Bostrom 2014; Harari 2015; Tegmark 2017; Kaku 2018; Bohan 2022). More on this later.

Biological life-forms elsewhere in the universe might have already moved beyond this early stage of inorganic transformation and/or may have evolved from an initial life platform that is not based on water, carbon, and nitrogen at all. These life-forms will be fundamentally different from those on Earth and, given enough time (and luck), may have developed superior intelligence to humans, along with the capacity for self-replication and evolution.

It is noteworthy for later reference that technologically advanced life has existed on Earth for only about the last 100 years of its 4.5-billion-year existence, or about 2×10^{-8} of the Earth's lifetime. Other life-forms in the galaxy/universe, if they managed to get started in spite of all of the challenges, are very likely to have had much more time to evolve (or become extinct!) than humans have had.

Life on Earth

The Earth's biosphere is much more ubiquitous than just on its outer, solid surface; it extends from around 20 kilometers below the surface to at least 80 kilometers into the atmosphere. It is found in soil, hydrothermal vents, and rocks kilometers below the sea floor; extends 900 meters below the ice of Antarctica; and is found in the deepest parts of the ocean. Studies conducted on the International Space Station show that bacteria can survive in the vacuum of outer space for at least three years.

Nonetheless, possessing a surface environment devoid of oxygen and high in methane for much of its early history, the Earth would not have been a hospitable place for the emergence of living things. The earliest direct evidence of life on Earth are mineralized microfossils of microorganisms in 3.465-billion-year-old Australian Apex chert rocks (Altermann and Pinti 2021) and in hydrothermal vent precipitates, considered to be about 3.42 billion years old. The evidence is preserved in the hard, mat-like layered structures ("stromatolites") these early creatures made, which date to 3.5 billion years ago, and living forms of which exist to this day (e.g., at Hamlin Pool, Western Australia).

Life-forms first appeared on Earth when the oceans formed, soon after the formation of the Earth itself some 4.54 billion years ago (figure 1.1).[1]

Photosynthesis began with the evolution of cyanobacteria in the oceans around 3.4 billion years ago, making food using water and the Sun's energy and releasing oxygen (O_2) as a by-product. This produced a sudden, dramatic rise in oxygen, creating an environment less hospitable for other microbes that could not tolerate oxygen. The oxygen was initially consumed by oxidation of dissolved iron in the water, ultimately precipitating out to create the massive and ubiquitous banded iron (oxide) formations. Once all the dissolved iron in the oceans was consumed (about 2.5 billion years ago), the oxygen was then able to escape into the atmosphere, where it initially reacted with primordial methane. It then gradually built up to

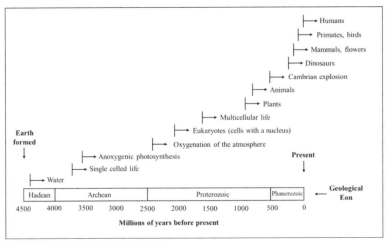

Figure 1.1. An approximate timeline of the emergence of various forms of life on Earth from its formation around 4.5 billion years ago to the present day. Technologically advanced humans have occupied the planet for around 10^{-8} of the time since its formation. (*Data extracted in part from https://en.wikipedia.org/wiki/Timeline_of_the_evolutionary_history_of_life.*)

its present proportion of around 21 percent, paving the way for the evolution of more-complex life-forms.

But microbes are just single cells and they do not have specialized cells called "organelles," the specialized subunits of cells that are needed for the construction of complex structures such as the skin, blood, and bone cells in animals. When microbes began living inside other microbes, functioning as organelles for them, the evolutionary path of life changed dramatically. Mitochondria, the organelles that process food into energy, evolved from these mutually beneficial relationships, and DNA began to be packaged inside nuclei. The new complex cells ("eukaryotic cells") boasted specialized parts playing specialized roles that supported the whole cell.

Cells "discovered" that by living together, certain critical benefits for survival could be obtained—they could feed more efficiently and/or gain protection by simply being bigger. Also, cells started to support the needs of the collective by each cell doing a specific job—some were tasked with holding the group together, while other cells made digestive enzymes that could break down food (Vesteg and Krajcovic 2008). Clusters of specialized, cooperating cells eventually became the first animals (sponges), which DNA evidence suggests evolved around 800 million years ago.

The assembly instructions for an animal's body plan are contained in the genes within its DNA, which control the action of other genes and proteins at specific locations and times to correctly assemble the special component parts of the animal.

At the beginning of the Cambrian period (541 to 485 million years ago) there was a massive burst of new life-forms over 15 to 25 million years, many with hard body parts like shells and spines that allowed these animals to radically engineer their environments, such as digging burrows. The animals became more active, with defined heads and tails for directional movement and the development of unique feeding styles. This subdivided the environment into feeding niches and led to even more diversification. The event was accompanied by major diversification in other groups of organisms as well.

Most of these new organisms were evolutionary experiments and quickly died out, but some groups thrived and dominated Earth for hundreds of millions of years before becoming extinct. By the end of the Cambrian nearly all existing animal types, or phyla (such as mollusks, arthropods, annelids, etc.), were established and food webs emerged, leading to the foundation for the ecosystems on Earth today.

The sudden rise in diversity of species during the Cambrian "explosion" has been attributed to:

1. A higher availability of oxygen that allowed the rise of large, complex animals, which need O_2 for chemical energy.

2. An increase in the amount of ozone (O_3) in the atmosphere, shielding Earth from biologically lethal UV radiation and enabling the evolution of complex life on land, as opposed to life being restricted to the water.

3. Massive global erosion caused by glaciers during the "Snowball" or "Slushball" Earth[2] depositing nutrient-rich sediments into the oceans, setting the stage for the diversification.

4. Mid-ocean ridge volcanoes causing a sudden increase in the calcium concentration in the oceans, making it possible for marine organisms to build skeletons and hard body parts.

5. Self-organization into higher morphological complexity, such as layers, segments, lumens, appendages, and even eyesight, through the mobilization of the products of genes that had previously evolved to serve unicellular functions. Eyesight, for example, would have precipitated the evolution of a wide range of defense mechanisms in prey creatures.

6. Widespread O_2 deficiency in the ocean seafloor giving mobile animals with the ability to seek out more oxygen-rich environments an advantage over fixed life forms. (Koonin 2007)

More than 99 percent of some 5 billion (but perhaps even up to a trillion) species that ever lived on Earth are understood to have become extinct, either by natural evolutionary processes (viz., Darwin's "Struggle for Life") or by massive single or multiple contemporaneous disruptive global events (discussed in chapter 2).

These species numbers pale into insignificance when it is considered that there are an estimated 10 nonillion (10^{54}) individual viruses and virions[3] on Earth, which some biologists consider to be life-forms. Indeed, there are more individual viruses than the estimated number of stars in the universe, which, in turn, are thought to be more numerous than all the grains of beach sand on Earth.

Figure 1.1 shows that while single-cell life began very soon (around 500 million years) after the formation of the Earth, multicellular creatures took another 2.4 billion years to emerge, and the complexity of plants and animals took a further 800 million years to appear. Primates then took almost another 800 million years to emerge, but technologically advanced life (i.e., modern humanity) has occupied the Earth for only about the last 100 years.

Clearly, the path to complexity and human intelligence required the operation of many critical steps in the Earth's chemistry and environment for we humans to emerge and flourish, and suggests that complex life is not easy to achieve.

The Central Role of Water

The study of life on Earth (the only example that we know about) suggests that the liquid solvent critical to life's origin and sustainment here (and potentially elsewhere) is likely to be water, because of the following:

1. Comprised of two hydrogen atoms and one oxygen atom, it is one of the most cosmically plentiful molecules, consisting of the most abundant atom (hydrogen, 74 percent) and the third-most-abundant atom (oxygen, 1 percent). Helium is the second-most-abundant atom in the cosmos, at 24 percent.

2. Its distinct physicochemical properties make it highly suitable for mediating macromolecular interactions through its action as a universal solvent for polar molecules and its critical role in protein folding, protein substrate binding, enzyme actions, the rapid transport of pro-

tons in aqueous solution, maintaining the structural stability of DNA/RNA, and the segregation of salt ions at cellular interfaces.

Carbon chemistry is likewise favored as a basis for life's chemical constituents because carbon has a high cosmic abundance (fourth, at 0.5 percent) and carries the ability to form an inordinate number of complex molecules. Alternative biochemistries to carbon and water may exist (including sulphur, which also presents a large number of elemental combinations) but their plausibility has not yet been convincingly demonstrated, so the search for life to date has focused on water as the key indicator for its past or current presence.

Life Elsewhere in and Beyond the Solar System

The development of spaceflight since Earth's first satellite, *Sputnik 1*, was launched by the Soviet Union on October 4, 1957, and the subsequent ability to orbit, land, and move around spacecraft on other bodies in the Solar System provided the ability to study the atmospheres and surfaces of these bodies for telltale direct and indirect signs of life there. In more recent times, the discovery of more than 5,000 planets in other star systems ("exoplanets") has included the identification of several hundred of these with environments that are like those of Earth and which might therefore harbor life.

In the Milky Way galaxy, it is expected that there are many billions of planets, with many more free-floating planetary-mass bodies that have been thrown out of their stellar system by one means or another and are now orbiting the galaxy directly. The characteristics of the eight planets in our Solar System range from the relatively small, inner rocky worlds of Mercury, Venus, Earth, and Mars to the outer giant gas planets of Jupiter, Saturn, Uranus, and Neptune, but the size, mass, orbital position, periodicity, density, and temperature diversity of exoplanets observed to date far exceed the variety in the planets found within our own Solar System. More on this variety later in this chapter.

Exoplanets where the surface temperature is "just right" for water to exist in liquid form are now referred to as being in the so-called Goldilocks

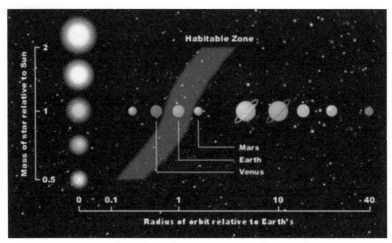

Figure 1.2. A schematic relationship between the distance of the Goldilocks/habitable zone (shaded panel) from its host star and the mass of that star relative to the Sun. The Solar System is depicted in the horizontal center panel, corresponding to a host star mass of 1. The more massive (i.e., brighter) the host star, the farther the Goldilocks zone is from that star. (*Wikimedia Commons; found at https:// www.e-education.psu.edu/astro801/content/l12_p4.html.*)

zone around a star. This name is the colloquial version of the term defined in astronomy and astrobiology as the "circumstellar habitable zone" (CHZ). It is the region around a star within which the incoming radiation and atmospheric pressure onto planetary-mass objects is appropriate for the sustainable presence of liquid water at their surfaces. The conditions in the CHZ are determined using the only example that we have of the requirements of life, namely our home planet's position in the Solar System and the amount of radiant energy it receives from the Sun (figure 1.2).

This figure demonstrates that the main determinant of the position and size of the habitable zone depends on the luminosity of the star, which dictates the equilibrium temperature of the planet. However, the exact boundaries of the zone also depend on subtle effects, like the influence of the carbonate-silicate cycle of the planet in regulating the level of carbon dioxide (and hence greenhouse gas effect—see the discussion in chapter 2) in its atmosphere, which can extend the zone farther from the host than its luminosity alone would dictate.

Exoplanets that have been detected in the Goldilocks zone are considered to provide the greatest prospects for life to evolve, at least in the form that would be familiar to us. Discoveries of confirmed Earth-like planets currently number less than 100, but the capability is emerging to measure the general composition of the atmosphere of these planets from the absorption of the light from their host star when the planet passes in front of that star. So far, no robust evidence has been found of the presence of gases that might indicate the presence of life on these bodies. What would be especially exciting would be the observation of industrial gases such as Freon (chlorofluorocarbons and hydrofluorocarbons that have been widely used as refrigeration gases) and some compounds of nitrogen that do not occur naturally and therefore would be indicative of the presence of advanced, industrial life-forms.

How Special Are the Circumstances Required for Life to Emerge?

The absence of evidence, so far, of life elsewhere in the Solar System, Milky Way galaxy, or the universe in general is not surprising because the circumstances required for life to begin and, in particular, for it then to flourish to the point where that life-form becomes "intelligent" are complex, numerous, and very demanding.

While very simple life-forms (prokaryotes) evolved on Earth after only around 500 million years after it formed, the evolution of complex cells (eukaryotes) took a further 1.8 billion years to emerge, and another 500 million years for multicellular life to develop (figure 1.1). The genus *Homo* emerged just 2 million years ago and *Homo sapiens* came into being only in the past 300,000 years, while technologically advanced life developed on Earth only in the past 100 years or so.

For complex (multicellular) life to form and flourish on a planet, the full set of the following conditions is likely to be required:

1. The parent star needs to be at least second generation to have condensed from a dust cloud that contained all of the elements heavier than iron that are essential to life. These heavy elements are not able to be produced by the fusion of hydrogen at the center of a star during its evolution; their formation requires the much larger energy generated

by a nova or supernova.[4] Thus a first-generation star needs to have "died" and its constituent material dispersed into the galaxy to enable these heavier elements to be incorporated into the building blocks of a second- or subsequent-generation solar system.

2. The gravitational aggregation of the star from an initial massive cloud of hydrogen and dust needs to have left a residual "accretion disc" of orbital material that is then able to condense into local accumulations that aggregate into one or more circumstellar planets.

3. The host star needs to have a mass, size, and luminosity that permits it to have a life expectancy of several billion years—that is, to allow enough time for life to emerge on one or more of the planets in its Goldilocks zone and to evolve into a complex form. In essence, this means that the star needs to be a red dwarf, like the Sun, or have a similar size. Being much more massive than this means that the additional gravitational pressure in the core of the star that arises from the larger size will produce more rapid "burning" of hydrogen fuel and thus greater outward "radiation" pressure from the core of the star. A balance is ultimately achieved between these two opposing forces, but the end result of the higher burn rate of hydrogen means that the life of the star is shortened significantly.

4. It is important that the star be stable in its energy output and that it not be prone to flares or other outbursts of intense X-rays and other dangerous parts of the electromagnetic spectrum that would extinguish any life that had formed on its planets. This requirement means that preferably the star's location is outside the more densely packed main arms of a spiral galaxy, where it is less likely to be subjected to similar massive bursts of radiation from stars that "go" nova or supernova in its immediate environment.

5. It is also preferable that the parent star is not part of a multiple-star system, thereby making it easier for a planet around that star to assume a stable orbit favorable to the emergence and evolution of life.

6. The planet needs to be at a particular distance from that star to receive an appropriate amount of energy (and therefore surface temperature)

for primitive life to begin, that is, for it to be in the Goldilocks zone, where it is not too hot and not too cold for water to exist in liquid form. Moreover, it needs to remain in the Goldilocks zone for billions of years, not just passing through or eventually pushed out by other planets, to allow life to mature and flourish long enough for intelligent life to evolve.

7. The planet needs to have a chemical composition (resulting from agglomeration of the residual disk of dust particles around the star) that allows it to have the wide variety of chemical elements required for life processes.

8. The agglomeration of particles must produce a body that is large enough to gravitationally clear its orbital neighborhood of other objects (thereby reducing potential extinction-level impact events) and to assume a size sufficient for it to acquire hydrostatic equilibrium (viz., a roughly spherical shape) under gravity, namely a diameter of more than about 500 kilometers (Hill 2022, and references therein). This allows impact kinetic energy to create heat sufficient to permit "differentiation" of the location of heavy and light compounds and elements to produce a solid/rocky crust (to act as a platform for life), a mantle, and a liquid or semiliquid metallic core. Convection of liquid metal in this core can then generate a magnetic field strong enough to repel most of the flux of cosmic rays from the sun that would otherwise destroy all life.

9. Convection in the liquid metallic core may also be needed to generate plate tectonics,[5] a mechanism for recycling elements crucial to life and also for sequestering (as carbonates) the carbon dioxide from the atmosphere to prevent the development of a runaway greenhouse effect (as is likely to have occurred in the case of Venus billions of years ago).[6] The presence of such plate tectonics is believed to have been critical for the so-called Cambrian explosion of life on Earth 500 million years ago by increasing the availability of essential nutrients in the oceans.

10. A significant atmosphere needs to be present on the planet that is thick enough to absorb harmful radiation, preserve water in the

liquid state on the surface, and generate rain on otherwise dry continental areas.

11. The planet's rotation period needs to be stable, with a significant inclination of its spin axis to the plane of its orbit around the star (23 degrees in the case of the Earth and 25 degrees for Mars) such as to provide seasons and thus a variable chemical and thermodynamic environment conducive to the genesis and evolution of life. On Earth, this stability/consistency of seasonal temperatures has been deemed necessary for the emergence of agriculture and permanent settlements and a transformation away from the hunter-gatherer lifestyle, allowing for the generation of food excess that enabled time for technology to develop and population to grow.

12. It would be advantageous (perhaps even essential) for the planet to have a satellite (e.g., our Moon) of sufficient size and mass to keep the rotation and axial orientation of the planet stable, and to produce tides and regular periods of wet and dry, conducive to the evolution of life. Not only did the impact of another Mars-size planet on the proto-Earth result in the formation of the Moon (see discussion below), but it tilted Earth's axis by about 23 degrees and likely introduced water and valuable elements into its mantle. The presence of the Moon then stabilized and slowed down Earth's rotation and helped to lock in the axial tilt that produces the four seasons of the solar year. Typically, small planets like Earth do not have a moon that is so large in a relative sense to its "parent" (we are in some senses close to being a binary planet in the Solar System) to enable it to play such an important role in the evolution of life. What, then, are the chances that a body the size of Mars would enter our orbit, collide with Earth, and have such a significant effect on our evolutionary history?

13. The solar system in which the planet exists needs to have other, much larger planets (like our own Jupiter and Saturn) big enough to clear the solar system (and especially the planet's neighborhood) of objects of such a size that their impact could destroy any life that had taken a foothold (as has nearly occurred several times in Earth's history—see chapter 2).

The conditions for complex (multicellular) life to emerge on the planet are likely to require many, if not all, of the above characteristics to be in play. If some of these conditions are missing or altered in some way, complex life is likely to have been extinguished, or it will have been extinguished before it had time to evolve fully. We have only one example from which to draw these conclusions, but we know that on Earth this process took nearly 3 billion years, or more than a quarter of the age of the universe.

Taking Earth's example once again, the conditions for technically advanced life to emerge must be even more demanding and/or "fortuitous" than the formation of simple and multicellular life, since it has been in existence here only over the past 100 years or so. This is a mere 2×10^{-8} of the age of the Earth, and around 10^{-9} of the age of the entire universe. It implies that while simple life-forms may be quite common, more-complex life-forms and especially intelligent life-forms are likely not to be so common (see the discussion in the next section).

Technical competence may be even more unlikely. For example, sharks have been around for more than 400 million years, yet they have not developed the ability to read or write, build spacecraft, or predict when eclipses of the Sun will occur, because these skills provide no evolutionary advantage to them.

Of all the complex species on Earth, only *Homo* has developed advanced intelligence and technologies. The survival of *Homo* may itself have been a matter of chance, since 98 percent of all currently documented species from the fossil record have been extinguished by one natural occurrence or another over the course of life's history (see chapter 2).

Even advanced species are not immune from extinction due to environmental causes and perhaps to their own hands. Over the 2 million years in which the genus *Homo* has lived on Earth, only one of nine species of *Homo* (i.e., *sapiens*) has survived from 1.2 million years ago to the present. At one point, no less than six of these species actually coexisted (*naledi, heidelbergensis, floresiensis, neanderthal, denisovans,* and *sapiens*), but only *sapiens* emerged unscathed (Stringer 2012). Indeed, the average time span for mammals is around 1 million years, so *sapiens* might have already passed its statistically expected span of existence.

The Drake Equation: What Are the Chances?

Many attempts have been made to calculate what the chances are of life existing elsewhere, based on a process of multiplying our best guesses for the likelihood of certain circumstances existing. One of the most popular of these algorithms, equation (1), is the so-called Drake equation (Drake 1961) devised by Cornell University astronomer Frank Drake (1930–2022). This provides an estimate, *N*, of the number of civilizations in the Milky Way galaxy with which direct communication might be possible:

$$N = R* \cdot f_p \cdot n_e \cdot f_l \cdot f_i \cdot f_c \cdot L \qquad \text{Eqn (1)}$$

where
- $R*$ = average rate of star formation in the galaxy
- f_p = fraction of those stars that have planets
- n_e = average number of planets that can be expected to support life per star that has planets
- f_l = fraction of planets that could support life that do develop life at some point
- f_i = fraction of planets with life that go on to develop intelligent life (civilizations)
- f_c = fraction of civilizations that develop a technology that releases detectable signs of their existence into space
- L = length of time for which such civilizations release detectable signals into space

Given the uncertainty in the estimates of the parameters, the original meeting in 1961 that gave rise to the Drake equation concluded that $N \approx L$ and that there were probably between 1,000 and 100,000,000 planets with civilizations in the Milky Way galaxy. Carl Sagan, using optimistic numbers for the parameters, produced estimates of *N* in the 12th episode of his program *Cosmos* (aired on December 14, 1980) ranging from "a pitiful few" to millions of communicating civilizations in the Milky Way. Modern astronomy has given us a better indication of likely values for $R*$, f_p, and n_e, but the others are still not known with any certainty and are relegated to guesswork relatable only to the single example of Earth. This has resulted in a massive variation in the estimates of *N*.

Worked examples using the maximum and minimum values assigned to each of the parameters in equation (1) and the resulting estimates of *N* are provided in table 1.1.

Clearly, with this huge range of possible outcomes for *N*, the Drake equation is of limited use in providing meaningful estimates of the number of contactable civilizations in the galaxy. In addition, it certainly says nothing about the possible existence of simple or indeed complex

Table 1.1. Worked examples of the parameters in the Drake equation using contemporary minimum and maximum estimates of each parameter.

Parameter in the Drake Equation	Minimum Estimate	Maximum Estimate
$R*$ = average rate of star formation in the galaxy	1.5	3
f_p = fraction of those stars that have planets	1	1
n_e = average number of planets that can be expected to support life per star that has planets	0.1	1
f_l = fraction of planets that could support life that do develop life at some point	10^{-4}	1
f_i = fraction of planets with life that go on to develop intelligent life (civilizations)	10^{-9}	1
f_c = fraction of civilizations that develop a technology that releases detectable signs of their existence into space	0.2	0.2
L = length of time for which such civilizations release detectable signals into space	300	1B
N = possible contacts	10^{-12}	600M

Source: https://en.wikipedia.org/wiki/Drake_equation#Range_of_results

life-forms that have not (yet) developed the ability to transmit radio or other signals of their presence. Then again, the equation was not meant to provide a specific number. Rather, it was the intention of Drake to focus the attention of astronomers on what attempts could be made/developed to *detect* such civilizations.

Whatever was the background objective, some 60 years of searches for extraterrestrial radio signals since the 1960s have failed to find anything (reproducible) that might have been generated by an alien civilization. This outcome (called the "Great Silence") might be considered to be an important indicator suggesting that the number of advanced civilizations "out there" must be nearer the lower end of the possible range than the upper more-numerous end, if indeed it is not zero.

A number of theories have been provided for the absence of the detection of such signals to date, namely that we are searching at the wrong wavelengths; our technology is not advanced enough to detect very, very weak signals from far away; these civilizations have disappeared before they could send signals out for a significant period of time; and/or they are not interested in declaring their presence in this or any other way.

Nevertheless, the conclusion that there may be no other intelligent life in our entire galaxy cannot be confidently eliminated (Sandberg et al. 2018).

Probabilities Calculated from Other Perspectives

Using the specific list of critical parameters provided above that are necessary for intelligent life to exist elsewhere in the galaxy, the probability can be calculated by a less formal process (though still with some very subjective steps) than is provided by the Drake equation, as follows:

1. The number of stars in the Milky Way is around 100 billion.

2. Half of these are single-star systems, which reduces the number of options to 50 billion.

3. Since only around 10 percent of these are like our Sun, there are now only 5 billion options.

4. Fifty percent of these have planets in the habitable zone, reducing the options to 2.5 billion.

5. Let's say that 10 percent of these planets have tilted axes, a large moon, plate tectonics, other large planets, etc., to allow the emergence of life; this drops the options to 250 million.

6. On Earth, technologically advanced life has only been present for 2×10^{-8} of the time since its formation, so the number of options therefore reduces to about 5.

This alternative calculation suggests that the number of technologically advanced civilizations in the galaxy is quite small, near the midpoint of the extreme values presented in table 1.1.

Another argument goes as follows: There are 100 billion stars in the Milky Way, most of which are billions of years older than our Sun. In the last 30 years more than 5,000 exoplanets have been discovered around stars not too far away from the Solar System, some of which are similar in size and temperature to our Sun. Some of these planets are orbiting in the so-called Goldilocks zone around their parent star, in which water might exist in the liquid state. As in the calculation immediately above, extrapolating these conditions to the entire galaxy, there are perhaps 250 million habitable planets in the Milky Way.

Some of these 250 million planets are likely to be Earth-like, which, if the Earth is typical, will have had the time to develop complex intelligent life. At least some, perhaps millions, of these civilizations (especially those on significantly older planets than Earth) will have been around long enough to develop interstellar travel (even if just by their autonomous uncrewed probes), a technology that Earth itself is developing right now. Thus, at any practical pace of interstellar travel, the galaxy should have been completely "colonized" within a few tens of millions of years, including the Earth, or at least visited, either by members of that civilization or its autonomous probes, multiple times in its long history.

This argument is not restricted by the term L in the Drake equation because these civilizations do not need to be in existence currently; all that is required is that intelligent life evolved sometime in the distant or recent past and that it had the capability of interstellar travel. That civilization itself may have died out millions or billions of years ago, but its signals and/or autonomous probes should continue to fly through the galaxy and be detected, eventually.

Indeed, no convincing evidence of visitation exists and no confirmed signs of intelligence elsewhere have been spotted, either in our galaxy or the more than 80 billion other galaxies of the observable universe (see discussion below).

It seems unlikely that, if they did indeed visit Earth, none of these civilizations chose to (1) make their presence known to us in any unambiguous way, (2) accidentally or deliberately extinguish life on Earth, or (3) settle the planet for themselves. Moreover, it seems unreasonable that they flew across the galaxy unscathed, yet crash-landed in a desert of the United States (or elsewhere) and/or were observed only by a handful of humans.

The Fermi Paradox: Where Is Everybody Then?

The alternative (to Drake) calculation provided above highlights the apparent contradiction between the high estimates of the number of advanced civilizations obtained from the Drake equation (or any other measures) and the fact that humanity has not yet had contact with or seen irrefutable evidence of such civilizations.

In response to this logic, Italian physicist Enrico Fermi (1901–1954) asked the now famous, and obvious, question: "Where are they?"

Many explanations have been put forward as a rationalization/explanation of this so-called Fermi paradox, namely why none of these civilizations have been detected, including that:

1. Our observations are incomplete/flawed, and it is just that we have not detected them yet.

2. Technically advanced humans on Earth have not been listening long enough.

3. The difficulty and cost (both financial and resource) of interstellar travel has been grossly underestimated.

4. The alien civilizations are so significantly different from us that contact is not possible.

5. They are unwilling to communicate with us.

6. Humanity exists only in a gigantic simulation experiment undertaken by a massively more advanced civilization and they want to see how we emerge and evolve in isolation from other life-forms.

7. Technologically advanced intelligent life is actually very rare.

8. It is the nature of intelligent life to destroy itself before it masters the technology of interstellar travel and/or communication (Hanson 1998; Bostrom 2008)—the so-called Great Filter.

The concept of the Great Filter originates from the Fermi paradox and the proposal (Hanson 1998) that the failure to find any extraterrestrial civilizations in the universe suggests that something is wrong with one or more of the arguments that advanced intelligent life is probable. Alternatively, or in addition, it suggests that there is a barrier that emerges after advanced life evolves that prevents it from reaching the stage of interstellar travel. It implies that there is a high probability of (1) an extinction catastrophe, (2) an underestimation of the impact of technology that "unburdens existence," or (3) resource exhaustion, perhaps especially energy (Baum 2010).

It is my contention that, of the suite of explanations listed above, the Great Filter is the most likely explanation for the absence of contact or evidence of contact with aliens, especially when considered in the light of the social and environmental problems that we on Earth are now facing on a number of levels.

The Search for Extraterrestrial Life

There are several methods at our disposal by which we might be able to observe evidence, or the implication, of extraterrestrial life. The evidence can be categorized into those searches targeting the observation of simple life-forms (bacteria, etc.) and those searching for complex, intelligent life-forms.

In the case of complex, technologically mature life, the relevant evidence might be:

- Direct observation—witness accounts of encounters with aliens, unidentified flying objects (UFOs), etc.

- Detection of electromagnetic emissions from extant civilizations as part of the Search for Extraterrestrial Intelligence (SETI), etc.
- Observation of large-scale engineering or other structures here or on other planets, including crop circles, the "canals" on Mars, Dyson spheres (see below) around other stars, etc.
- Observation of interstellar spacecraft or technological "junk" passing through the Solar System.
- Statistical implications from the sheer number of other stars and planets—Drake's equation, the Fermi paradox, etc.

In the case of simple life-forms, the evidence might include:

- Direct observation of living or fossil organisms on meteorites and from sample-return missions from Solar System bodies—the Moon, asteroids, comets, etc.
- Residual "trace fossils"[7] in data accrued from orbiters, landers, and rovers on Solar System bodies indicating the direct past or current presence of biota and/or environments conducive to their existence.
- Certain signature gases ("biosignatures") indicative of biological or industrial activity of life itself, detected in the atmospheres of exoplanets in the Goldilocks zone around other stars.

The evidence, or absence of evidence, of complex/advanced and simple life-forms is discussed in the following sections.

Complex/Advanced Life Forms
Unidentified Flying Objects (UFOs)

There have been close to 90,000 UFO (or unidentified aerial phenomena, UAP) sightings in the past 110 years—mostly in the United States and Europe. These sightings increased dramatically after the Roswell incident in 1947 (see below) and following the earlier "War of the Worlds" broadcast in 1938.

The "War of the Worlds" broadcast on Halloween night, October 30, 1938, is probably the most famous program in radio history. It was an adaptation of H. G. Wells's novel *The War of the Worlds* (1898)[8] and narrated by Orson Welles. Some (most?) listeners did not hear the disclaimer that it was a radio play and were convinced that a Martian invasion was taking place, in part due to the "breaking news" style of the storytelling.

The items in the broadcast began with relatively calm reports of unusual explosions on Mars followed by an account of an object falling on a farm in New Jersey. The crisis escalated dramatically when a "live" reporter described creatures emerging from an alien craft to incinerate police and onlookers with a "heat ray" until his audio feed abruptly ended. Subsequent updates detailed a devastating alien invasion and the military's futile efforts to stop it.

Although its overall impact has probably been exaggerated over the years, the broadcast raised concerns about Americans' susceptibility to believe what has become known as "fake news" as well as issues about the responsibility of the media.

To summarize, on close examination, when sufficient data is available to make a reasoned assessment of the event, the vast majority of UFO sightings have been explained by natural phenomena—but some remain classified as "open."

The Roswell Incident

The so-called Roswell incident is the July 8, 1947, recovery of balloon debris from a ranch in New Mexico by US officers from Roswell Army Airfield, and the accompanying press release stating that they had recovered a "flying disc." The army quickly retracted the statement, saying that the object was a conventional weather balloon.

The matter did not surface again until the late 1970s, when retired Lieutenant Colonel Jesse Marcel said he believed the debris he had retrieved was extraterrestrial. Immediately, ufologists began promoting a variety of increasingly elaborate conspiracy theories, claiming that one or more alien spacecraft had crash-landed and that the extraterrestrial occupants had been recovered by the military, which had then engaged in an elaborate cover-up.

Eventually, in 1994, the US Air Force released a report identifying the crashed object as a nuclear test surveillance balloon. A second report in 1997 concluded that stories of "alien bodies" probably arose from the admission that test dummies had been dropped from the balloon at high altitude. The incident has been described as "the world's most famous, most exhaustively investigated, and most thoroughly debunked UFO claim" (Gildenberg 2003).

While UFO sightings and conspiracy theories continue to be reported, the reality is that, at best, concrete evidence of visitation by extraterrestrials, or of artifacts left by them on Earth, has not been found to date.

Crop Circles

Crop circles are "unnatural" patterns that appear overnight in farmers' crop fields that are attributed to aliens or UFOs. They are mostly found in the UK but have spread to dozens of countries around the world in past decades. The mystery has inspired countless books, blogs, fan groups, and films and TV series such as *The X-Files*.

According to newspaper coverage at the time (Tim 2020), the worldwide crop circle phenomenon was energized by an event in January 1966, when a farmer from Euramo, near Tully in far north Queensland, Australia, said he saw a "large saucer-shaped object" rise up from a swampy area and fly away sideways "at terrific speed," leaving a roughly circular area of debris and apparently flattened reeds and grass, which he assumed had been made by the alien spacecraft. The incident was interpreted later as the effects of a local whirlwind, but without confirmatory evidence of this proposal, the alien explanation continued to have traction, as so often occurs in these circumstances.

Simple circles began appearing in the English countryside in significant numbers in the 1970s. The frequency and complexity of the circles increased dramatically, reaching a peak in the 1980s and 1990s when quite elaborate patterns were produced, including some illustrating complex mathematical equations. While there are countless theories ranging from the plausible to the absurd, the only known, proven cause of crop circles is humans. The origin of the crop circles in the UK remained a mystery until September 1991, when two men confessed that they had created the patterns for decades as a prank to make people think UFOs had landed.

SETI: Evidence from Outside the Solar System

The Search for Extraterrestrial Intelligence (SETI) is a collective term for scientific searches for intelligent life ex-Earth. It is primarily associated with monitoring electromagnetic radiation for signs of transmissions from civilizations on other planets, a suggestion originally made by Nikola Tesla in 1896. The film *Contact*, with a screenplay based on the book of the same name by Carl Sagan (1985), is perhaps the best-known example of this scenario. Scientific investigation of the presence of signals from outer space began shortly after the advent of radio in the early 1900s, and focused international efforts have been going on since the 1980s, many using very sophisticated hardware and software and the investment of significant financial support.

SETI Outcomes to Date

In 1924 the United States pronounced a National Radio Silence Day, urging citizens to keep their radios quiet for five minutes on the hour, every hour, so astronomers could use a powerful radio receiver strapped to a dirigible floating 2 miles above the Earth's surface to listen for any potential radio signals from Mars, which was at that time closer to Earth than it had been for about a century. Nothing!

In 1960 Frank Drake performed the first modern SETI experiment, using a 26-meter radio telescope to examine the stars Tau Ceti and Epsilon Eridani near the "water hole" 1.420 GHz marker frequency, in proximity to the hydrogen and hydroxyl radical spectral lines.[9] Nothing!

There have been many attempts to design a spectrum analyzer specifically to search for transmissions. Suitcase SETI was created in 1981 that had a capacity of 131,000 narrowband channels. It was expanded in 1985 to META with a capacity of 8.4 million channels, each with a resolution of 0.05 Hz, and the ability to use Doppler shift of frequency to distinguish between terrestrial and extraterrestrial signals. The follow-on to META in 1995 was named BETA, for "Billion-Channel ExtraTerrestrial Assay," that allowed the receipt of 250 million simultaneous channels with a resolution of 0.5 Hz. It scanned through the microwave spectrum from 1.400 to 1.720 GHz in eight hops, with two seconds of observation per hop. Nothing!

NASA's Microwave Observing Program (MOP) was funded in 1992 but drew the attention of the United States Congress, where the program was ridiculed and cancelled one year after its start.

In 1995 the nonprofit SETI Institute of Mountain View, California, resurrected the MOP program under the name of Project Phoenix, backed by private sources of funding. From 1995 through 2004 it studied roughly 1,000 nearby Sun-like stars, conducting observations at the 64-meter Parkes radio telescope in Australia; the 43-meter telescope of the National Radio Astronomy Observatory in Green Bank, West Virginia; and the 300-meter Arecibo Observatory in Puerto Rico. Nothing!

Many international radio telescopes continue to be used for SETI searches, including the Low Frequency Array (LOFAR) in Europe, the Murchison Widefield Array (MWA) in Australia, and the Lovell Telescope in the United Kingdom. Nothing!

The Wow! signal was a strong narrowband radio signal detected on August 15, 1977, by the Ohio State University's Big Ear radio telescope in the United States (figure 1.3). This signal was considered the strongest candidate for an extraterrestrial radio transmission ever observed but it had no detectable modulation (used to transmit information over radio waves), and innumerable attempts to reproduce the signal have failed.

Recent work has suggested that the signal may have been generated by the hydrogen clouds associated with one of two comets, 266P/Christensen and P/2008 Y2 (Gibbs), which were passing through the sector of sky the Big Ear Radio Observatory was surveying in 1977 (Breitman 2017).

On the 35th anniversary of the Wow! signal in 2012, Arecibo Observatory beamed a digital stream in the direction from which the signal was recorded. It was sent at 20 times the power of the most powerful commercial radio transmitter and contained around 10,000 Twitter messages solicited for the purpose and bearing the hashtag #ChasingUFOs. The sponsor also included a series of video vignettes featuring verbal messages from various celebrities. No reply yet!

In 2015 Stephen Hawking and Russian billionaire Yuri Milner announced a well-funded ($100 million) effort called Breakthrough Listen, described as the most comprehensive search for alien communications to date. Nothing yet!

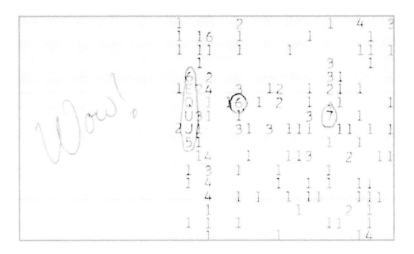

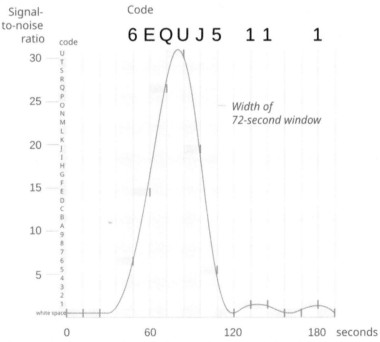

Figure 1.3. TOP: The original printout of the Wow! signal represented as "6EQUJ5" and including Jerry Ehman's handwritten exclamation. **BOTTOM:** "6EQUJ5" plotted as the signal's intensity variation over time, expressed in the measuring system adopted for the experiment. (*Public domain from http://www.bigear.org/Wow30th/wow30th.htm. and Wikimedia Commons.*)

In 2019 a narrowband signal at 982.002 MHz was intercepted that appeared to have originated from the direction of Proxima Centauri and showed shifts in its frequency consistent with the movement of a planet. No modulation was detected, and work is still underway to rule out terrestrial interference. Nothing yet!

Thus, around 60 years of SETI have failed to find any concrete and confirmed evidence of life-based radio signals arriving from outside the Solar System, even though radio telescopes, receiver techniques, and computational abilities have improved significantly during that time. The one thing that these searches have discovered, however, is that the galaxy is not teeming with very powerful alien transmitters continuously broadcasting near the 21-centimeter wavelength (frequency 1,420.4 MHz) of the neutral hydrogen atom.

This SETI version of the Fermi paradox is sometimes referred to as the "Great Silence." It represents a strong case for the absence of intelligent, advanced life elsewhere since (1) it does not require civilization to have mastered the technically difficult (perhaps ultimately impossible) challenge of interstellar travel and (2) any signals would have had the time to reach us from just about all parts of the universe by now. Explanations for the Great Silence run along the same lines as the answer to the Fermi paradox discussed above.

Massive Extraterrestrial Engineered Structures

There have been several cases of the purported observation of large-scale engineered structures on bodies within the Solar System which, if true, would indicate the presence of technically advanced life off-Earth.

The "Canali" on Mars

In 1877 Giovanni Schiaparelli (1835–1910), director of the Brera Observatory in Milan, Italy, started mapping the dark and light features on Mars. He named the larger ones "seas" and "continents" and dubbed the channel-like lineaments *canali*. This Italian word translates into English as "canal," "channel," "duct," or "gully," but it was, not surprisingly, taken to imply that they were nonnatural features that indicated the presence of technologically active life on Mars.

The construction of the Suez Canal and its opening in 1869 had become the engineering wonder of the era and supported the idea that if massive engineering structures could be constructed by an advanced civilization on Earth, why not on Mars? This triggered an extraordinary period of astronomical activity, excitement, and debate for several decades about the origin and implications of the "canals."

Percival Lowell, founder and director of the Lowell Observatory in Arizona, became very interested in Mars and studied it intensively from around 1893 to 1908. He was particularly interested in the surface features, many of which he characterized as "non-natural features." He became convinced that Schiaparelli's canals were real, and that they were built by intelligent beings to carry water from the polar caps (assumed to be composed of water ice, as on Earth) to the drier equatorial regions of the planet. Lowell mapped hundreds of them, for the most part matching Schiaparelli's observations (figure 1.4).

Objections from astronomers and many other scientists were raised, based on (1) the inability of some/most of them to observe the linear features, (2) the fact that calculations of the water content of the Mars atmosphere showed that it was so low as to be undetectable, (3) the extremely low atmospheric pressure of Mars (1 percent of Earth's) would not allow water to be present in the liquid state on the surface, and (4) massive fluctuations in the temperature of the surface (measured as having a high of about 20°C at noon at the equator to a low of about −153°C at the poles, with an average value of −63°C) that would make advanced life difficult to support.

Despite this, the presence of "canali" remained a popular proposition, including their representation in maps prepared as recently as 1962. They were still accepted right up to the flyby of Mars by NASA's *Mariner 4* spacecraft in 1965, when the surface was revealed to be a desolate cratered landscape with no signs of large engineering structures.

It is now suggested that the network of crisscrossing lines observed with the lower-resolution telescopes of the 19th and early 20th centuries was merely an optical illusion resulting from the tendency of humans to embellish patterns in dark and faint images, even when these patterns do not exist.

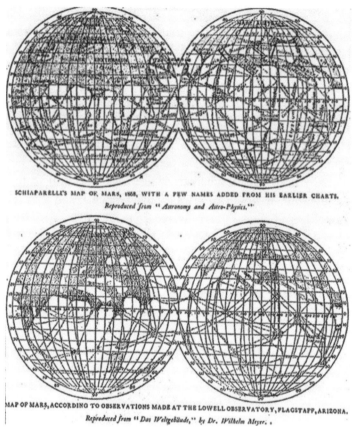

Figure 1.4. A comparison of some of Giovanni Schiaparelli's (upper) and Percival Lowell's (lower) drawings of the dark regions ("continents" and "seas") and linear features ("canals") of Mars. (*Image reproduced with permission from https://www.erbzine.com/mag14/1414.html.*)

The Face on Mars

The famous "Face on Mars" image captured by NASA's *Viking 1* spacecraft in 1976 (figure 1.5) was interpreted by some observers of the time as a deliberate artifact left by an alien civilization.

Equivalent terrestrial structures that might be visible to other civilizations looking at Earth from afar could include many large cities, especially when illuminated at night; the Great Wall of China; the pyramids

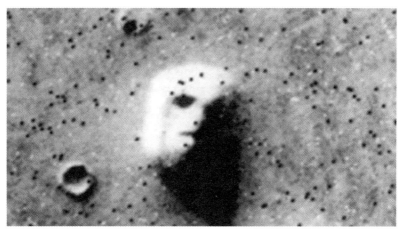

Figure 1.5. The "Face on Mars," captured by NASA's *Viking 1* spacecraft in 1976. Examination of this landform under different angles of solar illumination show that it is not an engineered creation of an advanced civilization on Mars. (*Public domain image by NASA/JPL.*)

of Egypt; and large mine sites. Later images of this feature, including a detailed study by the Mars Global Surveyor in 1997, revealed that it was a trick of low sunlight and just an illusion.

Advanced Life Outside the Solar System

Examples of engineered structures outside the Solar System have yet to be observed with current telescopes. However, it might be expected that evidence of exploration by other intelligent extraterrestrial life, such as their interstellar probes and information-gathering devices (if not occupied spacecraft) or detritus from spacecraft associated with these explorations, might appear passing through the Solar System from time to time, however remote the probability. Some of these possibilities are discussed in the following subsections.

Von Neumann Probe

Hypothetical exploration objects/craft, including von Neumann and Bracewell probes, have been proposed as being dispatched by other civilizations to observe and/or communicate with other civilizations. Even at the relatively low speeds achieved by Earth's fastest spacecraft (197 km/s in 2024, by the Parker Solar Probe during one of its ongoing flybys of the Sun to study the solar corona), these probes could explore exhaustively a galaxy the size of the Milky Way (diameter 105,700 light years[10]) in as little as 200 million years, with comparatively little investment in materials and energy relative to the significance of the potential results. If even a single civilization in the Milky Way attempted this more than 200 million years ago, such probes could/would be dispersed throughout the entire galaxy by now. An example of this in science fiction is the "monoliths" in the films *2001: A Space Odyssey* and *2010: The Year We Make Contact*.

A von Neumann probe is a spacecraft capable of replicating itself (exponentially), named after Hungarian-American mathematician and physicist John von Neumann (1903–1957). A self-replicating machine is a type of robot that can reproduce itself autonomously using raw materials found in its environment, thus exhibiting self-replication in a way analogous to that found in nature. For this reason, von Neumann machines are considered by some to represent an alternative form of life.[11]

Used in the context of interstellar exploration, it provides an ideal platform for the settlement of space since the machines can spread exponentially (Freitas 1980). The future development of such technology by humans may represent a neat solution to the safe, practical, and energy-efficient step-wise mining of moons and asteroid belt objects for ore, fuel, and other materials and, indeed, the off-Earth construction of a wide range of infrastructure. Being much less fragile than humans and much less vulnerable to the rigors of space, these robots could provide the solution to the manifest and manifold risks of long-distance space travel. More on this in later chapters.

Bracewell Probe

Another possibility that an alien civilization might use to contact human beings is a Bracewell probe (Bracewell 1960). Such a device would be a robotic interstellar space probe with a high level of artificial intelligence that is designed to seek out technological civilizations or to monitor worlds where there is a likelihood of technological civilizations arising. It carries all the relevant information that its home civilization might wish to communicate to another culture.

The proposition is that once the probe discovered a civilization that meets its contact criteria, it would make its presence known, initiate a dialogue with that culture, and communicate the results of its encounter to its place of origin. It would, in essence, act as an autonomous "local representative" of its home civilization, and thus would act as the point of contact between the cultures, avoiding the need for slow speed-of-light dialogue between vastly distant neighbors.

While a Bracewell probe is not a von Neumann probe, the two concepts are compatible, and a self-replicating device as proposed by von Neumann would be able to greatly speed up a Bracewell probe's search for alien civilizations. Some have claimed that probes of this sort may also be classified as "life" in their own right—more on this subject in chapter 13.

The monolith depicted in the film *2001: A Space Odyssey* appears to be a Bracewell probe placed on the Moon to ensure that only a civilization capable of spaceflight would be able to discover it. It is possible (and even very likely) that von Neumann and Bracewell probes might outlive the civilization that created and launched them. Though unfortunate to point out, the same outcome may come to pass in respect to the five spacecraft to date that humanity has launched into interstellar space, namely the two *Pioneer*, two *Voyager*, and the *New Horizons* vehicles. All of these craft are likely to be traveling through the galaxy for billions of years, long after the Sun has expanded to a red giant and destroyed Earth and all of the remaining life thereon.

Dyson Spheres

It has been proposed (Dyson 1960) that very advanced civilizations might need to construct a so-called Dyson sphere/ring/shell/swarm around their central star, to capture more of the host star's energy as the civilization's energy demands inevitably increased (figure 1.6).

Hypothetical megastructures like this would enable a civilization on one of the planets orbiting the star to harvest a far greater percentage of its solar power output than naturally falls upon the home planet of that civilization. The concept is a thought experiment that attempts to imagine how a spacefaring civilization of Type II on the Kardashev scale[12] would meet its ever-expanding energy requirements once it has exceeded what could be generated from the home planet's resources alone.

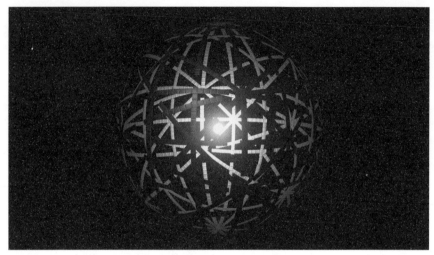

Figure 1.6. A representation of a hypothetical Dyson sphere, constructed by an advanced civilization to capture more of the host star's energy than the portion that falls on the civilization's planet. The scale is indicated by the bright object, namely, a star, located at the center of the sphere, demonstrative of the source of energy being captured. This conception of a Dyson sphere consists of massive independently orbiting arched panels, perhaps constructed with material extracted from a small planet or moon in the associated solar system. (*Wikimedia Commons.*)

The first modern imagining of such a structure was by Olaf Stapledon in his 1937 science fiction novel *Star Maker*. The concept was later explored by the physicist Freeman Dyson (1960), who speculated that such structures would be the logical consequence of the escalating energy needs of a technological civilization and would be a necessity for its long-term survival. Such spheres could be detected in astronomical searches through observed changes in the energy output and character of the light emitted from the central star and would be a strong indicator of extraterrestrial life, but evidence of this kind has not been observed as yet.

Oumuamua

An object subsequently named "Oumuamua" was observed traveling on a hyperbolic trajectory through the Solar System by the University of Hawaii's Pan-STARRS1 telescope in 2017. Coming from the direction of

Vega, a star 25 light years away, and traveling at 26.3 kilometers per second (km/s), Oumuamua would have taken 285,000 years to reach us from that stellar system. It was the first such interstellar object ever observed to be passing through our Solar System (identified by the hyperbolic character of its trajectory), but was followed shortly after, in 2021, by the object 21/Borisov (later identified as the first interstellar comet).

Oumuamua's unusual properties of no outgassing, steady rotation, very elongated/saucer shape (400 meters on its longest dimension, with an aspect ratio of 10), higher reflectivity than rocks or ice, and steady acceleration (up to 38.3 km/s) after passing by the Sun are not easily explained by it being an interstellar comet (like 21/Borisov). Loeb (2021) has suggested that the most likely explanation of these properties is that the object is a discarded/relic interstellar "light sail" (like the spacecraft proposed for the Breakthrough Initiative—see chapter 7) constructed by an advanced alien civilization for exploration, but this conclusion has been disputed.

Earth's Own Interstellar Probes
Humans have, so far, sent five spacecraft into interstellar space (figure 1.7):

- *Pioneer 10*, launched on March 2, 1972. Contact was lost in 2003, but it is estimated to have passed 134 astronomical units (AU).[13] It is heading toward the star Aldebaran (65 light years away) in Taurus.
- *Pioneer 11*, launched on April 6, 1973. Contact was lost in 1995, but it is now estimated to be at 111 AU. It will pass near one of the stars in the constellation Aquila in about 4 million years.
- *Voyager 2*, launched on August 20, 1977. It left the heliosphere[14] for interstellar space at 119 AU in 2018 and is still active. It is currently 136 AU from Earth and will pass by the star Sirius (86 light years away) in around 300,000 years.
- *Voyager 1*, launched on September 5, 1977. It passed the heliopause[15] at 121 AU in 2012 and is still active. It is the fastest of the five spacecraft with a velocity of 16.9 km/s and is currently 162 AU from Earth, the most distant human-made object. It

SPACE TO LIVE

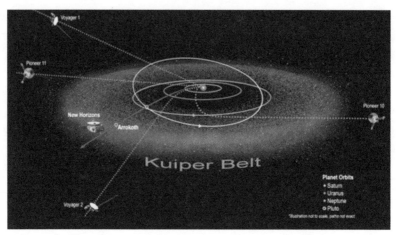

Figure 1.7. Map of the trajectories of the five interstellar spacecraft launched from Earth, namely, *Voyagers 1* and *2*, *Pioneers 10* and *11*, and *New Horizons*, relative to the plane of the orbits of the outer planets, Pluto, and the surrounding Kuiper Belt, as of 2023. All but *New Horizons* are now traveling in the interstellar medium. (*Public domain image from NASA/Johns Hopkins APL/Southwest Research Institute.*)

is headed toward a distant encounter with star Gliese 445, 17.1 light years from Earth.

- *New Horizons*, launched on January 19, 2006. It flew past the Kuiper Belt object Arrokoth on January 1, 2019, after its Pluto flyby in 2015 and is now about 58 AU from Earth.

Pioneer 10 was the most distant spacecraft from Earth, until *Voyager 1* caught and passed it in 1998. *Voyager 2* overtook *Pioneer 10* in 2023, but although *New Horizons* was launched far faster than any outbound probe before it, it will never overtake either of the *Voyagers*, thanks to gravity assists[16] they received from both Jupiter and Saturn. Unless anything unusual happens, each of these spacecraft will continue to travel through interstellar space for billions of years, perhaps to be detected eventually by another civilization, if they survive passage through the galaxy.

The fact that these spacecraft are a product of advanced technical competence on Earth demonstrates that if any other civilization in the galaxy exists and has had the time to develop equivalent (or much more advanced) technology, it might be expected that their own craft will ultimately arrive (or should have already arrived) near or in the Solar System. Indeed, as predicted by Enrico Fermi, there should have been innumerable examples of such encounters, not just the one proposed by Loeb (2021) in the case of Oumuamua, described above.

SIMPLE LIFE FORMS

Simple single-celled microorganisms like bacteria[17] and archaea[18] are numerically by far the dominant life-form on Earth. Bacteria were among the first life-forms to appear on Earth and are present in most of its habitats and in symbiotic and parasitic relationships with plants and animals. There are approximately 5×10^{30} bacteria on Earth, forming a biomass that exceeds that of all plants and animals.

Archaea also form a major part of life on Earth, but their cells have unique properties separating them from the other two domains of life: bacteria and eukaryote.[19] Classification is difficult, because the majority have only been detected by analysis of their nucleic acids in samples from their environment.

These two kinds of single-celled creatures evolved just 500 million years after the Earth was formed (figure 1.1) and have survived through 2 billion years of Earth's initial inhospitable past. Microbes known as "extremophiles" have been found in rocks up to 20 kilometers below the Earth's surface, up to 80 kilometers into the atmosphere, and in volcanic "black smokers" in the mid-oceanic trenches, in highly toxic, sulphurous fumaroles, with or without oxygen, and in highly saline or acidic environments (pH 0). Thus, it seems likely that these simple life-forms exist elsewhere in the galaxy and, indeed, may be quite common.

The extremes of pressure and temperature in space can quickly degrade most cell membranes and destroy its DNA. However, studies of clumps of *Deinococcus radiodurans* bacteria placed on the outside of the International Space Station and Skylab showed that they survived for three years, protected by the outer layers of organisms that shielded the ones below them from the extreme environment (Kawaguchi et al. 2020). Similarly, multilayers of *Bacillus subtilis* spores under space conditions beneath a perforated aluminum dome survived up to six years in the space missions of Spacelab/Shuttle.

Biosignatures

A "biosignature," "chemical fossil," or "molecular fossil" is any material, structure, or event that provides evidence of past or present life. To be useful, the signature(s) must be reliable, survivable, and detectable.

Some of the measurable indications of life include:

- isotope patterns, chemical features, or organic molecules that require or are formed by biological processes;
- minerals with composition or morphology that indicates biological activity;
- biologically formed cements, microtextures, microfossils, and films;
- macroscopic physical structures and textures (figure 1.8);

Figure 1.8. A biosignature, used to detect the presence of life off-Earth, is unlikely to be as obvious as this image of Buzz Aldrin's boot print left in the moondust about an hour into his and Neil Armstrong's lunar extra-vehicular activity during the *Apollo 11* mission on July 20, 1969. Aldrin took the photograph as part of studies into the soil mechanics of the lunar surface. (*Wikimedia Commons.*)

- time variations/disequilibrium of gases, reflectivity, or appearance;
- reflectance features due to biological pigments;
- gases formed by metabolic, aqueous, or industrial processes;
- signatures that are residues or outputs of a technologically advanced civilization.

Some of the telltale biosignature compounds are:

- methane (CH_4), alone or in the presence of oxygen (in disequilibrium) or carbon dioxide[20];
- dimethyl sulfide ($[CH_3]_2S$) and chloromethane (CH_3Cl);
- membrane lipids—a group of compounds (structurally similar to fats and oils) that form the double-layered surface of all cells;
- phosphene (PH_3)—recently suggested to be present in the clouds of Venus but now disputed;
- "chiral" compounds[21] (on Earth, the amino acids characteristic of life are all left-handed in shape, while all sugars characteristic of life on Earth are right-handed);
- magnetite (Fe_3O_4)—hexagonal platelet-like magnetite of a few hundred nanometers in size indicating the presence of the iron-reducing bacterium *Thermoanaerobacter*;
- pigments that are uniquely biologic in origin, from phototrophic and photosynthetic life-forms;
- chemical disequilibrium, necessary for life of any kind and exploited as a source of energy for metabolism—this is an "agnostic biosignature" and so is independent of the form of life that produces it.

EXPLORATION OF BODIES IN THE SOLAR SYSTEM

There have been a host of spacecraft missions to targets outside low Earth orbit, dating from the Soviet Union's *Luna 1* probe in 1959, the first artificial object to reach the escape velocity of Earth. Since then, semi-autonomous probes, rovers, and static equipment have survived and operated successfully on the surface of Venus (for short periods) and orbited and/or landed on the surface of Mars (for several years, and ongoing), on the surface of Saturn's moon Titan, on four asteroids (Ryugu, Eros, Itokawa, and Bennu), and two comets (Churyumov-Gerasimenko and Tempel). Six crewed Apollo missions landed on the Moon between 1969 and 1972, and 12 astronauts spent time on the surface on foot and (in three cases) riding on an electric Lunar Roving Vehicle to collect samples and set up a variety of equipment for remote monitoring of various physical parameters. More details of the history of these missions are provided in chapter 4.

Most of these probes had the dual objectives of (1) increasing scientific knowledge about the target bodies and their formation and (2) searching for evidence of past and/or present life. As discussed below, all these missions have failed, so far, to find direct evidence in the form of either "body" or "trace" fossils but, in many cases, have revealed the existence of physical and chemical environments that are conducive to life (at least as we know it).

Evidence from the Moon

To date, a total of 10 missions to the Moon by three nations have returned samples of lunar rocks and soils for direct analysis:

- Apollo (United States)—381 kilograms from six missions: 1969–1972;
- Luna (Russia)—0.301 kilograms from three missions: 1970–1976;
- Chang'e-5 (China)—1.731 kilograms from one mission: 2020.

None of this recovered material has shown direct or indirect evidence of life, but water or hydroxyl molecules have been found locked in glasses and minerals in the Apollo samples and have also been detected in spectroscopic analyses undertaken by lunar orbiter missions. Most of the water signals come from extremely cold regions on the floor of craters near the Moon's poles that are never warmed by sunlight. Outside the permanently shadowed regions, water molecules appear to have been expelled through micrometeorite impacts and then lost from the atmosphere due to the Moon's low gravity (17 percent of Earth's).

Today, the Moon is about as inhospitable to life as it gets, but the environment may have been host to simple life-forms shortly after the Moon formed around 4.5 billion years ago as a result of a collision between the proto-Earth and a Mars-size planetesimal, dubbed "Theia,"[22] soon after their formation in the early Solar System. The impact produced a ring of debris in orbit around the (new) Earth, mostly comprising material from the proto-Earth's mantle, which soon coalesced into the Moon.

There would have been enough water vapor associated with the collision to provide the Moon with a substantial atmosphere and possibly even liquid water on the surface. Volcanic activity would have been prevalent on the Moon at that time, which could have replenished the atmosphere with water vapor from deep in the interior. By the time volcanic activity settled down some 500 million years after the collision, the Moon's low mass would have been unable to retain this atmosphere and, much like Mars, the Moon would have then dried up except for water still trapped within rock and/or shaded from sunlight in the polar regions. Digging under the surface in future missions could yield evidence for lunar life, including the possibility of fossilized microbes.

Evidence from Orbiters, Landers, and Rovers on Non-Lunar Solar System Bodies[23]

The objective of recent missions to a wide range of objects in the Solar System has switched from searching for direct evidence of past and/or present life to looking for indirect evidence of life in the form of biosignatures and/or environments that are conducive to life, specifically "finding the water," given that this chemical is essential for the development and flourishing of life (as we know it).

Mars

Ever since 1877, when Schiaparelli dubbed the channel-like lineaments on Mars *canali*, this planet has been considered the most likely place to host life in the past and/or present and so has been the dominant non-lunar exploration target. However, it has been a difficult mission objective, with just 25 of 55 missions (45.5 percent) through 2021 being fully successful, with a further 3 partially successful. Eleven missions are currently active on the surface (both fixed and rover) and orbiting Mars, and several more are on the drawing board for future missions.

It has long been known from telescopic images that Mars has two permanent polar ice caps. The caps at both poles consist primarily of water ice, but carbon dioxide freezes out of the atmosphere (composed of 95 percent CO_2, 3 percent nitrogen, and 1.6 percent argon) during the relevant winter season when each lies in continuous darkness. The CO_2 accumulates as a layer of "dry ice" about 1 meter thick on the north cap in the northern winter, while the south cap has a permanent dry ice cover about 8 meters thick. When the poles are again exposed to sunlight during their summer seasons, the frozen CO_2 sublimes back into the atmosphere, giving rise to Earth-like frost and large cirrus clouds.

Several of the spacecraft missions have revealed unequivocal evidence that there is water in non-polar regions of the surface and that it has been present in massive amounts, potentially as rivers, open lakes, and oceans that have produced characteristic landforms in the distant past (figure 1.9).

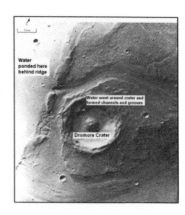

Figure 1.9. Pictorial evidence for the existence of water on Mars. **Left:** Photograph of the surface of Mars by the *Viking Lander 2* at its Utopia Planitia landing site on May 18, 1979, and relayed to Earth by *Viking Orbiter 1* on June 7. It shows a thin coating of water ice on the rocks and the soil. **Right:** Detail of landforms attributed to water flow around Dromore crater (14.8 kilometers) within the outflow channel Maja Valles, as seen by *Viking Orbiter 1*. (*Public domain images. Left: #P-21841 from NASA/JPL, https://www.nasa.gov/wp-content/uploads/2023/03/9464667429_d37c7d75de_o.jpg. Right: https://history.nasa.gov/SP-414/ch4.htm, modified by Jim Secosky.*)

In July 1976 NASA's *Viking 1* became the first spaceship to successfully land on the planet and operate effectively on its surface, with *Viking 2* following in September that year. These missions collected more than 50,000 photographs, but none showed signs of life anywhere in the vicinity of their landing sites. In 1997 NASA's *Mars Pathfinder/Sojourner* (the first, largely experimental, rover) and *Surveyor* (in orbit) collected detailed data on several specific minerals. In 2003 NASA launched two rovers as part of its Mars Exploration Rover Mission. The rover *Spirit* found rocks rich in key chemicals (magnesium and iron carbonates) indicating that they formed when Mars was warm and wet (with a thicker CO_2 atmosphere and near-neutral-pH water). It also found rocks with 90 percent pure silica, usually found in hot springs or hot steam vents, where life as we know it on Earth often exists, and discovered evidence of powerful steam eruptions from heated underground water, systems that support microbial life on Earth.

Spirit's sibling rover *Opportunity* found the mineral hematite, which typically forms in water, while near the rim of Endeavor Crater it found brightly colored veins of gypsum. These rocks likely formed when water flowed through underground fractures, leaving calcium behind. It also found other compelling signs of a watery past on Mars, specifically clay minerals formed in neutral-pH water, and other friendly conditions favorable to ancient microbial life.

In 2012 the rover *Curiosity* found smooth, rounded pebbles that (most likely) formed while rolling downstream for at least a few miles in a river that was ankle- to hip-deep. It detected a tenfold spike in methane in the atmosphere and found other potential biosignature organic molecules in rock-powder samples (collected by the robotic laboratory's drill) from Mount Sharp and the surrounding plains.

The methane is not necessarily evidence of life because cosmic dust can provide a source of methane, and it can also be produced by the interaction of water with olivine minerals. It is likely that intermittent winds on Mars disperse temporary accumulations of methane at some times of the day but leave it undisturbed in more concentrated form at other times, leading to these observations.

In total, the findings from *Spirit*, *Opportunity*, and *Curiosity* do not directly suggest past or present life on Mars, but they show that the necessary raw ingredients existed for life to get started there at one time.

NASA's car-size *Perseverance* rover landed inside the Red Planet's Jezero Crater on February 18, 2021, tasked with exploring an ancient lakebed in the crater for signs of ancient microbial life. This 45-kilometer-wide crater is a promising place to do such work on Mars because photographs taken from orbit suggest that it hosted a big lake and a river delta in the ancient past (figure 1.10). On Earth, river deltas like this can preserve carbon-containing organic compounds, the building blocks of life as we

Figure 1.10. Image of the ancient river delta in Jezero Crater on Mars, taken from NASA's *Mars Reconnaissance Orbiter*. The *Perseverance* rover landed in the region near the bottom right of the image in February 2021. The landforms in this image are convincing structural and topographic testimony for water-carved channels and transported sediments forming fans and deltas within lake basins. Spectral data acquired from orbit show the presence of clay and carbonates minerals resulting from chemical alteration by water. (*Public domain image from NASA/JPL-Caltech/MSSS/JHU-APL.*)

know it. *Perseverance* also found new evidence for ancient hot magma and abundant water and discovered various organic molecules that are present in all living things.

One of *Perseverance*'s primary objectives was/is to extract small rock cores and seal them in special sample tubes for retrieval by a future Earth-return mission. This will allow much more detailed chemical and physical analysis of the samples than that which can be achieved in situ by the instruments on the rover.

Venus

Venus, the planet that comes nearest to Earth and the one with the most similar size, lies near the inner edge of the Sun's Goldilocks zone (figure 1.2) and therefore might be considered suitable for hosting life. However, a runaway greenhouse effect that has produced a surface temperature of around 454°C and an atmospheric pressure 92 times that of Earth have created a nightmare scenario for life. The temperature is lower in the upper (highly acidic) atmosphere and so there is a possibility that reactions there between sulphur dioxide and carbon monoxide might serve as a food source for floating thermoacidophilic extremophile microorganisms (Clark 2002; Dartnell et al. 2015). More recently, Greaves et al. (2020) reported the presence of phosphine in the planet's atmosphere, a potential biosignature, although this finding has yet to be confirmed.

Saturn's Moon Titan

Of all the other bodies in the Solar System, spacecraft missions to Saturn's moon Titan have shown this moon to have the greatest promise of hosting life, although perhaps based on a fundamentally different chemistry from that on Earth. Rather than a mixture of around 80 percent nitrogen and 20 percent oxygen, Titan's atmosphere is composed of 95 percent nitrogen and 5 percent methane, with traces of acetylene, ethane, and propane, along with other organic molecules. This atmosphere is so dense/thick that humans would not need a pressure suit while moving around on the surface—but they would certainly need breathing apparatus and protection from the extreme cold of -179°C.

The *Huygens* lander parachuted down through the atmosphere from the orbiting *Cassini* spacecraft in 2005. It discovered that the climate is dominated by seasonal weather patterns as on Earth (with clouds and rain), and these have created surface features like those found on Earth, including dunes, rivers, deltas, lakes, and seas (probably of liquid methane and ethane). Its frozen surface is littered with rounded pebbles composed of water ice, while the ground surface consists of a mixture of water and hydrocarbon ice (figure 1.11). The probe survived 72 minutes on the surface of Titan, and it currently holds the record as the most distant landing from Earth on a celestial body permanently outside the orbit of Mars.

Although no evidence of life on Titan has been found, its complex chemistry and unique environments could potentially harbor circumstances

Figure 1.11. Image taken with the Descent Imager/Spectral Radiometer, one of two NASA instruments on the ESA's *Huygens* probe during its 2.5-hour descent from orbit to the surface of Saturn's moon Titan in January 2005. The two objects/pebbles just below the middle of the image are about 15 centimeters (left) and 4 centimeters (center) across, respectively, about 85 centimeters from *Huygens*. *(Copyright ESA/NASA/JPL/University of Arizona reproduced with permission.)*

suitable for life, both as we know it (in the extensive subsurface liquid water ocean) and perhaps as we don't know it (in the hydrocarbon-rich conditions on and above the surface). The surface of Titan is one of the most Earth-like places in the Solar System, albeit at vastly colder temperatures and with different chemistry, and despite these differences it is a destination worthy of continued exploration.

Jupiter's Moon Europa
Europa is slightly smaller than Earth's Moon and orbits Jupiter at a distance of around 670,000 kilometers every 3.5 days. It is believed to be a geologically active world, due to strong tidal flexing by the competing gravitational fields of Jupiter and the other Galilean moons that serves to heat its rocky, metallic interior and keeps it partially molten.

The surface of Europa is a vast expanse of water ice, beneath which is thought to be a global ocean of liquid water over 100 kilometers deep, prevented from freezing by the heat from gravitational flexing. Evidence for this ocean includes geysers erupting through cracks in the surface ice, the presence of a weak magnetic field, and the chaotic terrain on the surface. This icy shield insulates the subsurface ocean from the extreme cold and vacuum of space and helps to retain the heat generated from the flexing.

At the bottom of this ocean, it is possible that there are hydrothermal vents and ocean floor volcanoes. On Earth such features often support very rich and diverse ecosystems, but evidence of life will have to await closer examination by future dedicated orbiters and/or spacecraft on Europa's surface.

Saturn's Moon Enceladus
Like Europa, Saturn's moon Enceladus is ice-covered with a subsurface ocean of liquid water. Enceladus first came to the attention of scientists as a potentially habitable world following the surprise discovery (when backlit by the Sun) of enormous geysers near the moon's south pole. These jets of water escape from large cracks on the surface and are clear evidence of an underground store of liquid water. The geysers also contain an array of organic molecules and grains of rocky silicate particles that can only be present if the subsurface ocean water was in physical contact with the rocky ocean floor at a temperature of at least 90°C.

This is very strong evidence for the existence of hydrothermal vents on the ocean floor, providing the chemistry and localized sources of energy needed for life, but, like Europa, no evidence for life has been discovered to date.

Outside the Solar System

Exoplanets

The first confirmation of detection of an exoplanet came in 1992 from the Arecibo Observatory in Puerto Rico, with the discovery of several terrestrial-mass planets orbiting the pulsar PSR B1257+12 (Wolszczan and Frail 1992). The first confirmation of an exoplanet orbiting a main-sequence star (like the Sun) was made in 1995, when a giant planet was found in a four-day orbit around the star 51 Pegasi. As of April 18, 2024, there are 5,612 confirmed exoplanets in 4,170 planetary systems, with 948 systems having more than one planet (figure 1.12).

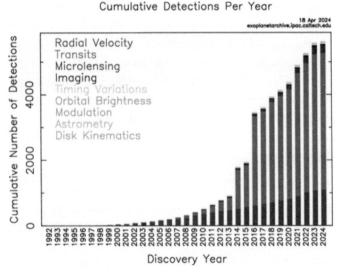

Figure 1.12. Cumulative detections of exoplanets per year from 1992 to April 18, 2024, distinguished by their method of discovery (see text for explanation). The "transit" detection method—observation of a periodic reduction of the light received from the exoplanet's host star as that planet transits across the face of the star—is the most successful of the exoplanet discovery methods so far. (*Public domain image from NASA Exoplanet Archive.*)

It is very likely that the number of discoveries of exoplanets, including those broadly/superficially similar to Earth, will continue to increase as the number of spacecraft and telescopes dedicated to the search increase in number and technical sophistication.

The nearest exoplanets to Earth are located 4.2 light years from Earth in orbit around Proxima Centauri, the closest star to the Sun (Anglada-Escudé et al. 2016), while the farthest from Earth (to date) is located 27,700 light years away, orbiting the star SWEEPS in the constellation Sagittarius. Of course, as the distance to a potential exoplanet stellar system gets larger, the difficulty of detection becomes greater because the light of a planet is a very small fraction of the diminishing amount of light received at Earth from the parent star.

The techniques used for exoplanet discoveries include the following:

- Transit photometry, or "looking for shadows" (4,171 discoveries to date). If a planet crosses (transits) in front of its parent star's disk, the observed visual brightness of the star drops by a small amount, depending on the relative sizes of the star and the planet. For example, an Earth-size planet transiting a Sun-like star produces a dimming of around 0.008 percent. When the planet is eclipsed by (passes behind) the star, there is a corresponding reduction in the total emission from the pair of bodies due to the loss of reflected starlight from the planet. This method can provide information about the transit depth, transit duration, ingress/egress duration, and orbital period of the exoplanet.

- Radial velocity, or "measuring wobble" (1,089 discoveries to date). A star with a planet will move around the center of mass (barycenter) of the star-planet system, which leads to small (depending on the relative mass of the planet) movement of the star toward or away from Earth. This "radial velocity" of the star with respect to Earth can be deduced from the displacement in the parent star's spectral lines due to the Doppler effect.[24] The method provides information about the planet's mass.

- Gravitational microlensing (214 discoveries to date). This occurs when two stars lie almost exactly in the observer's line of sight and the gravitational field of the closer star (and its planet) acts like a lens, magnifying the light of the more distant background star. A planet orbiting the closer star is detected when its own gravitational field makes an observable contribution to the lensing effect. Unlike other planet detection methods, lensing events are one-off (and therefore cannot be confirmed), lasting for weeks or days, as the two stars and Earth are all moving relative to each other.

- Direct imaging (69 discoveries to date). Since the light from planets tends to be lost in the glare from their parent star, it is usually very difficult to resolve the planet(s) directly, even if a coronagraph is used to block light from the star while leaving the planet visible. Planets orbiting far enough from stars to be resolved reflect very little of their host's starlight, but their infrared radiation can be detected through their thermal emission instead of the visible parts of the spectrum. The detection is more likely if the star system is relatively near to the Sun and the planet is especially large (a lot larger than Jupiter), widely separated from its parent star, and hot.

- Transit timing (29 discoveries to date). This method considers whether transits occur with strict periodicity or if there is a variation. If a planet has been detected by the transit method, then variations in the timing of the transit provide an extremely sensitive method of detecting additional non-transiting planets in the system with masses comparable to Earth's. This is useful in planetary systems far from the Sun, where radial velocity methods cannot detect them due to the low signal-to-noise ratio.

- Reflection and emission modulations (9 discoveries to date). During the orbit of a planet around a star, especially those in a close orbit, the combined light from the system will produce variations in reflected light as the planet goes through illumination phases from full to new and back again (just like the Moon). Corresponding changes in overall thermal emissions are also potentially detectable.

- Astrometry, or "miniscule positional changes" (3 discoveries to date). This method consists of precisely measuring a star's position in the sky and observing how that position changes over time. The technique has been used since the late 18th century when William Herschel, and many astronomers since, used it to infer the presence of binary star systems. If a star has a planet (or companion star), the gravitational influence of the planet will cause the star itself to move in a tiny circular or elliptical orbit around the mutual center of mass. Unfortunately, the changes in stellar position are so small that even the best ground-based telescopes generally cannot produce precise-enough measurements, and very long observation times are required. However, a potential advantage of the astrometric method is that it is most sensitive to planets with large orbits (i.e., greater distance from their host star), unlike other methods that are most sensitive to planets with small orbits.

By no means are all the exoplanets discovered so far capable of hosting life like that on Earth; most are too hot or too cold to support liquid water on the surface (i.e., they do not reside in the Goldilocks zone around the star), the surface itself is not solid/rocky, and/or they are orbiting a star that emits extreme levels of X-ray and ultraviolet (UV) radiation, which can be up to hundreds of thousands of times more intense than the amount that the Earth receives from the Sun. Many will be very unlikely to have any, let alone all, of the characteristics required for life to emerge.

Figure 1.13 provides a recent summary of some of the basic characteristics of the approximately 5,000 exoplanets discovered to date, including their mass, radius, orbital period, and incident radiation, plotted along with the equivalent characteristics of Earth, indicated by the large star on each part of the figure.

Only a few of the exoplanets in figure 1.13 are really "Earth-like" at least partly because the sensitivity of most detection methods favors larger planets and those that lie close to the host star, although this situation is rapidly changing as detection technologies and data analysis techniques improve. Not all of the techniques by which they were discovered allow measurement of all the characteristics of mass, radius, orbital period, and irradiation, so the separate parts of figure 1.13 do not all contain the same

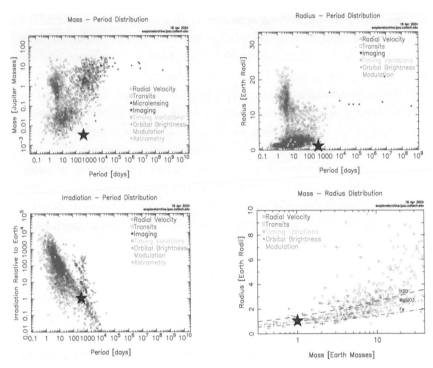

Figure 1.13. Plot of the relationship between the mass and revolutionary period of an exoplanet around its host star (top-left section), its radius and revolutionary period (top right), its irradiation level and revolutionary period (bottom left), and its mass and radius (bottom right) for exoplanets discovered up to 2024. For comparison, the corresponding position of the Earth in each diagram is indicated by the large star, indicating that Earth is an outlier in the general exoplanet distribution of all plots. (*Public domain image from NASA Exoplanet Archive.*)

data, and the exoplanets often cluster by detection method in the plot window. However, most of the known exoplanets, relative to Earth:

1. are much larger in mass and radius;
2. have a shorter revolution period around their star;
3. have a lower density (and therefore are less likely to be "rocky");
4. have a much higher irradiation influx.

The irradiation data indicates that most exoplanets discovered to date do not lie in the Goldilocks zone around their parent star and so are unlikely to have conditions that allow the presence of liquid water and, therefore, life as we know it. Other factors dictating the likelihood of the presence of life include whether the planet is large enough to sustain an atmosphere or is so massive that the atmospheric pressure at the surface is too large be conducive to life as we know it.

Another notable characteristic of exoplanets discovered to date is that 85 percent lie inside the so-called tidal locking zone[25] and may therefore have evolved to present the same face to their host star, either all the time or through a resonant relationship (Barnes 2017; Ballesteros et al. 2019). This may present challenges for the emergence of life on the exoplanet because the incoming starlight is not dissipated between the hot and cold hemispheres over the course of the planet's rotation (unless it has an atmosphere and wind). As detection methods improve, the proportion of tidally locked exoplanets is likely to decrease since we will be able to observe planets that are located farther from their host star, where the tidal forces that produce the locking are weaker.

A key factor for habitability is the estimate of equilibrium surface temperature, T_{eq}, of the exoplanet, namely, the theoretical temperature that a planet would be if it were a black body,[26] without an atmosphere, being heated only by its parent star. The equilibrium temperature is determined purely from a balance with incident stellar energy and ignores the influence of any greenhouse effect.

In figure 1.14, T_{eq} has been estimated for the same exoplanets as plotted in figure 1.13. Most of these 5,000 exoplanets have a much higher T_{eq} than Earth and so are unlikely to be habitable unless the presence and composition of an atmosphere is able to moderate the conditions to make them more suitable for life, perhaps by increasing the reflectivity (or "albedo") of the planet due to cloud formation.

Mediocrity Principle

The mediocrity principle[27] arises, in part, from the Copernican model (in which the Sun is at the center of the Solar System) and is based on the idea that if Earth-like planets are not unusual in the universe

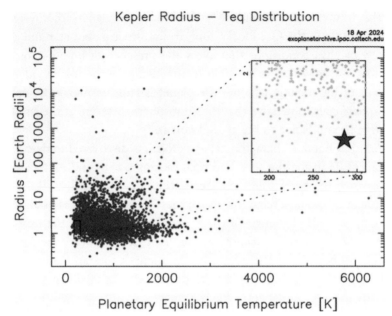

Figure 1.14. An estimate of equilibrium surface temperature T_{eq} (in "absolute" degrees Kelvin, or degrees Centigrade plus 273) of around 5,000 exoplanets plotted against their radius (in Earth radii). For comparison, the position of the Earth is indicated by the large star in the upper-right expansion of the small rectangular section of the extreme lower-left portion of the main plot. (*Public domain image from NASA Exoplanet Archive.*)

(as demonstrated by the large number of exoplanet observations), then, by extension, so is life. However, this view ignores the fact that since the steps from nonlife to life, especially complex life, are still largely unknown, they cannot be considered a straightforward consequence of Earth-like environments (Gleiser 2021). Indeed, a planet may have all the right characteristics for harboring life (as documented above) but there would still be no guarantee that life would exist there. Thus, the preconditions for life may exist on another world, but life must emerge, evolve, and then exist long enough for its intelligence, technology, and presence to be detectable from afar (by us).

The sequence of events is far from "mediocre" and perhaps has more in common with the "fine-tuning hypothesis." This asserts that the natural values of certain fundamental physical constants appear to be arbitrary and possibly random, making the conditions for intelligent life implausibly rare and the probability of the existence of a universe like ours vanishingly small (Lewis and Barnes 2016; Sawyer 2009).[28]

Physical and Chemical Properties of Exoplanets

Depending on a planet's position with respect to its host star, the total light collected by a telescope will include a varying fraction of light blocked by, or reflected off, the planet (figure 1.15). The planet reflects no light during the "secondary eclipse," when it is hidden from view behind the star, whereas it reflects the maximum amount (although still very small) of light shortly before and after this phase (see the lower section of figure 1.15). The planet blocks a small fraction of the star's light when it transits in front of the star in the "primary eclipse" part of its orbit.

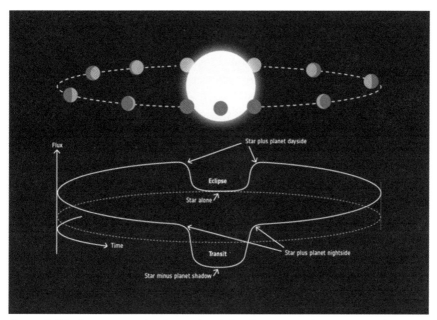

Figure 1.15. An annotated schematic of the light curve (lower section) observed for an exoplanet that alternately transits and is eclipsed by its parent star (upper section). Changes in the illumination of the Earth-facing side of the exoplanet are indicated as it revolves around the host star. (*Public domain from the Webb Space Telescope image gallery.*)

The changes in the intensity of light from the star during the transit of the planet across its face provide, as has been described above, critical information about the characteristics of that planet. Likewise, changes in starlight reflected by the planet as it orbits its star reveal details of the planet's atmosphere, including the presence of any clouds, and provide insight into the physical processes that drive the transport of heat from the hot day side to the cooler night side.

Spectral analysis of the filtered light of the host star that has passed through the atmosphere of a transiting exoplanet can provide information on the chemical composition of the atmosphere. It can reveal the presence of absorption lines due to potential gaseous biosignatures (e.g., methane), along with line-of-sight Doppler shifts of the planet's spectrum as it orbits the star that provide information about revolution speed.

This kind of analysis of exoplanet atmospheres is in its infancy, but as the technology improves it will no doubt provide very valuable information about biosignatures of any life that is present on the exoplanet.

Moons of exoplanets are no doubt as abundant in other star systems as they are in the Solar System, but they are even more difficult to detect than their host planets and none have yet been confirmed. As in our own Solar System (e.g., Saturn's moon Titan), these "exomoons" might be habitable worlds, but proposed future instruments are unlikely to have sufficient power to detect life on them, especially if they lack an atmosphere.

Exoplanet Life: Summary
- It is estimated by astronomers that there could be up to 40 billion exoplanets in the Milky Way, but the vast majority of these are too far away to be detected by observations from Earth, let alone examined in detail.
- The most common types of exoplanets in our galaxy that have been observed to date are "super-Earths" (up to two times Earth's size) and "mini-Neptunes" (between two and four times larger than Earth).
- Of the 6,000 or so exoplanets that have been discovered to date, only 55 have been confirmed to be Earth-like. While this is cer-

tainly a small proportion, Earth-like planets will not be as common as other types of planets because (at the moment) they are not as easily detected as are larger exoplanets.

- Life on other worlds might use vastly different chemistry, perhaps based on a solvent other than water (as may be the case for Titan in our own Solar System). Thus, limiting searches to exoplanets in the habitable zone of stars won't help to find life if its environmental requirements differ significantly from our own.
- Life on exoplanets might be hidden in a subsurface ocean encased in ice, invisible even to our most powerful space telescopes.
- The direct imaging of planets around other stars is in its infancy, but the new suite of telescopes, on and above Earth (including James Webb), will dramatically improve our chances of observing these planets and of finding biosignatures in their atmospheres.

Panspermia and Cross Contamination

Whether or not any life discovered off-Earth is the same as or distinct from life on Earth, the building blocks of life and/or life itself may have been transferred from their place of origin to other bodies through the agency of meteorites or other means. This process is known as "panspermia." Another complication is that past impactors, landers, and rovers sent to other bodies in the Solar System, despite our best efforts to sterilize them, may have contaminated the bodies they visited.

More than 300 meteorites have been exchanged in both directions between Mars and Earth, and nearly 400 between the Moon and Earth. Thus, the discovery of life or its biosignatures on the Moon or Mars does not necessarily mean that life emerged there separately from life on Earth. Only a DNA analysis of this (new) life-form would be able to provide concrete evidence that it was distinct from the DNA of life on Earth and had a separate and independent evolution pathway.

One of these Mars meteorites created a major stir around the world in 1996 when NASA announced that the 1.93-kilogram ALH84001 meteorite that had been found in 1984 in the Allan Hills of Antarctica contained mineralogical (carbonate grains) and structural (apparently biological in form) evidence of microfossils (Sawyer 2006). The claims immediately made headlines worldwide, including a formal announcement of the find by no less than the then-president of the United States, Bill Clinton, but they were controversial from the beginning. The wider scientific community ultimately rejected the hypothesis when all the supposedly unusual features in the meteorite could be explained without requiring life to be present. Despite this, the scientific and public attention caused by the intense analysis and discussion represented a turning point in the history of the science of astrobiology.

In a logical extension of potential contamination *within* the Solar System, given that organic molecules and microbes can survive for long periods in outer space, it has been proposed that bodies such as asteroids, dust, comets, etc., could have transported life-forms such as bacteria (complete with their DNA) through space to the Earth in the distant past. Thus, life on Earth (or anywhere else in the Solar System) might have been "seeded" by life present elsewhere in the galaxy. The possibility that the interstellar visitor Oumuamua contains DNA from another civilization in the galaxy immediately raises the possibility that other interstellar bodies at some time in the past could have collided with Earth in their passage through the Solar System, thereby spreading their "life seeds" onto the planet.

In a softer version of this hypothesis—"molecular panspermia"—it is suggested that the pre-biotic organic building blocks of life originated in space and became incorporated into the solar nebula from which planets condensed. These were then further distributed to planetary surfaces where life then emerged.

Summary and Conclusions from Chapter 1

1. The sheer number of stars and planets in the galaxy and the resilience of primitive (single-celled) life suggests that this form of life is likely to have evolved elsewhere, as it did on Earth.

2. Our many spacecraft observing from orbit and on the surface of several bodies in the Solar System have detected conditions conducive

to life as we know it (i.e., water and other necessary compounds), but no direct or indirect evidence of life itself has been found. Although unlikely in the author's view, primitive life may yet be found on other bodies in the Solar System but may have evolved there independently or by cross contamination (panspermia) of life from elsewhere.

3. The conditions and time required for complex life to emerge are much more special and took much longer than for simple life-forms, while the development of life with advanced technology took even longer to appear on Earth—more than 4 billion years, or nearly one-third of the lifetime of the universe. This latest stage in the evolution of *sapiens* represents only around 2×10^{-8} of the time since the Earth formed. As a consequence of this singular example, the probability of the existence of technologically advanced life existing elsewhere, contemporaneously to Earth, is likely to be very low.

4. There is no conclusive evidence of visitation by aliens on Earth, no sign of massive extraterrestrial engineering projects (e.g., Dyson spheres) despite a formidable array of telescopes in space and on the ground, and no convincing evidence of radio signals from advanced civilizations elsewhere, despite 60 years of comprehensive search protocols. The "galactic silence" and the absence of other visual evidence suggest that technologically advanced life is extremely rare and/or short-lived.

5. We have found Earth-like exoplanets in the Goldilocks zone of several other nearby stars but are yet to find evidence of life there, intelligent or not.

6. If technologically advanced life emerged elsewhere in the galaxy, even in only a few locations, and it lasted for only a few thousand years with rates of development similar to that on Earth, the galaxy should be overrun with concrete evidence of its existence by now (the Fermi paradox).

So, in the view of the author, advanced life is likely to be unique to Earth and we are indeed alone in the galaxy!

Thus, in our quest to search for and ultimately settle on another platform outside the Solar System, it is unlikely that we will be helped (or hindered) by contact with other advanced civilization(s). It is also unlikely that in so doing we will "trespass" on another civilization's territory. Thank goodness! However, we will need to be cognizant of our (high) potential to encounter "simpler" forms of life elsewhere and so we should tread very carefully and sensitively wherever we venture.

Chapter Two

Scenarios for the Demise of Humanity

We always thought the living Earth was a thing of beauty. It isn't. Life has had to learn to defend itself against the planet's random geological savagery.
—Arthur C. Clarke and Stephen Baxter,
in *The Light of Other Days*, 2000

The Fragility of Human Populations on Earth

Life on Earth for all species of plants and animals has been perilous and fragile from the very beginning of its existence around 1 billion years ago. The geological record reveals that natural disasters, including asteroid impacts, massive lava eruptions, and associated firestorms and atmospheric gaseous and particulate compositional changes, have significantly impacted plants and animals.

The oldest hominins are thought to have emerged as early as 7 million years ago and the earliest species of the *Homo* genus around 2 to 1.5 million years ago (Kaneda and Haub 2022). Although there does not appear to be complete consensus on the lineage relationship, status, and timing of all documented subspecies of *Homo* during the last 1 to 2 million years or so, there were perhaps up to four recognizable entities occupying the Earth at roughly the same time (although in different locations) during this period: *H. sapiens*, *H. neanderthals*, *H. denisovans*, and *H. heidelbergensis*. Several other subspecies appear to have died out earlier: *H. erectus*, *H. habilis*, *H. floresiensis*, *H. antecessor*, *H. rudolfensis*, and *H. ergaster*.

Homo sapiens ("modern humans") probably appeared around 300,000 years ago in Africa, though the exact location has long been debated. Diverse groups are thought to have lived in different locations across Africa for the first two-thirds of human history before migrating into Europe, Asia, and the rest of the world. It is likely that these early human populations grew or declined significantly in response to food and water availability, periods of peace or hostility among themselves and other species of *Homo* (most likely *neanderthalis*), and changing weather and climatic conditions.

When modern humans began moving into Europe about 45,000 years ago, *Homo neanderthalis* had been there for at least 100,000 years, but by 35,000 years ago the Neanderthals were extinct and *Homo sapiens* had the world largely to itself. It has been suggested (Balter 2012) that while initially small in number, *Homo sapiens* came to outnumber the Neanderthals by a factor of ten to one, allowing them to outcompete their physically larger rivals. During an extended period of coexistence, *sapiens* and *neanderthalis* mated, leaving the genomes of modern humans with 1 to 8 percent *neanderthalis* DNA.

About 25,000 years ago, human population numbers, at least outside of Africa, began to decline, as glaciers associated with a new ice age began to cover much of the northern part of Europe. When the ice age finally ended about 15,000 years ago, the population began to climb again, setting the stage for a major turning point in human evolution and the eventual emergence of advanced technological capability that has both enhanced and threatened the survival of the species.

The notion that humans or other organisms can become extinct gained scientific acceptance during the Age of Enlightenment in the 17th and 18th centuries. The groundbreaking comprehensive research undertaken by Charles Darwin for *On the Origin of Species* suggested that the extinction of species was a natural process and a core component of natural selection, although Darwin was skeptical of sudden extinctions, viewing these as more gradual processes.

In the 20th century, several existential threats to human survival emerged, including nuclear and biological weapons; increasing population with associated excessive resource depletion; a rise in global pandemics due to inevitable increased physical contact between more numerous humans; global warming and its associated changes in weather patterns and potential impact on human health, food, and freshwater availability; and the potential for runaway artificial intelligence and nanotechnology capability leading to, perhaps, the eventual displacement ("extinction") of naturally evolved humans through the emergence of a new species of *Homo* produced by genetic engineering or technological augmentation (see chapter 13).

On top of these "home-grown" anthropogenic threats, natural external risks to human population numbers (if not extinction per se) include supervolcano eruption, asteroid impact, nearby supernova or gamma-ray bursts, extreme solar flares, or much less likely, invasion by an alien civilization. In any event, the Earth will inevitably become uninhabitable and will eventually be destroyed by the increased size and luminosity of the Sun as it evolves naturally through its red giant phase.[1] This fate can only be avoided if humanity has already established a platform on which to live in another location in the outer reaches of, or actually outside, the Solar System.

Most experts opine that anthropogenic risks are much more likely than "natural" risks since the likelihood of many of the natural risks can be roughly estimated from the geological record and they have not yet resulted in human extinction throughout the 200,000-year history of the species. However, most of the anthropogenic risks have only emerged during the past 80 or so years, and they are dependent on a wide range of unpredictable social, political, and technological development scenarios. As a consequence, the jury is still out on their long-term impact.

We don't know exactly when humans might be snuffed out by any of these events, but a massive wave of deadly radiation from a nearby star "going supernova" could be a millisecond away from hitting us as you read this word. Alternatively, it could be that we have another 4 billion years left to solve our problems before the Earth and whatever then are its life passengers are engulfed by the inevitable expansion of our Sun. In between these two extremes in time is the plethora of other scenarios described above, any of which, alone or in combination, could wipe out all or a large part of life on Earth, or set back its development for thousands of years.

There have been events of these kind (some documented and probably others that we don't know about) that have already extinguished significant components of life on Earth. As a result, over 98 percent of all currently documented species from the fossil record are now extinct. Indeed, mass extinctions are relatively common throughout the Earth's history, and some of them may have played an important, some would say critical, part in the evolutionary pathway of life as we currently know it, including ourselves.

Based on the preserved record in geologic sediments, the background rate of extinctions on Earth is between two and five taxonomic families of marine invertebrates and vertebrates every million years, and about one species per every one million species per year (Lawton and May 1995). Most of these extinctions have been the result of natural evolutionary—"survival of the fittest"—trends. However, since life began on Earth, several major mass extinctions have occurred outside normal evolutionary processes that have significantly exceeded the background rate.

In the last 500 million years, five mass extinction events (Raup and Sepkoski 1982; Ritchie 2022) have resulted in the demise of more than 50 percent of the animal species then present on Earth over relatively short periods of time (figure 2.1). As many as 20 other events with lower impact, but well above the background extinction rate, have also been documented. The most recent of the five major events occurred around the boundary between the Cretaceous–Paleogene geological systems or periods,[2] approximately 66 million years ago.

Many researchers have posited that we have now entered a sixth period of mass extinction, resulting specifically from human activity (McCallum 2015). This outcome is being driven ultimately by growth of the global human population (quadrupling in the 20th century) and the associated increased consumption of natural resources, removal of natural habitats, changes in atmospheric composition, and accumulation of non-biodegradable waste.

At the end of the last ice age, 10,000 years ago, many North American animals went extinct, including mammoths, mastodons, and glyptodonts. While climate changes were a factor, paleontologists have evidence that overhunting by humans was also to blame. Indeed, the disappearance of macrofauna in every continent, including Australia, has been associated

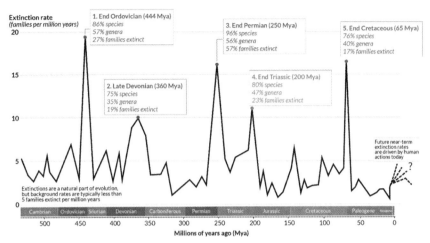

Figure 2.1. A timeline showing the extinction rates (families per million years) for life on Earth since the Cambrian period, 500 million years ago. Annotations describe the five largest extinctions over this period, defined as the loss of at least 75 percent of species within a period of around 2 million years. There is some evidence that ongoing massive human impact on the environment may be driving a future sixth extinction. (*Our World in Data*, Ritchie [2022]. Public domain, licensed under CC-BY by Hannah Ritchie.)

directly with the arrival of *Homo sapiens* on that continent (Dariment et al. 2015; Dembitzer et al. 2022). Early humans worked cooperatively to trap and slaughter large animals in pits, and the adoption of widespread farming and the creation of large towns and cities removed or drastically altered native habitats, with concomitant negative impacts on other animals and species.

The most devastating of the five major extinction events in figure 2.1 occurred over a period of up to 60,000 years around 250 million years ago during the time of the Permian and Triassic systems.[3] The precise causes of this extinction (the "Great Dying") remain unknown, but the main driver is generally considered to be the flood basalt volcanic eruptions that created the Siberian Traps. These eruptions released massive amounts of sulphur dioxide and carbon dioxide, both from the eruptions themselves and from the associated wildfires and burning of hydrocarbon deposits. The CO_2 content of the atmosphere is estimated to have risen from about 400 ppm to 2,500 ppm, with elevation of global temperatures of around 8°C and acidification of the oceans. There may have also been a contribution from an asteroid impact at about the same time, creating the 40-kilometer-wide Araguainha crater (impactor diameter of 2 to 3 kilometers) in Brazil and consequent release of huge amounts of methane, destruction of the ozone layer, and increase in harmful solar radiation.

Most of the world was rendered uninhabitable due to the immediate effects of the eruption and impact event, and around 96 percent of all species on Earth were removed. Marine creatures were particularly badly affected, and insects suffered the only mass extinction of their history. Incidentally, this latter point casts doubt on the claim that insects will be the only living organisms on Earth resilient and adaptable enough to survive in perpetuity.

All current species on Earth are derived from the 4 percent that remained after this event, so the prospect of humanity being destroyed by a future catastrophe may not be such a remote possibility. However, it appears that mass extinctions of this magnitude require the cooperative action of multiple, largely simultaneous causes and that the effect of individual events acting in isolation, while having severe regional impacts, may not always have global consequences.

Other potential extinction scenarios that might precipitate, individually or in combination, the demise of humanity (along with everything else) or that will at least significantly disable human civilization for long periods is provided in the following two lists:

1. Extinction scenarios of our own making (i.e., anthropogenic) that could all occur within the next century or so in isolation or in combination, but which we have a chance of limiting if we put our collective effort toward this end:

    - accidental or deliberate nuclear or chemical holocaust
    - global disease pandemics and/or weaponized microbes
    - genocide
    - artificial intelligence
    - self-replicating nanorobots consuming all resources on Earth ("grey goo")

- uncontained population growth and/or global-scale pollution
- massive, runaway anthropogenic climate change

2. "Natural" extinction scenarios over which we have no or very little control, presented in order of potential increasing time to incidence:
   - catastrophic impact of a large comet or asteroid (anytime, perhaps soon)
   - supernova explosion or massive gamma-ray burst[4] from a nearby star (anytime, perhaps soon)
   - explosion and mass flooding of basalt from a supervolcano on Earth (potentially in hundreds or thousands of years)
   - changes in the reflectivity of the Earth's atmosphere and surface (impacted by ongoing climate change and associated ice and cloud cover changes)
   - Milankovitch cycles (40,000 years to complete a full cycle)
   - plate tectonics (tens or hundreds of millions of years)
   - natural changes in the energy emitted by the Sun (gradual, due to aging of the Sun)
   - collision or near-collision with a star or other celestial body within our own galaxy (very unlikely and dependent on changes in other parts of the Milky Way)
   - Earth rendered unlivable by the natural expansion and increased luminescence of the Sun (in about 1.3 billion years)
   - the chaos produced by a merger with a neighboring galaxy (e.g., the Andromeda galaxy M31 in around 5 billion years)

As documented in the sections below, some of these scenarios have impacted Earth already (e.g., at least five mass extinctions arising from glaciation, volcanism, and/or major asteroid impacts), and we have come dangerously close (and may still be close) to annihilation or disablement by nuclear warfare (e.g., the 1962 Cuban missile crisis between the United States and the then–Soviet Union and the contemporary Ukraine-Russia war) and plagues (e.g., the "Black Death" bubonic plague in the mid-14th century). Still others are gradually building momentum (e.g., global warming and overpopulation).

It is imperative that humanity seeks to collectively and seriously do something to reverse and/or ameliorate the avoidable scenarios in the not-too-distant future or it could be too late to ensure its survival over the next few hundred years. Failure to do this could be our own manifestation of the "Great Filter" discussed in chapter 1 that has provided an answer to the Fermi paradox by suggesting that civilizations find a way of wiping themselves out soon after reaching a certain stage of technological development.

The following discussion deals with each of these extinction scenarios in detail.

EXTINCTION SCENARIOS OF OUR OWN MAKING
Accidental or Deliberate Nuclear Holocaust

Under this scenario, a significant part of the Earth is made uninhabitable due to the immediate and/or long-term effects of nuclear explosion fallout, the loss of most of modern technology due to damage from electromagnetic pulses, and the initiation of a long-term "nuclear winter"[5] (Toon et al. 2008). While the accuracy of computer models of such firestorms, especially in relation to the severity, longevity, and impact of the resulting climatic effects, has been disputed (Martin 1988), most recent models continue to suggest that some deleterious global cooling would result, assuming that many of the fires occurred in the spring or summer.

Notwithstanding these reservations, the prospect of the impacts of a nuclear holocaust has been a real concern since the development and deployment of the first atomic and hydrogen bombs in the early 1940s. Devastation to local life systems and food chains can result from even a single nuclear blast such as occurred at each of the cities of Hiroshima and Nagasaki in Japan in 1945, and from the accidental escape of radioactive solids, liquids, and gases from a nuclear power plant, as at Chernobyl in Russia in 1986 and at Fukagima in Japan in 2012. As mentioned above, the world came very close to nuclear war during the Cold War and particularly at the time of the Cuban missile crisis of 1962.

The so-called Doomsday Clock is a symbolic representation through the analogy of a clock counting down to midnight of how close the

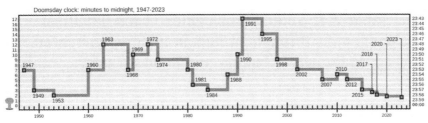

Figure 2.2. Doomsday Clock graph, simulating the estimated time to a global catastrophe, represented as occurring at midnight (the lower axis) on a clock with minutes running down on the vertical axis from 17 to 0 minutes for the years 1947 to the present. (*Wikimedia Commons.*)

world has been to global disaster (figure 2.2). It was initiated in 1947 and is maintained by the Science and Security Board of the *Bulletin of the Atomic Scientists*. It was originally designed to depict the threat of nuclear war,[6] but since 2007 it has also reflected other negative factors with a global impact on humanity. The clock was moved forward 10 seconds on January 24, 2023, to be 90 seconds from midnight (its nearest-ever location to date), due largely, but not exclusively, to the Russian invasion of Ukraine in 2022 and the increased risk of nuclear escalation stemming from that conflict.

Although the threat of a mutual-nuclear-destruction scenario itself has diminished over the 30 years or so since 1991, the clock has reflected a consistent increase in risk of global disaster since then because the Cold War nuclear threat has been replaced by the possibility of nuclear terrorism, nuclear proliferation into the hands of "rogue" states, and the inclusion of the effects of climate change, exacerbated by the inability of the governments of all of the major CO_2 emitters to reduce their emissions either individually or collaboratively.

Global Disease Pandemic

A pandemic is an epidemic of infectious disease that has spread through human populations across a large region, multiple continents, or even worldwide. Throughout history there have been several major pandemics, listed in table 2.1 in order of the overall number of human deaths. Recent major pandemics include the human immunodeficiency virus (HIV) from 1981, the 1918 and 2009 swine flu/influenza (H1N1) pandemics,

Table 2.1. Selected examples of the deadliest pandemics throughout human history, compared to the worst famine and flood disasters.

Pandemic or Natural Disaster	Estimated Number of Deaths	Impacted Region	Time Period (Modern Era)	Pathogen
Black Death	75M	Europe, Asia, North Africa	1346–1353	*Yersinia pestis* bacterium
Spanish flu	50M+	Worldwide	1918–1920	Influenza A virus H1N1
Plague of Justinian	30–50M	Europe, West Asia	AD 541–549	*Yersinia pestis* bacterium
HIV/AIDS	40.1M	Worldwide	1981–present	HIV-AIDS
COVID-19	7–29M	Worldwide	2020–2023	SARS-CoV-2
Third plague	12–15M	Worldwide	1855–1960	Bubonic plague
Famine	11–40M	China	1959–1961	—
Famine	25M	China	1906–1907	—
Flood	400K–4M	China	1931	—

Source: https://en.wikipedia.org/wiki/List_of_natural_disasters_by_death_toll

the short outbreak of severe acute respiratory syndrome (SARS) in 2003, and the related coronavirus disease (COVID-19) in 2020–2022 caused by the SARS-CoV-2 virus.

The so-called Black Death (or Bubonic Plague) started in the 14th century, although a previous outbreak of a closely related pandemic (the Plague of Justinian; see below) had occurred some 800 years earlier. The total number of deaths worldwide is estimated to have been up to 75 million people. Starting in Asia, the disease reached the Mediterranean and western Europe in 1348 and killed an estimated 20 to 30 million Europeans in six years, a third of the total population, and up to a half in the worst-affected urban areas. It was the first of a cycle of more than 100 European plague epidemics that continued until the 18th century. By the 1370s, England's population was reduced by 50 percent, culminating in the Great Plague of London in 1665–1666 that killed around 100,000 people, 20 percent of London's population.

The influenza pandemic from January 1918 to December 1920 was caused by the H1N1 virus and was unusually deadly (it came again in 2009). It was nicknamed "Spanish flu" not because it originated there, but because neutral Spain released illness and mortality figures whereas the then-warring nations of Britain, France, Germany, and the United States in the First World War imposed censorship on the reporting of its effects to maintain morale. The virus infected 500 million people across the world, including in remote Pacific islands and the Arctic, and probably ended up killing more than 50 million people, or some 3 to 5 percent of the world's population at the time, making it one of the deadliest natural disasters in human history.

The plague of Justinian, from AD 541 to 549, was the first recorded outbreak of the bubonic plague (arising from the *Yersinia pestis* bacterium). It started in Egypt and reached Constantinople the following spring, killing (according to the Byzantine chronicler Procopius) 10,000 a day at its height, and perhaps up to 40 percent of the city's inhabitants. Recurrences of the plague over the next two centuries eliminated one-quarter to one-half of the human population in the known world and caused Europe's population to drop by around 50 percent between AD 550 and 700.

Over the past century or so, there has been a dramatic increase in contact between humans and between humans and other creatures due to increased urbanization (now greater than 50 percent) and vastly increased travel and intercontinental food and goods transport. This has led to a significant increase in the potential for a global disease pandemic, despite major advances in medicine and health care.

Possible Future Candidates for Pandemics
Viruses causing viral hemorrhagic fever such as Lassa fever virus, Rift Valley fever, Marburg virus, Ebola virus, and Bolivian hemorrhagic fever are

highly contagious and deadly diseases, with the potential to become pandemics. Their ability to spread efficiently enough to cause a pandemic is limited, however, as transmission requires close contact with the infected vector, the vector only has a short time to reproduce before the death or serious illness of the host, and medical professionals can prevent patients from carrying the pathogen elsewhere.

Antibiotic-resistant microorganisms, or "superbugs," may contribute to the reemergence of diseases that are currently well controlled. For example, cases of tuberculosis that are resistant to traditionally effective treatments remain a cause of great concern to health professionals. Every year, nearly half a million new cases of multidrug-resistant tuberculosis (MDR-TB) are estimated to occur worldwide. The World Health Organization reports that approximately 50 million people worldwide are infected with MDR-TB, with 79 percent of those cases resistant to three or more antibiotics.

In recent years, common bacteria have developed resistance to various antibiotics, and to entire classes of antibiotics. These organisms have become an important cause of healthcare-associated infections, that is, ones developed in hospitals after or during treatment for other ailments and/or surgery. Inappropriate antibiotic treatment and overuse of antibiotics have been an element in the emergence of these resistant bacteria. The problem is further exacerbated by the nontherapeutic use of antibiotics as growth promoters in animal meat and plant production (Tang et al. 2017).

While not necessarily affecting significant numbers of other species on Earth, humans could well be wiped out if one of these pandemics took off in a big way.

Genocide

The term "genocide" was coined by Raphael Lemkin (1900–1959), a Polish Jewish legal scholar, in 1943 and is defined by the United Nations Convention on the Prevention and Punishment of the Crime of Genocide (CPPCG) Article II (1951) as:

> *Any of the following acts committed with intent to destroy, in whole or in part, a national, ethnic, racial or religious group, as such: killing members of the group; causing serious bodily or mental harm to mem-*

bers of the group; deliberately inflicting on the group conditions of life calculated to bring about its physical destruction in whole or in part; imposing measures intended to prevent births within the group; and forcibly transferring children of the group to another group.[7]

Despite being perhaps the world's only universal taboo, genocide appears to be a regular and widespread event in human history. Justifications for past instances include claiming a right to preserve the well-being/existence of a group, claiming that others are inferior, difficulties in assimilating with the majority population, contributing nothing of value, and/or threatening the smooth functioning of society.

Examples have been variously quoted as including the disappearance of the Neanderthals around 30,000 years ago; biblical genocide (Canaanites, Babylon); atrocities perpetrated in the Roman Empire (Carthage), the Americas (Beothuk in Canada, Native Americans in the United States), the African Congo and Rwanda (Tutsis and moderate Hutus), and Germany (Jews, Roma); the Armenian genocide by Turkey in the Middle East; and the disappearance of the Tasmanian Aborigine.

Some consider that the expansion of the British and the Spanish Empires, and the subsequent establishment of settlements[8] on indigenous territory, frequently involved acts of genocidal violence against indigenous groups in the Americas, Australia, Africa, and Asia (Adhikari 2021).

Not taking anything away from the seriousness of these examples of genocide (perhaps except for the Neanderthals, where the cause of their demise, often attributed to *Homo sapiens*, remains uncertain), they relate to relatively small subpopulations of humanity, rather than to the elimination of the human species as a whole. Nevertheless, the prospect remains that large sections of humanity could meet their end through rampant acts of genocide that have massive knock-on acts of vengeance and retribution. The chances of this happening may be amplified in the future through desperate actions by individual groups and/or nations to acquire scarce resources (food, freshwater, etc.) as population increases put further pressure on these resources and their "fair" distribution.

It is also possible that an extraterrestrial species at a certain level of technological capability might seek to destroy another intelligence that it encounters out of expansionist motives, paranoia, or simple aggression. In Harrison (1981) it is argued that such behavior would be an act of prudence since an intelligent species that has overcome its own self-destructive tendencies might view any other species bent on galactic expansion as a kind of virus. It has also been suggested that a successful interstellar alien species would, by definition, be a super predator, as is *Homo sapiens* in its domain on Earth.

Artificial Intelligence (AI)

As reported in the *Wall Street Journal* in September 2023 (Schechner and Seetharaman 2023), despite the recent attention that AI's existential risk ("x-risk") has been getting, until now this was confined to a fringe of philosophers and AI researchers. That all changed after OpenAI released ChatGP freely to the general population in late 2022. Based on learned statistical occurrences (including "Markov chains"), this so-called generative AI has the capability of generating text, images, or other media, using learned patterns and structures in its input training data and then generating new data that has similar characteristics. It delivers humanlike responses that have precipitated warnings that such systems could soon gain superhuman intelligence.

Prominent researchers, including Geoffrey Hinton, considered one of the godfathers of AI at Alphabet Google, have contended that generative AI "contains a glimmer of humanlike reasoning." In a talk given at a Collision Conference in Toronto in 2023 (Siddiqui 2023), Hinton outlined six potential risks posed by the development of current AI models: bias and discrimination, unemployment, online echo chambers, fake news, "battle robots," and existential risks to humanity. David Krueger, a machine learning professor at Cambridge University, helped organize a statement in May 2023 signed by hundreds of AI experts saying that "extinction risk from AI was on par with the dangers of pandemics and nuclear war" (Krueger 2023).

Some researchers are concerned that AI systems trained to seek rewards could end up with hidden power-seeking urges, inadvertently harming humans while carrying out our wishes, or that they could simply outcompete humans and take control of our destiny. Research in this community aimed at ameliorating this perceived threat focuses largely on (1) "alignment," or how to make sure tomorrow's computer "minds" have goals that are consistent with ours, and (2) being able to understand how AI thinks, or might think, referred to as the "interpretability problem."

Other specialists in AI ethics and fairness are concerned about how the generative AI tools are accidentally or intentionally exploiting workers and deepening inequality for millions of people by using systems that are trained on historical data and can, therefore, entrench past discrimination into future high-stakes decisions.

Chapter 13 delves into this complex scenario not so much from the perspective of human extinction outcomes from out-of-control AI, but in terms of its implications for the evolution and definition of future humanity itself, and whether recent advances in AI can provide a different (inorganic) platform for the preservation of biological human civilization.

Self-Replicating Nanorobots ("Grey Goo")

"Grey goo" is a hypothetical end-of-the-world scenario involving molecular nanotechnology ("tiny machines") in which out-of-control self-replicating robots consume all matter on Earth while building more of themselves, a scenario that has been called "ecophagy," or "the literal consumption of an ecosystem." The term "grey goo" was first used by molecular nanotechnology pioneer Eric Drexler in his book *Engines of Creation* (1986) to emphasize the difference between "superiority" in terms of human values or, alternatively, "competitive success."

The original idea assumed machines would be designed by humans to have ecophagy capability, while popularizations of the concept have suggested that machines might somehow gain this capability by accident. Self-replicating machines of the macroscopic variety were originally described by mathematician John von Neumann and are sometimes referred to as von Neumann machines (see chapter 1).

A simple example in Drexler (1986) proposes such a replicator floating in a bottle of chemicals, making copies of itself every 1,000 seconds. The exponential (power of 2) rate of replication means that after 10 hours there are over 68 billion replicators. In less than 24 hours their weight would be one ton and in less than two days they would outweigh the Earth, and so on, if there were enough chemicals in the bottle to support ongoing replication.

In another hypothetical doomsday scenario, billions of nanobots[9] are released to clean up an oil spill off the coast of Louisiana. However, due to a programming error, the self-replicating nanobots lose their discrimination/preference for oil-based carbon and devour all carbon-based objects, destroying everything on the planet within days.

Bill Joy, one of the founders of Sun Microsystems, discussed some of the challenges with pursuing this technology in his now-famous article in *Wired* magazine, titled "Why the Future Doesn't Need Us" (Joy 2000). In direct response to Joy's concerns, the first quantitative technical analysis of the ecophagy scenario has been published in Freitas and Merkel (2004). This work suggests that the maximum rate of global ecophagy by self-replicating nanorobots is fundamentally restricted by (1) the replicative strategy employed, (2) the maximum dispersal velocity of the mobile replicators, (3) the operational energy and chemical element requirements, (4) the resistance of biological ecologies, (5) thermal pollution limits, and (6) most importantly, the determination of humans to stop them.

In time, we will see if these proposed (hopeful) limitations are enough to prevent this extinction scenario from coming to pass, by unfortunate accident or design.

Population Explosion and/or Global-Scale Pollution

A related scenario to "grey goo" is one in which humans themselves gradually overpopulate the Earth to such an extent that resources essential for the maintenance of life—namely, clean water and air, food, and materials for shelter—are no longer available in sufficient quantities to support the population and/or are unevenly distributed to such an extent that this causes massive destructive conflict between subpopulations.

The extent to which humans impact the earth's biosphere can be measured with a so-called ecological footprint analysis. This type of analysis assesses the area of biologically active land and ocean required to provide the resources a population consumes and to absorb the corresponding waste. The footprint values are often stated in terms of the "total number of Earths" needed to sustain the world's population at that level of consumption. The consumption of energy, biomass, building material, water, and other resources are converted into a normalized measure of land area ("global hectares per capita," "ecological footprints," or "number of Earths required") and compared to a measure of the Human Development Index (HDI)[10] (figure 2.3).

The figure demonstrates clearly that the number of Earths required to support a population rises steeply with the relevant HDI—in other words, the more "affluent" a society is, the more resources it consumes.

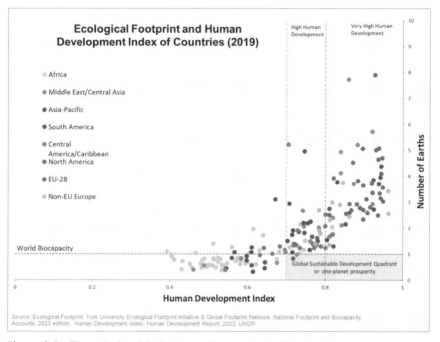

Figure 2.3. The relationship between the ecological footprint per person ("Number of Earths"—vertical axis) and the corresponding national Human Development Index (HDI—horizontal axis), a measure of quality of life, for eight world regions (2019). The regions are designated by dots with different shades of gray in the legend on the left of the figure. Most of the nations with a high HDI lie above the World Biocapacity of "1 Earth," shown as the horizontal dashed line, and only three nations fall within the Global Sustainable Development shaded region at the bottom right. (*Ecological Footprint: York University Ecological Footprint Initiative & Global Footprint Network. National Footprint and Biocapacity Accounts 2023 Edition, Human Development Index: Human Development Report, 2023, UNDP.*)

Very few countries (i.e., those in the shaded portion of the figure on the lower right-hand side) come close to achieving these basic conditions for global sustainable development. This global comparison highlights the inequalities of resource use between nations on Earth at the beginning of the 21st century and can be used to argue that most current lifestyles are not sustainable unless they can be supported by other populations where the ecological footprint is much lower.

The HDI has been criticized on several grounds, including alleged lack of consideration of technological development, positive contributions to human civilization, and overall global (rather than exclusively national) development. However, at the very least, it exposes the inequity between countries and the fact that resource use everywhere is increasing at the expense of the natural environment's ability to support it.

Lin et al. (2021) estimated that the average biologically productive area per person worldwide in 1961 was approximately 3.1 global hectares per capita and the average ecological footprint of each human was 2.3 hectares; in other words, the world was living within its means. However, they estimated that by 2015, the biocapacity and footprint had changed to 1.6 and 2.8, respectively, meaning that humanity was consuming resources 1.7 times faster than the Earth's biocapacity to regenerate these resources.

Prospects for the future look mixed. On the one hand, the world's population has grown continuously from near 370 million since the end of the Great Famine and the Black Death in 1350 to its current level of 7.9 billion. In recent years, the birth rate (fertility) has been declining faster than the reduction in death rate, so the rate of population *increase* has fallen. A 2023 analysis by the United Nations Population Division (figure 2.4) shows that annual world population growth peaked at 2.3 percent per year in 1963 and dropped to 0.9 percent in 2023, equivalent to about 74 million people each year, and that it could drop even further to −0.1 percent by 2100 (Roser and Ritchie 2023).

Even with a (slightly) declining population of around 10 billion after 2100, the previous 75 years or so to that date are likely to have witnessed massively increased demands on energy and other physical and biological resources (and thus a much-increased ecological footprint) as increasing numbers of humans in all countries endeavor to increase living standards

Scenarios for the Demise of Humanity

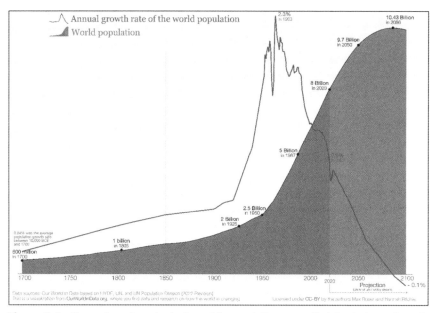

Figure 2.4. Current and projected world population growth (shaded area) and population growth rate (line) from 1700 to 2100. The projections indicate that the world population will reach a maximum of around 10.4 billion in 2086 and then begin to decline. The maximum growth rate of 2.3 percent was reached in 1963. (*Our World in Data, Roser and Ritchie [2023]. Public domain, licensed under CC-BY-SA by Max Roser.*)

and achieve increased longevity. It would not be surprising if this decline in the availability of resources, and especially their inequitable distribution around the globe, leads to significant conflicts between nations.

Climate Change

If the Earth had no atmosphere, its "black body" temperature due to absorption of the Sun's radiation would be −13°C. The fact that the average surface temperature is a much more pleasant +14°C is due to the presence of greenhouse gases (GHGs), primarily water and CO_2, in its atmosphere. Without this warming, the Earth would not have produced life as we know it, and certainly not technologically advanced human life, since most of its water would be permanently frozen. Fortunately, several natural mechanisms exist to control the abundance of the GHGs in the atmosphere, including plate tectonics, volcanism, ice ages, and photosynthesis, and warming due to the greenhouse effect has been relatively stable at non-life-threatening temperatures over most of Earth's history.

Earth's climate system is driven by solar radiation. Annual changes in the level of radiation received at any point on the Earth's surface arise from changes in the distance to the Sun due to the elliptical shape of the Earth's orbit and the tilt of its axis (giving rise to the seasons). These specific seasonal changes in climate result from differences in surface heating taking place between summer and winter that drive storm tracks and pressure centers and changes in cloudiness, precipitation, and wind. Seasonal responses of the biosphere (especially vegetation) and cryosphere (glaciers, sea ice, snowfields) also feed into atmospheric circulation and climate through the sequestering of CO_2 in leaves, etc., and changes in the albedo (reflectivity) of Earth's surface.

Multiyear climate variations, including droughts, floods, and other events, are caused by a complex array of factors and Earth system interactions. The periodic change of atmospheric and oceanic circulation patterns in the tropical Pacific region, collectively known as the El Niño–Southern Oscillation variation and its opposite effect, La Niña, has an important role in these variations.

This relative overall stability during the last 500 million years or so of Earth's 4.5-billion-year history allowed Earth's complex life to emerge and evolve but, nevertheless, the climate has undergone significant changes. Figure 2.5 shows that excursions of average global surface temperature over the past 500 million years have ranged from +14°C to -6°C relative to current temperature.

These changes in global climate have involved the following diverse mechanisms, magnitudes, rates, and consequences:

- During "Snowball Earth" (about 700 million years ago) the Earth was almost completely covered with ice sheets many kilometers thick.
- During some ice ages, sea level was so low (due to ice formation) that people could walk from Indonesia to Tasmania.
- At other times it was so warm that the polar ice caps disappeared, and the sea level was 120 meters higher than today.

Scenarios for the Demise of Humanity

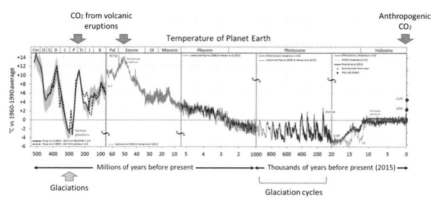

Figure 2.5. Global temperature variation in °C of the Earth over the past 500 million years, relative to the 1960–1990 average. Major volcanic eruptions and anthropomorphic events likely to have produced the rises in temperature through their associated CO_2 emissions are annotated, along with intervening periods of lower temperature associated with known periods of glaciation. Note the five changes in the scale of the horizontal time axis. The small dots on the right-hand axis indicate observations and predictions of global average temperature from the present to the year 2200. (*Glen Fergus, with data sources provided in https://en.wikipedia.org/wiki/File:All _palaeotemps.png#Summary [not including the annotations added by the current author outside the main plot rectangle]. Licensed under the Creative Commons Attribution-Share Alike 3.0 Unported license.*)

- During the last billion years or so, plate tectonics have resulted in a very different distribution of continents and changes in ocean and atmospheric circulation.
- There have been occasional asteroid impacts that have produced years-long "global winters."
- During the last 8,000 years, which includes most of recorded human history, the climate has been relatively stable at the warmer end of this temperature range (right-hand panel of figure 2.5).

These massive (relative to contemporary observed and projected) changes in temperature and climate have been variously and/or collectively triggered by:

- The Sun's changing light output (the luminosity of the Sun has increased by 25 percent since around 3.8 billion years ago).

- Weathering of rocks, which serves to sequester CO_2 from the atmosphere into carbonate minerals.
- Variations in the orbit and tilt of Earth (Milankovitch cycles—see discussion below).
- The configuration of continents and oceans (plate tectonics—see discussion below).
- Large volcanic eruptions (e.g., Krakatoa—CO_2/particulates).
- Evolution and spread of life (bacteria, plants, and animals).
- Land cover changes due to ice/water and vegetation (influencing reflectivity, or "albedo").
- Asteroid impacts ("global winters").

Over the period between 5 and 1 million years ago, these excursions of temperature have been more "modest," with a pattern of low-amplitude climatic variation superimposed on a generally declining average value from +4 to −3°C relative to the 1960–1990 average (figure 2.5). The excursions alternate between warmer ("greenhouse") phases characterized by higher temperatures, higher sea levels, and an absence of continental glaciers and cool ("icehouse") phases marked by lower temperatures, lower sea levels, and the presence of continental ice sheets.

Over the past million years, there have been higher-amplitude variations of between +1°C and −5°C, in which cool periods are embedded within warmer greenhouse phases and warm periods are embedded within icehouse phases. After a period of relatively stable temperatures around the 1960–1990 average over the past 11,000 years, the temperature has been rising over the past 200 or so years due to ongoing anthropogenic CO_2 emissions. This rise is projected to increase significantly over the next 120 years (shown in the far right of figure 2.5).

The evolution of human life has been strongly influenced by these changes. Recent research on human genomes (Hu et al. 2023) showed that our human ancestors went through a severe population bottleneck with only about 1,300 breeding individuals between around 930,000 and 810,000 years ago. The bottleneck lasted for about 120,000 years and brought humans close to extinction. The population decline is consistent with a substantial chronological gap in the available fossil record and with severe cooling that resulted in the emergence of glaciers, a drop in ocean surface temperatures, and perhaps long droughts in Africa and Eurasia.

Life itself (especially plant life) is implicated as a causative agent of some of these climate changes, as the processes of plant photosynthesis and respiration have largely shaped the chemistry of Earth's atmosphere, oceans, and sediments, especially during its early history. As the population of humans increases, we are starting to witness our own impact on the environment increasing in a major way, and on several fronts.

Since the beginning of the Industrial Revolution around 1750, the global average temperature has increased by 1.8°C, largely ascribed to the impact of significant increases in CO_2 and CH_4 emissions due to contributions from increased human numbers and industrial activity (so-called anthropogenic inputs), arising primarily from the burning of fossil fuels. A long-term warming trend started shortly after 1900 (figure 2.6), but with

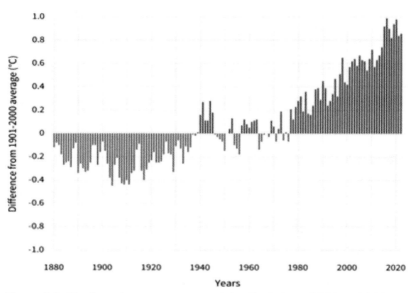

Figure 2.6. Yearly surface temperature of the Earth from 1880 to 2022, compared to the 1901–2000 average (horizontal line). The relatively consistent rises in annual surface temperature from 1910 to 1940 and from 1970 to 2020 are very clear. (*Public domain from NOAA Climate.gov, based on data from the National Centers for Environmental Information.*)

many short-term fluctuations and a significant leveling-off from 1940 to 1975 and another period of leveling starting in 2000. The net effect is around a 1.4°C rise in temperature over this 120-year period.

Unless this trend can be halted (or reversed) over the next century or so, the Earth will experience average temperatures up to 5°C higher than it has experienced at any time in the past 5 million years.

Impact on Humanity

Projections of future increases in average global temperature vary greatly, depending on the effectiveness of existing and proposed CO_2 and CH_4 emissions reduction processes. Rises even as low as 2°C above late-nineteenth-century levels are projected to be accompanied by a sea level rise of around 0.5 meter, ocean acidification, loss of polar and continental ice sheets, changes in rainfall patterns and natural habitats, and probably increased extreme weather events (Guivarch et al. 2022).

On time scales of centuries and millennia, previous populations of humans have experienced climate changes that (together with other events) have often had profound effects on their location, culture, and society. Such changes are believed to have impacted the initial cultivation and domestication of crop plants, as well as the domestication and pastoralization of animals. Human societies have mostly adapted in response to climate variations, although there is evidence that certain ancestral societies and civilizations have collapsed, at least in part, due to associated rapid and severe climatic change; to wit, the Pueblo Indians who lived in Utah, Arizona, New Mexico, and Colorado, from the 8th to the 12th centuries. The Puebloans left their very sophisticated infrastructure and cliff communities in the 12th and 13th centuries due to a complex mixture of causes, including regional climate change, environmental degradation (topsoil erosion or deforestation), hostility from new arrivals, religious or cultural change, and influence from Mesoamerican cultures (Diamond 2005).

The capacity of humans in the 21st century to adapt effectively to rapid changes in climate is much more limited than these earlier societies because (1) their numbers are vastly higher, (2) their societies are much more compact (urbanized), (3) their ability to relocate to more favorable parts is more limited because of the impenetrability of national borders, and (4) they are more dependent on complex and interdependent food and water supply chains/systems. Furthermore, the climate changes in recent times appear to be taking place on much shorter time frames, except for those occasions in the past when the temperature changes resulted from an asteroid impact or massive volcanism. Thus, while earlier humans in dispersed, smaller groups were probably able to adapt to the more gradual changes in climate at that time by incremental migration, modern societies are not so fortunate.

Contemporary climate "alarmists," and even large and respected institutions like the United Nations, are suggesting that if actions to eliminate CO_2 emissions altogether (and ultimately to begin to extract the CO_2 that is already in the atmosphere) are not accelerated significantly, by 2100 the average global surface temperature will have increased by 4 to 6°C above the 1901–1960 average temperature (Wuebbles et al. 2017). While a temperature rise of this order is less than that which life on Earth has survived in the distant past (figure 2.5), the contemporary conventional wisdom is that unless effective action is taken, humanity is doomed—for all the reasons outlined in the previous paragraph.

Runaway Climate Change

This is a scenario in which a planet's climate system passes a threshold, or tipping point, after which internal positive feedback effects cause the climate change to accelerate. An example of this playing out on Earth is that when the temperature of the ocean rises due to the GHG effect of increased atmospheric CO_2, the solubility of this gas in the ocean decreases (as it does for all gases), thereby expelling further CO_2 into the atmosphere, causing a further temperature rise, and so on.

Venus is a relevant case in point within the Solar System. This planet has only a slightly smaller diameter than Earth, and it orbits at about two-thirds of Earth's distance from the Sun. Its atmosphere is composed of 96.5 percent supercritical CO_2,[11] with the remainder being mostly nitrogen. The surface temperature is 467°C, and the pressure is 93 bar, about the pressure found 900 meters underwater on Earth.

It is thought that early Venus may have had a global ocean, possibly derived from the original rocks that formed the planet or delivered later from comets, although this has been disputed, based on the unlikely ability of volcanoes to replenish water at the rate at which it was being lost to space due to the high surface temperature. Even if oceans had existed in the past, at some point in the evolution of Venus, a runaway greenhouse effect is believed to have occurred, leading to the current CO_2-dominated atmosphere, probably about 4 billion years ago. The process began with an increase in the brightness of the early Sun, leading to an increase in the amount of water vapor in the atmosphere. This produced an increase in temperature (with the water acting as a GHG in the same way as CO_2) and consequently an increase in the evaporation of the ocean (a "runaway effect"), leading eventually to the situation in which the oceans boiled and all the water vapor entered the atmosphere. As explained in chapter 1, the surface temperature continued to rise to nearly 500°C before equilibrium was reached between incoming and outgoing radiation.

Venus has not recovered from this event, and the danger is that Earth might be headed for the same runaway fate if emission abatement processes are not successful, and the CO_2 (and subsequently H_2O) content of the atmosphere continues to increase. On the other hand, Goldblatt and Watson (2021) have suggested that the addition of anthropogenic greenhouse gases to the atmosphere is unlikely, by itself, to trigger a full runaway greenhouse on Earth.[12]

Although most of the attention of climate change protagonists is on CO_2, this gas contributes (only) between 9 and 26 percent of the greenhouse warming on Earth, second to water vapor which, although less intense as a GHG, has a much higher abundance in the atmosphere than CO_2 (0.1 percent vs. 0.004 percent) and contributes 66 to 85 percent of the GHG effect. Methane gas (CH_4) is even more potent but has a much lower abundance than CO_2. Unlike CO_2 and CH_4, increased evaporation of H_2O at the higher temperatures associated with climate change is a greenhouse mechanism over which humanity has virtually no leverage (Held and Soden 2000).

The historical abundance of CO_2 in Earth's atmosphere is provided in figure 2.7, along with some of the projected increases up to the year 2500 that would arise under four models of anthropogenic emission abatement (RCPs, or Representative Concentration Pathways), together with the "Wink12K" line that shows the effect of burning all of the Earth's avail-

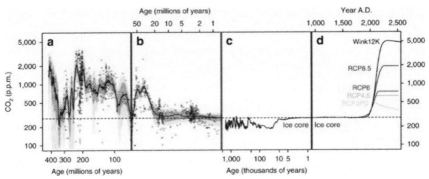

Figure 2.7. CO_2 content (ppm, log scale) of the Earth's atmosphere over the past 400 million years. Note the four changes in the scale of the horizontal time axis. Parts (a) (b) are proxy-based atmospheric CO_2 on a log timescale determined with best-fit LOESS (loosely compacted windblown sediment) data and associated uncertainty envelope, (c) is ice core–based atmospheric CO_2 on a log timescale, and (d) is atmospheric CO_2 on a linear timescale in years AD, taken from ice cores and the observation record together with several future emission scenarios. The near-vertical lines in the extreme right-hand panel indicate predictions of global CO_2 content up to the year 2500 under four RCP8 (Representative Concentration Pathway) models and the Wink12K scenario that reflects the case in which all of Earth's fossil fuels are burned. (*Open access publication by Foster et al. [2017], licensed under a Creative Commons Attribution 4.0 International License.*)

able fossil fuel reserves (if this was possible). This figure demonstrates that CO_2 abundance in the atmosphere has varied over the range of 200 to nearly 3,000 ppm over the last 400 million years. Indeed, the worst-case RCP scenario of future CO_2 abundance (in around 2100) is less than the levels recorded around 400 and 200 million years ago.

The level of CO_2 in the atmosphere is determined by the balance between sequestering of this gas in sediments (mainly carbonates), dissolution in the oceans and incorporation into the biosphere, and its emission through decomposition of organic matter and volcanic activity. As documented above, the recent relatively rapid rise in CO_2 has been ascribed to the increased burning of fossil fuels from the industrialization of human civilization over the past two centuries.

In terms of the impacts of *decreasing* temperature, recent work by Margari et al. (2023) suggests that an ancient European population of

Homo erectus may have been wiped out by an "extreme cooling event" about a million years ago. The previously undocumented temperature downturn coincides with the disappearance of fossils and stone tools during the period from 1.1 to 0.9 million years ago, following the arrival of *Homo erectus* in Europe from Africa between 1.8 and 1.4 million years ago.

Several other factors have caused, and are causing, changes in Earth's climate, over which humans can have no impact, including plate tectonics, Milankovitch cycles, and volcanic activity. These are discussed in the sections below, dealing with potential extinction mechanisms that are beyond our control.

Summary

- The Earth has experienced many periods in the geological past when the temperature was much warmer (+14°C) and colder (-6°C) than the present.
- Most of the big changes in Earth's temperature have been caused by Milankovitch cycles and are associated with glaciations.
- Very large volcanic eruptions occurred over short time periods and may have acted in concert with other influences (e.g., firestorms) to change climate locally and to put pressure on certain species at the time.
- Plate tectonics has played a significant role in changing global climate through modification of ocean currents, wind patterns, CO_2 content of the atmosphere, and (possibly) the nutrient content of the oceans.
- CO_2 in the atmosphere has increased by around 50 percent (from 280 to 417 ppm) since the Industrial Age began in the mid-18th century, but over the past 500 million years it has ranged between eight times higher and half as high (figure 2.7) and living things still survived and thrived.
- Water vapor in the atmosphere causes between 36 and 85 percent of the GHG effect on Earth, with CO_2 only 9 to 26 percent.

- Natural emissions of CO_2 and CH_4 may be similar to anthropogenic emission levels when the full effects of volcanoes, submarine basalts, faults, etc., are considered (Plimer 2021).
- With more CO_2 in the atmosphere, the biosphere mass increases.
- Changes are happening more rapidly now than in the past (except for unforeseen asteroid impacts and volcanic eruptions!), so we and other species (some essential to our well-being) haven't got a lot of time to mitigate, adapt, or evolve in response.
- There are billions more human beings than there were previously on Earth, many living near the sea or in marginal land, with little ability to move (unlike the days of much less numerous and largely nomadic hunter-gatherer societies).
- Finally, our species is now constrained by national boundaries and is much more dependent on complex, interconnected supply chains for food, water, and everything else than it has ever been in the past. This also limits significantly options for dealing with climate change impacts.

"Natural" Extinction Scenarios over Which Humans Have No, or Very Little, Control

Impact of a Large Comet or Asteroid

Despite the benign appearance of the night sky, the Solar System is a very active place. Most readers will have seen the occasional meteor or perhaps a meteor shower or "storm" such as the Leonids (returning to Earth's vicinity every 33 years or so) light up the sky at night for several seconds or minutes, or a bright comet do the same for weeks or months. These events demonstrate that by no means all the small and medium-size objects in the Solar System have been gravitationally captured by its planets, dwarf planets, and asteroids.

It is now widely accepted that collisions of asteroids and comets with the Earth in the past have played a significant role in shaping the geological and biological history of the planet (see the discussion above). Bodies as small as 20 meters in diameter can cause significant damage to the local environment and human populations (Rumpf et al. 2017). This can occur through the creation of large craters and the "splashing" of debris far and wide if they impact on land or massive tsunamis if they impact the sea.

On the morning of February 15, 2013, an asteroid the size of a semi-trailer (20 meters) came from the direction of the rising sun and exploded in a fireball over the city of Chelyabinsk, Russia (Specktor 2023). It disintegrated around 22 kilometers above the ground with 30 times more energy than the bomb that destroyed Hiroshima, briefly glowing brighter than the Sun. The blast shattered windows on more than 7,000 buildings, temporarily blinded some individuals outside, and inflicted ultraviolet burns and otherwise injured more than 1,600 people. This asteroid is thought to be the biggest natural space object to enter Earth's atmosphere in more than 100 years, but no one saw it coming because it arrived from the direction of the Sun, hidden in our biggest blind spot, until it was too late. Who knows what else lurks in unknown orbits around the Sun?

Even now, occasionally, a massive chunk of solid material left over from the formation of the Solar System impacts on a planet with dramatic and devastating effect. The comet Shoemaker-Levy 9, which had been captured by the gravitational pull of Jupiter some time at the beginning of the 20th century, got too close to Jupiter on one flyby in 1992 and broke up into around 20 large chunks, each a few kilometers in size. In July 1994 they fell onto the planet, exploding in the upper atmosphere with the energy of many millions of tons of TNT.[13] The resultant discoloration of the clouds of Jupiter associated with several of these chunks of material were roughly the size of the Earth (figure 2.8).

This figure provides irrefutable evidence that the Solar System remains a very active and dangerous place in contemporary times, even 4.5 billion years after its fiery formation. Had it collided with the Earth, the Shoemaker-Levy comet impact would have been devastating to plant and animal life, not least because of the extended absorption of sunlight due to similar plumes of dust and debris as those that appeared in the atmosphere of Jupiter.

Because of its large size and mass, Jupiter is a regular target for impacts. As discussed in chapter 1, without Jupiter's ability to "sweep up"

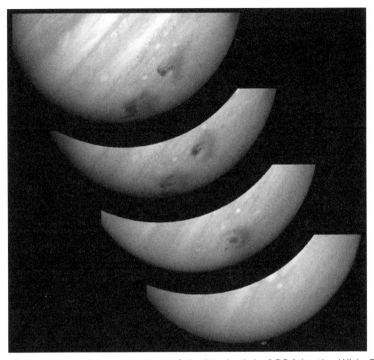

Figure 2.8. Mosaic of images taken of Jupiter in July 1994 by the Wide Field and Planetary Camera 2 on the Hubble Space Telescope (HST). They show the evolution over five days of the impact site (dark splashes) on Jupiter, created by the largest fragment (*G*) of the Shoemaker-Levy 9 comet. From lower right to upper left, the images show (on the planet's lower edge): the impact plume about 5 minutes after the impact; the fresh impact site 1.5 hours after impact; 3 days later after evolution by the winds of Jupiter (together with the new *L* impact after 1.3 days); and 5 days later, showing further evolution of the *G* and *L* sites due to winds and an additional impact (*S*) in the vicinity of *G*. (*Copyright R. Evans, J. Trauger, H. Hammel, and the HST Comet Science Team and NASA/ESA. Permission granted under Creative Commons Attribution 4.0 International License.*)

debris in the Solar System, it is very likely that life on Earth would not have evolved very far before being eliminated, or severely set back, by the impact of large bodies left over from the formation of the main planets.

On the other hand, mass extinctions have sometimes accelerated the evolution of life on Earth when the dominance of particular ecological niches passes from one group of organisms to another. This is rarely because the newly dominant group is "superior" to the old one, but usually because an extinction event eliminates the old, dominant group and makes way for the new one, a process known as "adaptive radiation" (van Valkenburgh 1999).

For example, mammaliaformes (or "almost mammals") and then mammals existed throughout the time of the dinosaurs but could not compete in the large terrestrial vertebrate niches that dinosaurs monopolized. The Cretaceous-Tertiary boundary mass extinction removed the non-avian dinosaurs and made it possible for mammals to expand into the large terrestrial vertebrate niches, ultimately making possible the evolution of humans. The dinosaurs themselves had been beneficiaries of a previous mass extinction at the Triassic-Permian boundary, which eliminated most of their chief rivals, the crurotarsans (or reptiles).

As evidence of the prevalence of these ancestral collisions, the surfaces of all the rocky planets—Mercury, Venus, Earth, and Mars—along with numerous asteroids and most moons of the planets, are pockmarked with craters arising from the impact of small and large solid bodies (figure 2.9). Indeed, the Earth's Moon is thought to have been created from the accretion (in Earth orbit) of debris thrown up by the impact of a Mars-size planet (dubbed "Theia") on Earth in the very early stages of the formation of the Solar System. Likewise, the retrograde rotation[14] of Venus is thought to have arisen from the impact of another large object early in its history, while Uranus's 97.8-degree tilt to the ecliptic and retrograde rotation is considered to be the result of an impact of a body about the size of the Earth many "moons" ago.

The very large impacts that produced most of the craters on the Moon and other planets and moons occurred during the so-called Late Heavy Bombardment[15] period in the Solar System's history, about 4 billion years ago. Since then, the number of impacts has declined as the number of residual small bodies has decreased due to gravitational "cleanup" by the larger objects, but they still occur occasionally, as evidenced by the Shoemaker-Levy collision on Jupiter in 1994.

Visual evidence of massive impacts on Earth has been largely eroded away over the eons, but some prominent examples remain, such as the

Figure 2.9. A photomosaic of the southern hemisphere of Mercury, taken by the *Mariner 10* spacecraft from a distance of 48,069 kilometers during its second flyby of the planet in September 1974. It looks very similar to the surface of the Moon, with an abundance of impact craters and splash rays of all sizes. (*Public domain from NASA/JPL.*)

famous 1.2-kilometer-diameter Meteor (also known as Barringer) Crater in Arizona, thought to have formed some 50,000 years ago from the impact of a 50-meter-wide asteroid, and the less well known 65-kilometer-diameter Manicouagan Impact Crater in Quebec, Canada, formed by the impact of a 5-kilometer-wide asteroid around 214 million years ago.

Indirect evidence of major impacts on Earth is occasionally provided by trace-element "signatures" of the unique chemical composition of impacting bodies. The best example of a signature of this kind is the anomaly of the abundance of the element iridium in sediments all around the globe at the boundary between the Cretaceous and Tertiary geological eras. As discussed above, it was this impact that is believed to have been responsible for the extinction of the dinosaurs and ammonites, some 66 million years ago, due to the lengthy "nuclear winter" that arose from the debris thrown up into the atmosphere at that time.

Every 2,000 years or so, a meteoroid the size of a football field hits Earth and causes significant damage to the area. Objects larger than this could be sufficient to wreak havoc with the Earth's climate and the food chains of life, but they would occur only every 100,000 years or so. An even larger object, big enough to threaten Earth's civilization, is thought to occur once every few million years or so.

Supernova Explosion or Gamma Ray Burst from a Nearby Star

Supernovae are huge explosions of high-mass stars that occur when the star collapses to form a "white dwarf," a "neutron star," or a "black hole" or when two neutron stars merge. A particularly energetic form of these explosions is known as a gamma-ray burst (GRB), which produces a narrow beam of very intense radiation.

Supernovae and GRBs cause a level of emitted radiation that often briefly outshines an entire galaxy, before fading from view over several weeks or months. Explosions of this type expel much or all of a star's material at a velocity of up to 30,000 km/s (10 percent of the speed of light), driving a shock wave into the surrounding interstellar medium and generating massive quantities of deadly radiation that would be fatal to any life-forms on planets close by. Although no supernova has been observed in the Milky Way since 1604, the number of documented supernovae remnants (identified historically as "planetary nebulae") indicates that on average the event occurs about three times every century.

There are several stars in our neighborhood that look like they may "go supernova" in the near future. Primary among these is Eta Carina, one of the most massive and luminous stars in the galaxy, located between just 6,500 and 1,000 light years from Earth. The Homunculus Nebula that surrounds Eta Carina is believed to have been ejected from the star in a massive outburst in 1841, and briefly made Eta Carina the second-brightest star in the sky. Given this inherent instability, the star is expected to go supernova at any time, with an impact on the Earth that cannot be predicted with confidence but is certainly going to much bigger than was its outburst in 1841.

Another candidate for "going supernova" soon is the red supergiant star Betelgeuse around 640 light years from Earth in the constellation of Orion, the Hunter. Betelgeuse is the star that represents one of the shoulders of Orion and is the ninth-brightest star in the night sky. It is one of the largest and most luminous stars known. If located at the Sun's position, its surface would extend out to between the orbits of Mars and Jupiter. It is expected to go supernova sometime within the next 1,000 years and is close enough for the resultant radiation to provide a significant risk to life on Earth.

Although rare (Podsiadlowski et al. 2004), a gamma-ray burst from a nearby object (less than 6,000 light years away) would be powerful enough to destroy the Earth's ozone layer, leaving organisms vulnerable to ultraviolet radiation from the Sun. It has been suggested that a gamma-ray burst caused the Ordovician-Silurian boundary extinction about 435 million years ago (Melott and Thomas 2009), while Fields et al. (2020) have suggested that a supernova contributed to the Hangenberg extinction about 360 million years ago. This latter event was precipitated by a sea-level rise followed swiftly by an episode of severe global cooling and glaciation that occurred at the end of the Devonian period, marking the dawn of the Late Palaeozoic ice age.

On the positive side, the expanding shock waves from supernova explosions help to provide the conditions conducive to the formation of new stars (by assisting the agglomeration of gas and dust) and play a significant role in enriching the interstellar medium with higher mass elements that provide the foundation for the molecules and materials upon which complex life depends. Thus, somewhat ironically, we may have needed a supernova to get us started, but we don't want anything to do with them after that.

Volcanic Eruptions

Although volcanoes are not events of our own making, they are certainly part of the Earth's extended carbon cycle. Over very long (geological) time periods, they release CO_2 from the Earth's crust and mantle, counteracting the CO_2 uptake by sedimentary rocks and other geological sinks.

Volcanic activity also releases massive amounts of SO_2 and particulates (ash) into the atmosphere. The ash particles fall rapidly from the stratosphere and most are removed within several days to weeks, with little impact on climate change, but on the other hand, SO_2 can have a significant effect on climate. This arises through its conversion to sulphuric acid aerosols which then increase the reflection of radiation from the Sun back into space and partially block the transmission of solar radiation to the Earth's surface, cooling the Earth's lower atmosphere, or troposphere. The eruption of Mount Pinatubo in 1991, the second-largest terrestrial eruption of the 20th century (after the 1912 eruption of Novarupta), affected the climate substantially, with global temperatures decreased by about 0.5°C for several years.

The eruption of Mount Tambora in 1815 produced an aerosol cloud the size of Australia in the stratosphere, blocking sunlight and reducing the global temperature by 0.4 to 0.7°C. Crops failed across Europe and the United States due to the cold or lack of sunshine, torrential rains flooded crops in Ireland, novel strains of cholera killed millions of people in India, and people starved in many countries. It was described as causing the "Year without a Summer" in 1816 (Stothers 1984).

Studies of global volcanic CO_2 emissions indicate that present-day subaerial (land-based) and submarine volcanoes release less than a percent of the CO_2 released currently by human activities, but this has been disputed by others (Xi-Liu and Qing-Xian 2018; Plimer 2021), who suggest that the amount of CO_2 emitted by undersea volcanoes, mid-ocean ridges, and earthquakes has been massively underestimated and may rival anthropogenic contributions. While it has been proposed that extreme volcanic release of CO_2 in the deep geologic past did cause global warming, and possibly some mass extinctions, this is also subject to debate at present.

For example, the Toba eruption in Sumatra, Indonesia, around 74,000 years ago was one of the largest known explosive supervolcano eruptions in the Earth's history. It is thought that this event caused a severe global volcanic winter of 6 to 10 years and contributed to a 1,000-year-long cooling episode of 3 to 5°C, or even as much as 10°C in the northern hemisphere in the first year after the event, leading to a genetic bottleneck in humans (Rampino and Ambrose 2000). Several genetic studies have revealed that 50,000 years ago, the human ancestor population greatly expanded from only a few thousand individuals (Rogers 1995). However, some physical evidence disputes the links with the millennium-long cold event and genetic bottleneck, and some consider the theory disproved (Ge and Gao 2020).

Explosion and Mass Flooding of Basalt from a Supervolcano on Earth
Large igneous provinces, such as Iceland, the Siberian Traps, the Deccan Traps, and the Ontong Java Plateau, are regions of continental scale basalts resulting from flood basalt eruptions. When created, these regions often occupy several thousand square kilometers and have volumes on the order of millions of cubic kilometers (Bryan 2010; Keller 2014). In most cases, the lavas are laid down over several million years and release large amounts of gases, many of which are potent GHGs.

Events of this magnitude are thought to occur only every 100 million years or so, but more than 100 documented eruptions have produced between 1,000 and 9,000 cubic kilometers of lava and tephra.[16] These are at least a thousand times larger than the 1980 eruption of Mount St. Helens and at least six times larger than the 1815 eruption of Mount Tambora, the largest eruption in recent history, which produced 160 cubic kilometers of volcanic deposits.

It is likely that there have been many more such eruptions during Earth's history beyond those able to have been observed directly or inferred from the geologic record. Each of these large eruptions has impacted massively on the composition of the Earth's atmosphere. The largest of these supervolcanoes typically cover huge areas with lava and volcanic ash and cause a long-lasting change to weather and climate (such as the triggering of a small ice age) sufficient to threaten species with extinction (see section above on causes of climate change).

The Campanian Ignimbrite supereruption in the Mediterranean around 40,000 years ago, attributed to the 12-by-15-kilometer-wide caldera of the Archiflegreo volcano, 20 kilometers west of Mount Vesuvius, has been suggested as having contributed to the demise of the Neanderthals.

Another notable example occurred 65 million years ago and produced the Deccan Traps; this event spread 9,000 cubic kilometers of lava across much of what is now India. It was roughly coincident with the Cretaceous–Paleogene extinction event and may have caused additional environmental stresses on extant species. The largest flood basalt event, the

Siberian Traps, occurred around 250 million years ago and was coincident with the largest mass extinction in history, the Permian–Triassic extinction event, and may have been the primary cause of that extinction.

Evidence for even larger-scale eruptions on Mars has been observed by satellite imagery from the *Mars Global Surveyor* and in ash deposits detected by NASA's *Opportunity* rover. Such catastrophic eruptions on ancient Mars could have drastically altered the planet's climate by initiating long periods of chilling in a very lengthy "volcanic winter."

Changes in the Reflectivity of the Earth's Atmosphere and Surface

Albedo is the reflecting power of a surface, expressed as the ratio of reflected radiation from the surface to incident radiation falling upon it. The average overall albedo of Earth ranges from 30 to 35 percent due to variations in cloud cover, but it also varies widely across the surface because of different geological, biomass, and environmental features.

All weather is a result of the uneven heating of the Earth caused by different areas of the planet having different albedos. When the planet's albedo increases, more incoming sunlight is reflected into space, creating a cooling effect on global temperatures, and vice versa. Ice reflects far more solar energy back into space than the other types of land cover or open water. This "ice-albedo" feedback (Budyko 1969) plays an important role in global climate change since it provides a strong positive feedback mechanism, like oceanic CO_2 levels and their dependence on temperature.

For example, if warm temperatures decrease the ice cover (in polar, sea, and/or continental locations) and this melt area is replaced by water or land, then the albedo decreases. This, in turn, increases the amount of solar energy absorbed, leading to more warming, and hence further melting of the remaining ice. If additional global warming of 4 to 6°C occurs in the future, this could prevent Arctic Sea ice from reforming at all during the northern hemisphere winter, possibly representing an irreversible tipping point for Earth's climate (Armstrong McKay et al. 2022).

Milankovitch Cycles

Serbian scientist Milutin Milankovitch (1879–1958) hypothesized that the long-term, collective effects of changes in Earth's position relative to the Sun are a strong driver of Earth's long-term climate, especially in relation to triggering the beginning and end of glaciation periods (ice ages) and the advance and retreat of the Sahara Desert. Specifically, he examined how variations in three types of Earth orbital movements— shape, tilt, and wobble—affect how much solar radiation reaches the top of Earth's atmosphere and what impact it then has (Kerr 1978). These orbital movements occur on cycles of 26,000 to 100,000 years and cause variations of up to 25 percent in the amount of incoming insolation at Earth's mid-latitudes. They are not responsible for short-term climate changes and are not responsible for recent rises in temperature; these are due to increased CO_2 in the atmosphere.

Plate Tectonics

As discussed in chapter 1 in reference to exoplanet habitability, plate tectonics is the scientific theory that Earth's lithosphere is composed of several large tectonic plates that have been slowly moving since about 3.4 billion years ago, in response to convective movements in the semi-viscous mantle layer below the outer "crust." The model emerged from the concept of "continental drift" developed during the first decades of the 20th century.

At so-called "convergent" plate boundaries, a process of subduction (one plate moving under another) carries the edge of one plate down under the other plate and into the mantle, thereby often serving to create mountain ranges in the upper plate. This loss of total crustal surface area is balanced by the formation of new oceanic crust along "divergent" plate margins in a process known as "seafloor spreading," thereby conserving the total surface area of the lithosphere. A third type of plate boundary is called a "transform boundary," where existing plates slide/grind past each other, along transform faults, often giving rise to strong earthquakes along the fault line (e.g., the San Andreas Fault in California).

Over the course of millions of years, the motion of the Earth's tectonic plates reconfigures global land and ocean areas, generates topography (i.e., mountain ranges and oceanic trenches, as the most extreme examples), and can affect both global and local patterns of atmosphere-ocean circulation. The locations and movements of the seas are important in controlling the transfer of heat and moisture across the globe, and therefore in determining global climate.

A recent example of tectonic control on ocean circulation is the formation of the Isthmus of Panama about 5 million years ago, which shut off direct mixing between the Atlantic and Pacific Oceans. This strongly affected the ocean dynamics of what is now the Gulf Stream and may have led to northern hemisphere ice cover. During the Carboniferous period, about 300 to 360 million years ago, plate tectonics may also have triggered large-scale storage of carbon through subduction of oceanic crust and increased glaciation due to the resultant lowering of the GHG content of the atmosphere.

Plate tectonic processes such as the redistribution of continents, growth of mountain ranges, formation of land bridges, and opening and closing of oceans provide a continuous but moderate environmental pressure that stimulates populations to adapt and evolve (Stern 2016), one of the strongest defenses against extinction. Periods of mountain building associated with plate tectonics also provide increased erosion, raising the nutrient content of the oceans (Large and Long 2015). This may have been one factor contributing to the species explosion witnessed at the beginning of the Cambrian period around 540 million years ago, when most major groups of living animals appeared. In contrast, the nutrient-poor periods lead to a depletion of plankton in the oceans and a slowdown in rates of diversification. This process could have played a role in three major mass extinction events at the end of the Ordovician, Devonian, and Triassic periods.

Natural Changes in the Energy Emitted by the Sun

The Sun is the predominant source of energy input to the Earth; its long-term (power output) and short-term (11-year sunspot cycle) variations in solar intensity are both known to affect global climate.

The early Earth was molten during its accretion from the remnants of the solar nebula surrounding the Sun around 4.5 billion years ago. The heating was due to frequent/incessant collisions with other major and minor bodies, and it led to extreme, global volcanism. The resultant outgassing most likely created the primordial atmosphere, which would have contained almost no oxygen and would have been toxic to humans and most modern life. Over time, the Earth cleared its orbital region of these residual bodies and the bombardment reduced, allowing the planet to cool and form a solid crust. Even though the early Sun emitted only 70 percent of its current power (Sagan and Mullen 1972), the temperature did not cool far enough to solidify water on the surface due to the buildup of GHGs that kept the surface temperature above the freezing point of water.

Over the following approximately 4 billion years, the energy output of the Sun increased, the volcanism settled down, and the atmospheric composition became dominated by nitrogen and CO_2 (Kasting 1993, 2014). After that, the most notable alteration was the "great oxygenation" of the atmosphere between 2.4 and 2.1 billion years ago (Kasting 1993). It was likely caused by cyanobacteria which evolved porphyrin-based photosynthesis, producing oxygen gas as a by-product. The sudden injection of highly reactive free oxygen, toxic to the then–mostly anaerobic life, may have caused the extinction of many organisms, including the ubiquitous microbial mats (stromatolites) around the globe. The subsequent adaptation of surviving archaea may have led to the rise of eukaryotic organisms (cells with a nucleus) and the subsequent evolution of multicellular life forms and ultimately plants and animals (Schirrmeister et al. 2013).

Over the next 5 billion years of the Sun's life, it will continue to increase gradually in brightness by about 1 percent every 100 million years. After about a billion years from now, the luminescence will have increased such that liquid water will no longer be able to exist on Earth (if climate change has not already achieved this outcome) and the last remaining organisms on the planet will suffer a final, complete mass extinction.

The Sun will continue to expand for around 4 billion years to reach its red giant phase and will then perhaps be large enough to have engulfed the Earth in its outer atmosphere (see below for more details of this extinction scenario).

Collision or Near-Collision with Another Star Within the Milky Way Galaxy, or a Merger with a Neighboring Galaxy

Galaxies collide, but they are so sparsely populated with stars that modeling (at the University of Toronto) has demonstrated that it is extremely rare that any two individual stars would collide or come close to colliding. What does happen is that there is massive disruption to the gravitational equilibrium of both galaxies and when the clouds of dust and gas that form most of the matter in galaxies collide, the resultant pressure wave triggers a burst of star formation (Schneider 2010).

The Milky Way's "sister" galaxy in the so-called Local Cluster[17] is the Andromeda Galaxy. It is one of only 10 galaxies that can be seen from Earth's surface with the naked eye, and it is the farthest away of these objects. Andromeda is slightly larger than the Milky Way, with a diameter of around 260,000 light years and around 1 trillion stars, but it is thought not to be as massive. It is 2.5 million light years away (only 10 times its diameter), so its apparent size in the sky is about three times the size of the Moon, but its surface brightness is very low and so its full extent is only properly observed in photographs. It is moving toward the Milky Way at a speed of around 100 to 140 km/s and will "collide" with the Milky Way in about 4.5 billion years, at about the same time as the Sun will have expanded to encapsulate the Earth (see the next section).

As two galaxies approach each other, the tidal interaction will de-structure the galaxy and millions of stars will be driven from their home galaxy. Of course, any planets orbiting a star that has a close encounter with another star from the colliding galaxy, or indeed just a star from the home galaxy that has wandered into its vicinity, will not fare well. It is thought that there is a very large number of "orphaned" planets (indeed, probably as many as there are still attached to stars) "floating" around the galaxy in isolation after being torn from their host in stellar near-collisions that occurred in the long-distant past. The bad news is that stripping a planet from its primary, if not sole, source of energy[18]—its host star—would lead to the demise of all life on that planet and render it incapable of supporting any new life-forms.

It doesn't even have to be as severe as this "orphaning" scenario. Many of the comets that we witness in the Solar System are likely to have been extracted from the Oort Cloud[19] in the outer reaches of the Sun's influence because of the gravitational disturbance from passing stars. Some of the ejected objects may have a size that is sufficient (if they get close enough) to throw Earth into an unstable orbit that takes it out of the Goldilocks zone. This eventuality would also be disastrous for life on Earth.

Earth Engulfed by the Natural Expansion of the Sun

The evolution of stars from their origins in a collapsing molecular cloud of hydrogen gas and dust particles is complex and depends in large part on the mass of the initial cloud. The size of the star that results from the collapse of the dust cloud depends on the mass of accumulated material and the equilibrium that is reached between the outward pressure of the radiation emitted by the thermonuclear reactions in its core and the inward pressure of gravity.

So-called main sequence stars,[20] with masses in the range 0.3 to 8 times the mass of the Sun, go through a red giant phase. In the case of the Sun, this phase of evolution will occur at an age of around 10 billion years. Prior to this, the Sun will become gradually hotter (i.e., more luminous) for several billion years, putting severe, and ultimately terminal, pressure on the viability of life on Earth.

But this isn't the end of the story. After an old star burns through the last of its fuel, it balloons into a red giant, engulfing any inner solar system planets. Soon after, the star sheds all its outer layers in an enormous eruption of hot gas that sweeps across any solar system attached to the star, creating an expanding "planetary nebula" of debris and leaving its core as a hot white dwarf, which will survive for trillions of years before fading to a hypothetical brown or black dwarf. About 97 percent of all stars in the Milky Way will end their lives as one of these white or brown/black dwarfs with shriveled, crystalline cores that pack about a Sun's-worth of mass into a sphere no larger than Earth.

Scientists suspect that any surviving (outer) planets, or whatever is left of them after this cataclysm, can continue to orbit around a white dwarf in the surrounding debris disk. Sometimes, those planetary remnants will spiral into the dead star's surface and become part of the dwarf, revealing its elemental composition in the spectrum of the star as traces of heavy metals like iron and magnesium are mixed into the dead star's atmosphere.

This is the extinction event of life on Earth with the longest delay, and it is the one that is the most certain. It will happen in around 3 billion (from the luminosity increase) to 5 billion years (from engulfment) and there is nothing that we can do about it.

Whatever else has happened in the billions of years between now and then, unless humans have left Earth for other parts, they will necessarily be extinguished as this inevitable process in the Sun's history runs its course.

Summary and Conclusions from Chapter 2

The roughly 4-billion-year history of life on Earth is replete with very long periods of time when not much happened (or appeared to happen) as the accretion process and the solid and liquid surfaces of the planet "settled" down, if you like, from its tumultuous beginning and simple, single-celled life emerged from the maelstrom. Then, about 500 million years ago, near the commencement of the Cambrian geological period, life suddenly became a lot more complex and there was an explosion of new and far more complex species, culminating in the last few million years with the emergence of *Homo sapiens*.

These new multicelled species were and are much more fragile than the very resilient single-celled organisms that preceded them. Many of their number were periodically decimated by a range of global-scale extinction and near-extinction events, including periods of worldwide glaciation, large-scale changes in the composition of the atmosphere, the emergence of dominant species that outcompeted their rivals for resources, collisions with asteroids and comets causing medium- to long-term massive climate change, continental-scale volcanic eruptions, and combinations of all of these.

Over the past century or so, *Homo sapiens* has begun to contribute to these threats to life on Earth through its (until recently) exponentially increasing numbers and resulting often-profligate need for resources; its increasing urbanization and travel, which brings more of us into close contact with our fellow humans; its advanced technology; the unstable and competitive relationships between nations; and the impact of all of these factors on all species and the environment.

The potentially most destructive and immediate of these impacts are associated with increasing temperature and climate instability due to the increasing amount of greenhouse gases in the atmosphere. The associated threat to the availability of freshwater and sustainable food supplies, along with our predilection to "live beyond the means" that the Earth is capable of supplying to us, implies that unless there is a major change in these circumstances and our behavior, our future as a species will surely be very limited.

We are also under significant threat from events that are not within our control, including massive volcanic eruptions and cyclical changes in the characteristics of the Earth's orbit around the Sun. The Solar System is a dynamic place in which the planets continue to "sweep up" other objects that come under their gravitational influence at short notice, some of which could wipe out all life on Earth if our paths were to cross. Other stars nearby in the galaxy will inevitably explode in supernovae and/or suddenly emit massive quantities of lethal radiation in our direction.

On much longer time frames (billions of years), we could suffer from near collisions with other stars in our own galaxy and the merger of galaxies in our Local Cluster. Even in the unlikely event that none of these things happen, our days will surely end when the Sun passes into its red giant phase of evolution and expands to approach us ever more closely to destroy our food supplies, boil off our oceans, and then, ultimately, engulf our planet entirely.

We need to have found a way of forestalling all the other unfortunate mechanisms for our demise and then to have found and settled into another place to live in a remote location in the outer Solar System, or outside it completely, well before that inevitable event starts to unfold.

CHAPTER THREE

How Urgent Is the Imperative to Search for Other Locations for Humans to Live?

It's important that we attempt to extend life beyond Earth now. It is the first time in the four billion-year history of Earth that it's been possible, and that window could be open for a long time—hopefully it is—or it could be open for a short time. We should err on the side of caution and do something now.
—ELON MUSK, FOUNDER AND CEO OF SPACEX

There are three reasons, apart from scientific considerations, mankind needs to travel in space. The first . . . is garbage disposal; we need to transfer industrial processes into space so that the earth may remain a green and pleasant place for our grandchildren to live in. The second . . . to escape material impoverishment: the resources of this planet are finite, and we shall not forego forever the abundance of solar energy and minerals and living space that are spread out all around us. The third . . . our spiritual need for an open frontier.
—FREEMAN J. DYSON, IN *DISTURBING THE UNIVERSE*, 1979

THE FRAGILITY OF LIFE ON EARTH AND THE CHANCES OF ITS EXTINCTION

Many of the catastrophic events described in chapter 2 will be dismissed by some as too improbable; for example, the impact of a large asteroid, a collision with another star, or a supernova explosion nearby. Others will be considered so far in the future to be of no concern; for example, a chance collision with another star or engulfment of Earth during the red giant phase of the Sun. Others might claim that technology developed in the future will be able to deal with many of these events even if it isn't already able to do so, and render them relatively harmless, for example, runaway climate change, a global disease pandemic, or the adverse resource requirements of population growth.

Be that as it may, there can be no doubt that complex life on Earth is fragile and may consider itself to be somewhat lucky to have survived this long. It has suffered massive reductions in the numbers of species in the past, due to volcanic eruptions, asteroid impacts, and climate change, and combinations of these. The human population has itself suffered substantial reductions in numbers due to long periods of extensive ice coverage and pandemics such as the Spanish flu and the Black Death, and these at times when the population was much more dispersed with a much lower overall level of congregation in settlements and individual-to-individual contact.

If life elsewhere is not so fragile as Earth's, enabling these other civilizations to survive for millions, if not billions, of years and in so doing develop technology far, far greater than that achieved by *Homo sapiens* so far, the question raised by Fermi remains unanswered: "Where is everybody then?" Since there is no evidence of such life existing, perhaps it is as fragile as we are and has been snuffed out in the Great Filter, just as we may be.

Primitive life on Earth has been in existence for about 4 billion years, around a third of the age (13.6 billion years) of the universe. Intelligent life is defined here, somewhat arbitrarily, as beginning with the earliest existence of *Homo sapiens*, perhaps 750,000 years ago, or around 0.006 percent of the age of the universe. As discussed in chapter 2, for complex life to evolve, it needs building blocks of elements that are heavier than iron. These elements are not produced in a first-generation star; their synthesis requires the extreme temperature and pressure conditions that are produced only in the nuclear reactions that occur in various forms of supernova.[1]

The Earth's host star, the Sun, is a third-generation star, so for planets orbiting second-generation stars in the Milky Way and in other galaxies, there has been ample time, indeed billions of years, for intelligent life to develop and to contact Earth. The fact that no verifiable contact has been made with other civilizations (chapter 1) might be an indicator that the size of the term L in the Drake equation, namely, the length of time such civilizations are detectable, and/or the number of times they reemerge on the same platform, is quite small. Perhaps L is tiny because life has become extinct on these other platforms for any one or other, or all, of the extinction scenarios described in chapter 2. Perhaps, therefore, life on Earth and anywhere else is destined to take the same path. If so, we need to get on with the challenge of doing what we can to ensure our survival.

The following sections of this chapter examine the *likelihood* of each of the potential causes of the extinction of life happening, starting with the list of events over which we think we might have some control, and then those over which we almost certainly have no control.

Events over Which We Think We Might Have Some Control
Accidental or Deliberate Nuclear Holocaust
If the Doomsday Clock is anything to go by (figure 2.2), humans have not achieved much over recent years in reducing the likelihood of a nuclear holocaust. The reductions produced by the Nuclear Non-Proliferation and Disarmament Treaties between the superpowers Russia and the United States following the end of the Cold War have been neutralized by the increased risk posed by the outbreak of war between Russia and Ukraine in 2022, the accidental release of radioactive material into the environment by leak or explosion/meltdown, a deliberate terrorist act by a "rogue" state or individual group, and the prospect of chemical and/or biological warfare which would impact directly on humans and/or on their supplies of uncontaminated food, water, and air. In all these examples, we can only hope that their greatest effects are felt locally rather than globally (unless things really get out of control) and the risk to life in general remains relatively low due to political intervention at the national and global levels, advances in technology, and/or the sober realization that this is the only "home" we have to play in, at least at the moment.

So we might, after all, have a degree of control over this particular extinction scenario.

Global Disease Pandemic

So too, with disease pandemics. Nobel Laureate Peter Doherty (2013) has noted that the term "pandemic" refers not to a disease's severity but to its ability to spread rapidly over a wide geographical area, especially when the disease is a respiratory virus. Extremely lethal pathogens are usually quickly identified and confined, but the rise of high-speed transportation networks and the globalization of trade and travel have radically accelerated the ability of diseases to spread. Furthermore, global warming, increasing population density, the rise of more-efficient disease vectors, and growing antibiotic resistance all complicate efforts to control pandemics.

On the other hand, research is advancing at a fast rate and interventions, including improvement of health practices in hospitals, and the development of more effective inoculation and antiviral treatments can be made much more quickly, so that a pandemic of catastrophic proportions is unlikely to take hold. Doherty also notes that research into animal reservoirs of pathogens, from SARS in bats to HIV in chimpanzees, show promise for the development of new preventative strategies.

In all the deadliest examples of pandemics (table 2.1), a significant proportion of the global population survived either through natural or acquired immunity and/or because they remained out of contact with the pathogen. For example, when the Black Death broke out in 1346, the world's population was around 440 million; six years later the population had dropped by 20 to 45 percent, but it did not disappear completely.

Nevertheless, it is possible that there is a disease lurking out there somewhere (perhaps more deadly and virulent than the Black Death) that is so powerful that it will resist (by very rapid self-modification to produce new strains) our attempts to wipe it out through treatment or immunization. Its effect may be more catastrophic than pervious pandemics due to the much closer proximity and greater interaction between humans than in previous eras. In that case, humans may well be doomed, or at least those sections of humanity without access to treatment and/or unable to escape contact with affected individuals.

Genocide and Deadly Competition for Resources

The genocide scenario, by itself, is probably not going to affect humanity in its totality but, rather, will impact subsets that fall victim to mindless aggression and/or prejudice from another group. Perhaps its greatest impact may be felt in a complex, multicausal extinction scenario where several factors act in deadly combination, such as with a global disease pandemic, exacerbated by conflicts over diminishing resources and/or the impacts of climate change (see discussion on these scenarios below).

In any event, we can aim to reduce the probability of this scenario by removing/lowering existing inequities in living standards and resource access within and between all nations, along with improvements in education and intergroup relationships. Our history of success in this domain is not good, but perhaps with the shared enemy of mutual, assured destruction, we might have a chance.

Self-Replicating Nanorobots ("Grey Goo") Consuming All Resources on Earth and/or Reaching the "Singularity" of Artificial Intelligence

Discussions on these scenarios often imply the inevitability of reaching a "tipping point" at which the nanomaterials and/or robots suddenly and unexpectedly develop a runaway capacity (viz., superior intelligence and uncontrolled self-replication, or "grey goo") to act and replicate independently of their creators. The result might then be the exponential consumption of resources or the expropriation of control of the world from humans.

While the ecophagy scenario described in chapter 2 is particularly attractive as a doomsday scenario in science fiction, it is unlikely that an event of this sort would occur irreversibly and on a scale that would impact the entire globe.

The intelligence and control scenario has been explored, in part, in popular movies like *I, Robot* and is revisited in chapter 13 in respect to the definition of humanity itself. Its importance as a doomsday/extinction scenario has been much used in science fiction and has been highlighted by recent developments in artificial intelligence (AI), for example, the generative large language ChatGPT software. These developments move way beyond "machine learning/training" to encompass capability that displays some elements of human reasoning, the "holy grail" of AI. We can only hope that appropriate checks and balances can be incorporated into the operating codes of these machines and robots to prevent them from taking over, as the software system HAL (nearly) did in *2001: A Space Odyssey*.

However, who can predict what might happen to these checks and balances when the insight and aptitude (and perhaps ego, desire for control, and power) of these machines begins to emerge and to approach the capability of human intelligence—the so-called AI "singularity"? If this point is reached, it might be "game over" for all of us and we may have as much control over our futures as do the cattle in the paddocks of today.

Population Explosion and/or Global-Scale Pollution
The positive side of this scenario is that its effects develop relatively slowly and become increasingly obvious and can be predicted (and therefore demand attention and, hopefully, action) long before they reach the point of no return. Furthermore, increased education and affluence, especially among women, are serving to reduce the birthrate in developing countries. On the other hand, some of the developed countries (such as Monaco, Singapore, Italy, Japan, and South Korea) have such low birth rates already that they are starting to struggle to generate and maintain enough economic activity among working sectors of the community to sustain historical economic growth levels and living standards and to support those members of the society who are no longer productive by choice or circumstance.

On the negative side, even static or declining population does not guarantee reduced resource use and reduction of waste and pollution. Indeed, figure 2.3 shows that the challenges of resource availability and waste disposal generally increase as the standard of living in a population increases and demands for material goods and energy rise.

Who can say that the projected increase in population of around 2.5 billion over the next 80 years (figure 2.4) will not soon exceed the availability of naturally occurring freshwater or access to abundant energy for purification or desalination of seawater? It will also present major challenges for the area of arable land upon which to grow enough food for this population. Despite spectacular success in increasing the productivity of cropland, stimulated by the introduction of mechanization to farming and the use of artificial fertilizers following the Industrial Revolution (Smil 2022), enough food has been able to be generated to feed an ever-increasing world population. Unfortunately, due to the unequal distribution of this food (and other resources) among countries and regions, there remains the likelihood of "resource wars" of global proportions. There are already many disastrous examples in history, well documented in Diamond (2005), where societies have chosen to fail rather than adjust their behavior to survive.

The related but distinct issue is the so-called tragedy of the commons (Hardin 1968), an economic and social theory in which individuals, acting independently and rationally according to each one's self-interest, behave contrary to the whole group's long-term best interests by depleting some common resource, on the assumption that if they do not exploit the resource, someone else will anyway. This argument also applies if another group has already preemptively diminished the resource on their own, such as in the case of deforestation and industrialization by developed countries. "Commons" in this context can include the atmosphere, oceans, rivers, fish stocks, and any other shared resource.

It has been argued that overexploitation of the common resource is not inevitable since (1) the individuals concerned may be able to achieve mutual restraint by consensus (e.g., limits on commercial whaling and the use of ozone-depleting chemicals) or (2) the "right" to exploit common land is generally controlled by law. Thus, it should be well within humanity's power to address issues of overpopulation, waste, and pollution, even if remedial action only begins in earnest when individual humans and communities experience the effects personally and are forced into action by their own inability to access water, food, or energy. We must hope that it isn't too late when this remedial action eventually takes place.

Massive, Runaway Anthropogenic Climate Change

Many climate scientists predict that Earth is about to reach a "tipping point" at which the warming effects of increased levels of atmospheric

CO_2 enter a "runaway state" because of the creation of positive feedback loops. As discussed above, an example of this kind of loop is the melting of sea ice that decreases the albedo of the Earth, thereby creating more warming and more sea ice melting. Another example is the warming of the oceans, which decreases their ability to dissolve CO_2 and, at the same time (through evaporation), creates a higher atmospheric content of water. Both of these gasses produce a greenhouse effect, which then creates more atmospheric CO_2 and more warming.

The Intergovernmental Panel on Climate Change (IPCC) was established in 1988 under the auspices of the United Nations. It has 195 member states and releases a comprehensive report every six or so years, documenting the outcomes of complex climate model projections of the global temperature several decades into the future, under a range of anthropogenic CO_2 emission scenarios. The latest report contained dire warnings about the impact of a temperature rise of more than around 2°C in the years since the Industrial Revolution (1760–1840). These impacts include significant rises in sea level, increased severity of extreme weather events and their frequency, changes in rainfall patterns affecting food production, threats to animal and plant life, and increased frequency and intensity of wildfires.

Despite passionate recognition of the dangers of climate change and heartfelt messages of intent by individuals, companies, and nations, including the setting of strong emission reduction targets and investment in new technologies, current experience is that national governments are unable to execute meaningful emissions reductions that slow down the current increase in atmospheric CO_2 level, let alone decrease or stop it altogether. This is because humans in developed nations have become very attached to their current high standard of living, dependent as it is on the consumption of ever-increasing amounts of energy and a reliance on the burning of fossils fuels as the primary source of this energy.

Moreover, developing nations push out the scale and timing of their reduction targets to give them a chance to increase the standard of living of their citizens to the same standard of the developed nations, as many would consider they are entitled to do. The trouble is that, if the vast numbers of humans in the developing nations achieve higher living standards, the feeble attempts of the developed nations to reduce their emissions will be swamped by the increased emissions from everyone else.

Furthermore, a molecule of CO_2 remains in the atmosphere for up to 100 years, so even if all CO_2 emissions were to cease today, it is unlikely that global temperature will decrease before 2080, around two human generations from now (Smil 2022). Delayed outcomes/rewards of this magnitude do nothing to encourage action today!

To make matters worse, it is not just energy production that creates CO_2 emissions. Deforestation (~20 percent of total GHG emissions), livestock methane (~17 percent), and concrete[2] and iron/steel production for infrastructure (each ~7 percent) contribute around half of man-made global CO_2 emissions annually (Fennell et al. 2022), and this figure is unlikely to decrease in the near-term future.

But all may not be lost. New and emerging technologies for emission reduction and direct extraction (from the atmosphere), if implemented on a massive scale across all major nations on Earth, have the potential to gradually reduce the CO_2 content of the atmosphere and in so doing limit the rises that are already embedded in the system in the short term and ultimately start reducing the global temperature. In parallel with these efforts, attention should also be focused on acclimatization/adaption strategies to combat the inevitable increased temperature and other climate change effects that will be experienced over the next few decades.

Stepping back to take a long-term perspective, humans are inadvertently reengineering (terraforming) Earth through the continued generation of anthropogenic GHG emissions, the deforestation of huge areas of the planet to make way for food production and population expansion, the construction of massive infrastructure developments (dams, mines, factories, and ever-expanding cities), the diversion of waterways, and the production of enormous amounts of nonbiodegradable waste. All these actions have an impact on global climate, weather, sea levels, landscapes, and the biosphere in general.

Terraforming a planet, moon, or other body (Fogg 1995) will be discussed in chapter 6 as a future strategy for off-Earth settlement, but this process is taking place right in front of our eyes on-Earth. The trouble is that this current transformation is making Earth *less* habitable for life. Perhaps the unintentional "success" we are having in terraforming our home planet in a negative sense can inform ways in which we might instead turn these learnings into a positive process that provides a more amenable and sustainable environment for future habitability. If successful, the turnaround would take some of the pressure and urgency off the search for another settlement platform to ensure humanity's survival and at least buy us some more time.

The economic cost of this transformation would be huge and the technical difficulties enormous, and given the requirements for a turnaround in cooperation among nations of the Earth, it is probable that the difficulties are insurmountable—certainly even more difficult to achieve than our current feeble efforts to slow and reverse the GHG emissions that are triggering the climate change in the first place. Time will tell.

Events over Which We Almost Certainly Have No Control

Explosion and Mass Flooding of Basalt from a Supervolcano on Earth

The Yellowstone Caldera in Yellowstone National Park in the western United States lies over the Yellowstone hotspot where molten rock from the mantle rises toward the surface. This volcanism is relatively recent, with calderas created by three supereruptions that took place 2.1 million, 1.3 million, and 640,000 years ago (Matthews et al. 2015).

If another large, caldera-forming eruption were to occur at Yellowstone, it would have regional effects including ash falling over a very wide area and short-term (years to decades) changes to global climate. Fortunately, there is currently no evidence of unusual seismic or other worrying activity in the vicinity, so the chance of this sort of eruption at Yellowstone in the next few thousands years is very small.

Supernova Explosion or Gamma-Ray Burst from a Nearby Star

There is nothing that humans can do to stop a star from exploding in our neighborhood of the galaxy. We can keep a watch on nearby stars to assess their likelihood of exploding anytime soon (viz., Eta Carina and Betelgeuse, as discussed previously), but this will be of no help in preventing the event from occurring. With enough warning, and if humanity can marshal the necessary physical, financial, and energy resources, it may be possible to construct some sort of radiation shield (electromagnetic and/or physical) large enough to protect at least a part of the surface of the Earth from the deadly incoming radiation. However, the task would be akin to the construction of a Dyson sphere, discussed in chapter 2 in relation to the search for very advanced extraterrestrial life, and our capability of doing this in the medium-term future seems improbable, at best.

So, we are very likely to be stuck with this particular extinction scenario.

Catastrophic Impact of a Large Comet or Asteroid
We already discussed that a collision of this kind 66 million years ago caused 96 percent of species on Earth to be eliminated, including the dinosaurs. The impacting object is estimated to have been about 10 kilometers in size, with an energy of 3×10^6 megatons of TNT. How likely is it that an event of this kind, or even more devastating, will occur sometime soon?

An object in the Solar System with an orbit that brings it into proximity with Earth is known as a "near-earth object" (NEO), defined as one with a closest approach to the Sun (perihelion) of less than 1.3 AU, or 1.3 times the Sun-Earth distance. The category includes a few thousand near-Earth asteroids (NEAs), a few near-Earth comets, and anything else large enough and with an orbit that has been characterized with sufficient reliability to suggest that it will strike the Earth. Figure 3.1 shows that the number of observed NEOs is increasing rapidly as detection methods improve and commitment to the search increases in recognition of the importance of the danger to humanity.

The risk that any NEO poses takes into account its size and density and the (related) effect that a collision would have on the culture and technology of human society (Binzel 2000; Chesley et al 2002). Naturally enough, public awareness of the risks arising from NEOs rose significantly after the impact of the fragments of comet Shoemaker-Levy 9 on Jupiter was witnessed in detail in July 1994 (figure 2.8), but this was/is not the only such "wake-up call."

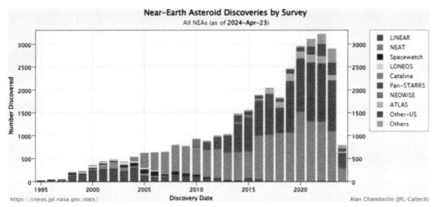

Figure 3.1. Annual discoveries of near-Earth asteroids from 1995 to April 2024 and the detection instruments used for these observations. (*Alan B. Chamberlin—Discovery Statistics Center for Near-Earth Object Studies, Jet Propulsion Laboratory, NASA. Public domain.*)

An asteroid discovered in 2012 called 2012 DA14 was described in February 2013 by NASA as the biggest asteroid ever seen passing so close to Earth (Yeomans and Chodas 2013). From its brightness, its size was estimated at around 45 meters with a mass of about 130,000 tons. It made its nearest approach to Earth on February 15, 2013, when it was about 27,700 kilometers above Earth's surface, or less than one-tenth the distance between Earth and the Moon and only twice the diameter of the Earth. It passed well outside Earth's atmosphere, but inside the belt of satellites in geostationary orbit at 35,800 kilometers above the surface. This was indeed a close call!

The main problem with estimating the number of NEOs and their associated risk is that detecting each of them is determined by its size, the characteristics of its orbit (and its evolution over time), and an estimate of the reflectivity of its surface (Bottke et al. 2000). Clearly, the easier it is to detect an object, the more frequently ones with these characteristics will be counted. Thus, there will be observational biases when trying to calculate the number of bodies in a population from the list of its detected members. Search protocols have discovered around 900 NEOs with a diameter greater than 1 kilometer, representing more than 90 percent of the expected number with this size in the Solar System, but there are around 15,000 objects that have been observed with a diameter between 30 and 1,000 meters and almost certainly many more in this size range that have not yet been detected.

The average time between the impact of "stony" asteroids[3] with a diameter of 7 meters, which impact with an energy of approximately 15 kilotons of TNT (about the same as the atomic bomb dropped on Hiroshima), is estimated to be five years. For an object 60 meters across (10 megatons of impact energy), the interval is around 1,300 years; one 1 kilometer across will impact only every 500,000 years; and one 5 kilometers across (equivalent to the object that wiped out the dinosaurs) can be expected to arrive every 18 million years or so.[4]

For another comparison, the fragments of the Shoemaker-Levy comet that hit Jupiter were up to 2 kilometers in diameter, not so different from the object that decimated the dinosaurs. Taken in total, the Shoemaker-Levy fragments would therefore have wreaked very significant havoc had any one of them, or all, landed on Earth. As a result, popular culture is replete with fictional stories about the destruction or forced deviation of objects determined to be on a collision path with Earth, including the films *When Worlds Collide* (1951), *Armageddon* and *Deep Impact* (both in 1998), and *Don't Look Up* (2021).

A total of 17 minor planets (asteroids, dwarf planets, and Kuiper Belt objects) and 8 comets have been visited by space probes to date, mostly with the goal of obtaining a greater understanding of their composition as an indicator of the nature of the primordial material that came together to produce the planets and moons in the Solar System. The most recent of these was NASA's DART (Double Asteroid Deflection Test) mission to the binary asteroid in the near-Earth Apollo group[5] 65803 Didymos (780 meters in diameter) and its moonlet Dimorphos (160 meters). The asteroid pair was about 11 million kilometers from Earth when the encounter took place, and it was/is not on a collision course with Earth. Unlike earlier missions to asteroids, the aim was not to spend time at the object making observations of its composition and/or to collect and return sample to Earth for analysis, but rather to undertake the first attempt to see whether a high-velocity impact with a spacecraft could materially alter the orbit of an asteroid of this size.

The impact occurred on September 26, 2022, at a speed of around 23,000 km/h. The event was monitored by DART's small companion craft (provided by the Italian Space Agency) that had been deployed from the 570-kilogram "mother" craft 15 days earlier. The companion craft was unaffected by the impact and captured images of the impact itself and of the resulting cloud of ejected matter. Follow-up radar and telescopic evidence confirmed that the spacecraft's impact altered Dimorphos's orbit around Didymos by 32 minutes, shortening it from 11 hours and 55 minutes to 11 hours and 23 minutes, with a margin of uncertainty of around 2 minutes.

DART's successful, largely autonomous targeting of a small (but quite large on the extinction scale) asteroid, with limited prior observations, represented a critical first step on the path to develop and demonstrate "kinetic impactor" technology as a viable operational capability for planetary defense. However, the method requires warning of an incoming body at least several years in advance, not just to mount the mission in time but also to execute what will inevitably be a relatively minor path deflection far enough away from Earth to avoid the anticipated collision. Nevertheless, it gives some realistic hope that this extinction scenario might not be so inevitable after all, at least for objects of this order of size.

SUMMARY AND CONCLUSIONS FROM CHAPTER 3
Intelligent, technologically adept life on Earth required the passage of more than 4.5 billion years to develop, and only in the past half-century has this life-form been capable of sending messages and objects into interstellar space. Given the vast number of stars and their associated planets in the galaxy and the passage of 13.8 billion years since its formation (three times the life of Earth), it seems that there has been enough time for technologically advanced life to have emerged elsewhere. The fact that we have not detected such life suggests that: (1) what life there may be "out there" is not yet able to be detected by our sensors, (2) there is actually no other technologically advanced life elsewhere, and/or (3)

any such life has, by one means or another, been extinguished before it could make its presence known.

The extinguishment scenario is not so far-fetched, given the raft of existential threats to life on Earth that we are facing, or could soon or eventually face. Some of these threats are more urgent than others and some, if not most, will need to act in concert if they are to succeed in extinguishing life completely, while others are completely beyond our control, at least for the foreseeable future.

Of the dangers to life on Earth that might present themselves in the short term, the most likely is the impact of a hitherto undetected largish asteroid or comet that has managed to avoid capture by the other large bodies in the Solar System. This clear and present danger to life was exemplified by the catastrophic impact of the fragmented Shoemaker-Levy 9 comet on Jupiter in 1994. We also know that the threat is real because there have been at least five cases over the past 450 million years in which up to 95 percent of all life on Earth has been extinguished by collisions of this kind.

Comprehensive efforts to document all of the large objects in the Solar System have been in progress for some time, but the possibility remains that one of those not-yet-detected bodies, or a large-enough body deflected from the Ort Cloud or Kuiper Belt in the outer reaches of the Solar System, could appear "out of nowhere" and wreak havoc if it collided with Earth. Current technology is being developed to divert smaller objects from a collision course with Earth, but a really large object will not be able to be disarmed in this way.

A massive continental-scale volcanic eruption might occur sometime in the next few hundred years in or near one of Earth's crustal "hot spots," but its effect is unlikely to be global. Radiation from a supernova explosion of a star could also occur suddenly (even our relatively quiescent Sun has frequent unexpected large coronal mass ejections that wreak havoc on satellites and electronic networks), but the supernova would have to be in our close neighborhood for the radiation dose to be fatal to life. We have a reasonable understanding of which local stars represent a danger of this kind, and none of these are likely to present a real threat to us in the next few million years.

All of the other external threats (Milankovitch cycles, plate tectonics, collisions with other stars and galaxies, expansion of the Sun) are on long time frames (the latter, very long), predictable, and out of our control.

The threats that we pose to ourselves provide different dangers to life, and are potentially much more urgent. Among these may be counted: (1) the increase in Earth's population, with associated often profligate and unsustainable resource consumption; (2) increased contact between humans, leading to amplified chances of disease pandemics; (3) political and religious instability in an age of nuclear, biological, and chemical weapons; (4) out-of-control technological developments (artificial intelligence, "grey goo"); and (5) massive, runaway anthropogenic climate change. All of these issues and events have the potential to be very serious threats to our well-being and/or to our existence on Earth, but all of them are within our capability to ameliorate and control.

But we need the concerted will and cooperative commitment to take action at the global as well as local level.

Conclusions to Part I

The development of complex, technologically advanced life requires the juxtaposition of many individual circumstances that, taken together, imply that this form of life is likely to be very rare, if not absent, everywhere in the galaxy (other than on Earth). Indeed, despite massive efforts over many decades, no evidence for the presence of even simple life-forms has been obtained in the Solar System or elsewhere.

The existence of life on Earth is fragile and subject to a host of existential threats. Depletion of resources that are critical to human life—clean water and air, food, shelter, and energy—is very likely to be accelerated by future increases in population, contamination, and/or the desire for increases in living standards across all developing nations. This depletion remains an issue of considerable concern (and moderate probability) because of the potential link to massive intra- and international social and political unrest.

There is clear scientific evidence of the widespread and accelerating negative impacts of climate change on the biosphere of Earth if current anthropogenic emissions of CO_2, exacerbated by additional population, are not abated dramatically over the next few decades.

If history is any guide, our increasing population density and human-to-human contact could provide a catalyst for a new fatal disease pandemic that spreads too rapidly to stop and/or cannot be controlled by advances in medical science.

We still live under the threat of deliberate or accidental nuclear, chemical, and biological holocaust that could render large parts, if not all, of the surface of the Earth inhabitable by humans for very long periods of time.

Advances in artificial intelligence and robotics have the potential to improve the quality of life for humans, but there is significant risk that their development will lead to a point where they are capable of "taking over" in their own right, perhaps with goals and values that are in conflict with our own.

The Shoemaker-Levy comet impact on Jupiter in 1994 demonstrated that large-scale impacts can, and still, occur in the Solar System at a scale that provides a clear and present danger to life on Earth. We have no way of knowing if disturbances in the Ort Cloud have already sent, or will in the future send, massive cometary objects into the inner Solar System on potential collision courses with Earth. We also have no idea what future collisions or near-collisions of asteroids will perturb their orbits to bring them into a similar intersection with Earth, or whether deflection methods will be adequate.

And finally, we have no control over the possibility, however remote, of a devastating, global-scale volcanic eruption or explosion of a nearby star, and we cannot alter the natural evolution of the Sun into its expansive and much more luminous red giant phase.

The documented history of species extinction on Earth suggests that each species has a limited lifespan. When linked with the "Great Silence"—the absence of direct or indirect evidence of intelligent life elsewhere in the galaxy (the Fermi paradox)—it suggests that our remaining time as a species may be quite limited. Perhaps we are about to usher in our own version of the "Great Filter" that has been proposed as an explanation of this lack of evidence—that it is the fate of technologically advanced civilizations elsewhere in the galaxy to become extinct before they can develop the means for interstellar travel or advanced communication.

Quite aside from the possibility that one or some of these extinction scenarios will come to pass, the imperative for humans to find another platform to support their existence may also be argued from the perspective of maintaining "pressure to innovate" (Zubrin 2019). At various stages in human history, Zubrin argues, the catalyst for the innovation that has made us the most successful species on Earth has been the need to deal with the challenges associated with our expansion into unfamiliar geographical frontiers (to wit, the migration of *Homo* out of Africa and the settlement of the Americas) and the associated generation of diversity through physical separation from the original cultural and technological norms.

As the world becomes ever more connected, and with its cultures and gene pools blending, Zubrin posits that the risks of technological malaise may increase, enhancing the likelihood of collapse, not with a "bang" (i.e., extinction), but with a "whimper" (i.e., decay). Opening a new frontier on another celestial body, remote from Earth, with its own unique set of innovation and evolutionary pressures might then be just the thing that humanity needs to avoid collapse.

All these near- and medium-term extinction uncertainties and pressures to innovate impact the future of humans on Earth and create a strong imperative to prepare for the prospect of the settlement of other worlds. It is based on this premise that the remaining sections of this book are written.

Part II
Potential Locations for Human Settlement

Man must at all costs overcome the Earth's gravity and have, in reserve, the space at least of the Solar System. All kinds of danger wait for him on the Earth.... We have said a great deal about the advantages of migration into space, but not all can be said or even imagined.
—Konstantin Tsiolkovsky, in the essay "The Aims of Astronautics," 1929

Part II examines options for the settlement and sustainable survival of the human species at locations beyond Earth. This does not mean that major efforts should not continue to be devoted to reversing and/or adapting to the negative impacts of humans and natural forces on our planet. Rather, it acknowledges that there is no guarantee that there is either the concerted will or the technology to reverse these impacts in a viable time frame. It means that the medium- to long-term survival of the human species may rest on providing it, at the same time, with another option to deal with things that are going badly wrong on Earth by finding another "space to live."

Putting aside the very significant (some would say insurmountable) human physical, social, and psychological challenges implicit in long-distance space travel for the moment (these are covered in detail in part IV), large-scale, long-term formal settlement of a platform off-Earth presents an enormous technological challenge.

In our Solar System there are no planets or moons that, with current technology, can support human life for periods of more than a few days. The maximum time spent on the Moon by any of the manned Apollo missions was three days, due to limits on the oxygen, electricity, and food supplies that could be transported with the crew on a single mission. Longer stays would require multiple supply missions and the creation of a local environment that could provide safe shelter from cosmic rays, micrometeorites, and other dangers; a continuous and ultra-reliable energy supply to maintain life and technical support systems (temperature, humidity, communications, etc.); and, ultimately, the ability to generate food, water, and a breathable atmosphere and to recycle and dispose of waste.

For semipermanent and/or large populations in residence at locations that are too far away from Earth for regular resupply missions to be viable (e.g., the outer planets and their moons), it would also be necessary to have the capability for local resource extraction and small-scale manufacturing capabilities.

At the time of writing (December 2024) the nearest star outside of our Solar System with a known planet that may have an environment close to that of Earth is the red dwarf Proxima Centauri. Evidence suggests that one of this star's planets, Proxima b, is likely to be a rocky body in the habitable zone of its parent star, with a mass 1.3 times the mass of Earth and with an orbital period of roughly 11 Earth days. The planet was discovered on August 24, 2016, and is 4.22 light years from Earth.

At the maximum speed (41,000 km/h) that the third stage of the Saturn V rocket could impart to the *Apollo 11* spacecraft on the way to the Moon, the journey to Proxima b would take around 110,000 years.[1] The *New Horizons* spacecraft that was sent to Pluto (and beyond) on January 19, 2006, with a speed of around 58,600 km/h was the fastest spacecraft to date, but this was only about 50 percent faster than the maximum speed of *Apollo 11*. *New Horizons* took nine and a half years to reach Pluto and even then, were it not for a planned and carefully designed "gravity assist" from Jupiter on the way, it would have taken three years longer to get to Pluto. Even a journey to Mars requires around seven months with current technology, and other targets for settlement within the Solar System require multiyear journeys. One would need to travel at around 4,000 times this speed, or some 14 percent of the speed of light, just to make the journey to Proxima Centauri in one human lifespan! Of course, speeds of this magnitude are, and will always be, way beyond the capabilities of chemical rocket technology.

The four chapters in part II summarize the record of space travel to date and then explore arguments for and against specific options for what might be new locations/platforms on which humanity might seek to live, namely, in Earth's immediate environment, elsewhere in our Solar System, and outside our Solar System.

Chapter Four

The Record of Space Travel So Far

The exploration of space will go ahead, whether we join in it or not, and it is one of the great adventures of all time, and no nation which expects to be the leader of other nations can expect to stay behind in the race for space.
—US President John F. Kennedy in a speech at Rice University, September 12, 1962

Houston, we've had a problem here.
—Command Module Pilot John L. "Jack" Swigert on the *Apollo 13* mission to the Moon on April 13, 1970

The definition of when "outer space" begins, namely, where aerodynamics stops and astronautics begins, is the subject of continuing debate. The so-called Kármán line (von Kármán and Edson 1967; McDowell 2018) is defined as the point beyond which the lift provided by an aircraft's wings is insufficient to sustain flight and Earth orbit cannot be achieved without additional input. This definition is not exact, because the geometry of a stable orbit can be circular, with a constant height above Earth's surface, or elliptical, in which the orbit may take the spacecraft much closer to the surface than its average distance. Most regulatory agencies (including the United Nations) accept the Fédération aéronautique internationale (FAI) Kármán line definition of spaceflight as any flight over 100 kilometers, or something close to it. The United States

Air Force prefers to differ, defining this achievement for any craft that travels above an altitude of "only" 80 kilometers.

As of October 2023, 619 people have reached the Kármán line limit and most have progressed into orbit, spread over 377 launches and comprising people from 44 countries (Wikipedia 2024a). Citizens of the United States represent slightly over 60 percent of these space travelers (374 astronauts), with 22 percent (126 cosmonauts) from the Soviet Union and Russia and 4 percent (22 travelers) from China. Of the total number, 24 (all from the United States) have traveled beyond low Earth orbit to circle, orbit, or walk on the Moon. These space travelers have spent a total of more than 29,000 person-days (or over 79 years) in space, beginning with the 108-minute flight of cosmonaut Yuri Gagarin in *Vostok 1* in 1961.

What follows is a summary of the major highlights in the history of spaceflight, both uncrewed and crewed, organized by country of origin. The data was extracted primarily from the internet (Wikipedia 2024b; Logsdon 2024).

Soviet Union

The Soviet Union was the leading nation in spaceflight during the 1950s and '60s, but lost motivation and momentum following the successful US National Aeronautics and Space Administration (NASA) *Apollo 11* Moon landing in July 1969. With the abandonment of NASA's Space Shuttle program (135 missions from 1981 to 2011), NASA lost its launch capability and began to rely on Russian *Soyuz* vehicles to ferry its astronauts (along with Soviet and other nationalities) to the International Space Station (ISS). Over the 60-year history of spaceflight by the Soviet Union and later Russia, it launched more than 1,900 spacecraft (157 of them crewed) and was responsible for a great number of pioneering accomplishments, including the first:

- intercontinental ballistic missile (1957)
- satellite (*Sputnik 1*, 1957)
- animal in space (the dog Laika on *Sputnik 2*, 1957)
- moon impact (*Luna 2*, 1959)

- photographic image of the far side of the moon (*Luna 3*, 1959)
- human in space and Earth orbit (Yuri Gagarin on *Vostok 1*, 1961) just 25 days before the first US suborbital flight (Alan Shepard on *Mercury/Freedom 7*)
- woman in space and Earth orbit (Valentina Tereshkova on *Vostok 6*, 1963)
- spacewalk (Alexey Leonov on *Voskhod 2*, 1965)
- lunar soft landing (*Luna 9*, 1966)
- lunar orbit (*Luna 10*, 1966)
- analysis of the environment of another planet (Venus, *Venera 4*, 1967)
- living creatures (including two tortoises) to circle the Moon and return safely to Earth (*Zond 5*, 1968)
- robotic rover (*Lunokhod 1* on the Moon, 1970)
- soil samples extracted and returned to Earth from another celestial body (*Luna 16*, 1970)
- soft landing and first data received from the surface of another planet of the Solar System (Venus, *Venera 7*, 1970)
- space station (*Salyut 1*, 1971)
- probe to soft-land on another planet (Mars, *Mars 3*, 1971)
- spacecraft to orbit another planet, soft-land, and take photos of the surface (Venus, *Venera 9*, 1975)
- international manned spaceflight (NASA's *Apollo "18"* spacecraft docked with a Soviet *Soyuz* spacecraft in low Earth orbit, 1975)

UNITED STATES

Likewise, the United States has a very commendable record of space flight. As of October 2023, NASA had successfully launched 166 crewed flights during its space exploration program.[1] It has also made more than 1,000 unmanned missions into Earth orbit or beyond, including several very successful long-term visits to all planets out to Neptune (Pluto was the target of the *New Horizons* NASA mission, but it is now classified as a "dwarf planet") and several soft landings of surface robotic vehicles on Mars, Venus, Titan, and several asteroids and comets. Five of the outer Solar System missions involved trajectories that allowed them to escape the Solar System and enter interstellar space.

NASA's achievements include the first:

- weather satellite (*TIROS 1*, 1960)
- manual control of a crewed spacecraft (Alan Shepard on *Mercury/Freedom 7*, 1961)
- successful interplanetary probe, with data returned (flyby of Venus, *Mariner 2*, 1962)
- geostationary satellite (*Syncom 2*, 1963)
- Mars flyby (returned pictures, *Mariner 4*, 1965)
- crewed spacecraft to change orbit (*Gemini 3*, 1965)
- rendezvous in space and docking with another crewed spacecraft (*Gemini 6A* and *Gemini 7*, 1965)
- crewed mission to leave Earth orbit, orbit the Moon, and return to Earth (*Apollo 8*, 1968)
- fastest speed ever traveled by a human—39,897 km/h, or 11.08 km/s (*Apollo 10*, 1969)
- crewed landing on the Moon (*Apollo 11*, 1969)
- spacecraft to orbit another planet (Mars, *Mariner 9*, 1971)
- flyby of Jupiter (*Pioneer 10*, 1973, followed by *Pioneer 11*, 1973, and *Voyager 1* and *Voyager 2*, 1979)
- use of "gravity assist" by an interplanetary spacecraft (flyby of Mercury, *Mariner 10*, 1974)
- pictures transmitted from the surface of Mars (*Viking 1*, 1976)
- probe to fly by two planets and first probe to Saturn (*Pioneer 11*, 1979, followed by *Voyager 1*, 1980, and *Voyager 2*, 1981)
- probe to fly by Saturn's moon Titan (*Pioneer 11*, 1979)
- reusable spacecraft (space shuttle *Columbia*, 1981)

- flyby of Uranus (*Voyager 2*, 1986)
- probe to penetrate a comet's coma and take close-up images of its nucleus (*Giotto* at Halley, 1986)
- flyby of Neptune (*Voyager 2*, 1989)
- asteroid flyby (Ida and its moon Dactyl, *Galileo*, 1989)
- large optical telescope in earth orbit (Hubble Space Telescope, 1990, with the European Space Agency [ESA])
- spacecraft to orbit Jupiter (*Galileo*, 1995)
- rover on another planet (*Mars Pathfinder* lander, with rover *Sojourner*, 1997)
- proof of concept for airbag-mediated touchdown and automated obstacle avoidance (*Mars Pathfinder*, 1997)
- use of ion-powered rocketry (*Deep Space 1*, to asteroid 9969 Braille, 1998)
- sample return of cosmic dust and material from a comet's coma (*Stardust* at comet Wild-2, 1999 and 2006)
- spacecraft to orbit and land on an asteroid (*NEAR*, at Eros, 2000 and 2001)
- solar wind sample return probe from Sun-Earth Lagrangian point L1 (*Genesis*, 2001 and 2004)
- spacecraft to orbit Saturn (*Cassini-Huygens*, 2004, with ESA and Italy)
- impactor to study the interior of a comet (*Deep Impact*, 9P/Tempel, 2005)
- landing on a moon other than Earth's (*Huygens*, on Saturn's moon Titan, 2005, with ESA and Italy)
- mission(s) to orbit the Moon's Lagrangian points L1 and L2 (*Artemis-P1* and *-P2*, 2010)
- spacecraft to orbit Mercury (*Messenger*, 2011)
- human-made object to enter interstellar space (*Voyager 1*, 2012, launched 16 days *after* its twin *Voyager 2*).

- spacecraft to orbit a dwarf planet (*Dawn* on Ceres and Vesta, 2015)
- spacecraft to fly by dwarf planet Pluto and its moons (*New Horizons*, 2015)
- test of an asteroid deflection method through a high-velocity impact (DART spacecraft on 65803 Didymos's moonlet Dimorphos, 2022)

NON-GOVERNMENT ("PRIVATE COMMERCIAL") SPACEFLIGHT

From the earliest days of spaceflight, nation-states were the only bodies that developed and flew spacecraft above the Kármán line. The US and Soviet space programs were operated using mainly military pilots as astronauts and cosmonauts, and no commercial launches were undertaken by private companies.

The first non-government space operations were launches of commercial communications satellites after passage of the US Communications Satellite Act of 1962. This allowed commercial consortia to own and operate satellites, but these were still deployed on state-owned launch vehicles.

In 1980 the European Space Agency created Arianespace, a commercial company operated after hardware and launch facilities were developed using government funding. In 2006, in anticipation of the termination of the Space Shuttle program in 2011, NASA created the Commercial Transportation Service (COTS) to tender for the development and coordination of vehicles to deliver crew and cargo to the ISS on a regular basis. The vehicles developed under these public-private partnerships would allow NASA to cease its dependence on Russia's *Soyuz* spacecraft to ferry its astronauts to the ISS. More than 20 companies tendered for contracts under the program, and in 2013 contracts were awarded to Elon Musk's Space Exploration Technologies Corporation (SpaceX) and the Orbital Sciences Corporation (now subsumed within Northrop Grumman). More recent successful commercial spaceflight projects include the suborbital flights of Virgin Galactic (founded by Richard Branson) and Blue Origin (founded by Jeff Bezos) and orbital flights of SpaceX.

Achievements by these non-government individuals and corporations are the first:

- privately funded rocket to reach the boundary of space, launched by Space Services Inc. on a suborbital flight (1982)
- private company to reach orbit—Orbital Sciences Corporation's (founded by David Thompson, Bruce Ferguson, and Scott Webster) *Pegasus*, an air-launched rocket (1990)
- crewed mission to the ISS—SpaceX's *Crew Dragon Demo-2* (2020)
- successful suborbital flights into space by Virgin Galactic and Blue Origin (2021)
- orbital spaceflight with only private citizens aboard—SpaceX's *Crew Dragon Resilience Inspiration 4* mission (2021)

OTHER NATIONS

Achievements by nations other than the United States and the Soviet Union/Russia are the first:

- non-American and non-Soviet citizen in space (Vladimír Remek, a Czechoslovakian, on *Soyuz 28* in 1978, as part of the Interkosmos program)
- sample return probe to launch from an asteroid (*Hayabusa*, 2010, Japan; 15 micrograms returned)
- spacecraft to orbit a comet (*Rosetta*, 2014, ESA)
- spacecraft to land on a comet (*Philae*, 2014, ESA)
- probe to orbit the Moon, visit Sun-Earth L2 Langrangian point, and make a flyby of asteroid 4179 Toutatis (*Chang'e 2*, 2018, China)
- rovers released on an asteroid (*Hayabusa*, 2019, Japan, on Ryugu)
- soft landing on the Moon's far side (*Chang'e 4*, 2019, China)
- soft landing near the south pole of the Moon (*Chandrayaan-3*, India, 2023)

RECORDS FOR HUMAN PRESENCE IN SPACE

All of the following records apply as of October 2023.

Russian cosmonaut Valery Polyakov holds the record of 438 consecutive days aboard the Mir space station, from January 1994 to March

1995, but Russia's Gennady Padalka currently holds the record for a collective 878 days (more than two years) spent in space over the course of five missions.

The International Space Station has the record for the longest period of continuous human presence in space, from November 2, 2000, to the present (more than 22 years). This record was previously held by Mir, from September 5, 1989, to August 28, 1999, a span of 3,644 days, or almost 10 years.

As of August 2, 2023, the ISS has been home to a total of 269 individuals from 21 countries, including 163 from the United States and 57 from Russia, with the next-most number, 11, from Japan. The visits were spread over more than 400/100 spaceflights/missions. Many of the occupants have made multiple visits (two have made five trips), and eight were "tourists" who paid a great deal of money for the privilege.

The greatest number of people who have gathered in the one place in space is 13 people in the ISS, with the first of these gatherings occurring at the time of the docking of the NASA shuttle *Endeavour* in July 2009.

The longest time spent on the Moon was 75 hours by the *Apollo 17* crew of Harrison Schmidt and Eugene Cernan in December 1972. They spent a total of 22 hours on the Moon's surface, spread over three excursions, and ventured the greatest distance from a spacecraft during their extravehicular activity outside the lunar module *Challenger* (7.6 kilometers).

As the above commentary demonstrates, the achievements in individual and collaborative human spaceflight have been very significant and are made even more remarkable considering that the first powered flight of a heavier-than-air vehicle only became a reality on December 17, 1903, just 66 years before the first humans landed on the Moon. This first flight of the *Wright Flier* (figure 4.1) was piloted by Orville Wright at Kill Devil Hills near Kitty Hawk, North Carolina. It lasted just 12 seconds, traveled 37 meters, and reached a top speed of 11 km/h. Orville and his brother Wilbur completed three other flights that day, taking turns piloting, the longest traveling 260 meters in 59 seconds. The highest altitude reached in any of the flights was about 3 meters.

The feat was made possible when the brothers combined their mechanical experience from their bicycle shop business with the

Figure 4.1. Orville Wright, lying prostrate on the lower wing of the *Wright Flier* during the first powered flight of a heavier-than-air aircraft in December 1903 at Kitty Hawk, North Carolina. His brother Wilbur is shown standing on the ground at the right of the aircraft. The photograph was taken by one of five unnamed witnesses to the event. (*Public domain from National Park Service and NASA.*)

breakthrough invention of three-axis control to enable them to steer the aircraft and maintain its equilibrium. In acknowledgment of the achievements of the Wright brothers, the *Apollo 14* lunar module that landed on the Moon on February 5, 1971, piloted by Alan Shepard and Edgar Mitchell, was named "Kitty Hawk."

SUMMARY AND CONCLUSIONS FROM CHAPTER 4

There is little doubt that one of the most significant events in human history occurred on July 20, 1969, during the *Apollo 11* mission, when Buzz Aldrin and Neil Armstrong became the first humans to step onto another celestial body. The launch of cosmonaut Yuri Gagarin into orbit on a 108-minute flight high above Earth on April 12, 1961, was also a historic event since it was on that occasion that humanity first became a spacefaring species, but this might be considered a natural extension of previous suborbital flights.

On the other hand, some have claimed (Collins 2019) that the most momentous achievement of NASA's Apollo program or, for that matter, the entire space program, if not the history of mankind, occurred on December 23, 1968, when *Apollo 8*, or more particularly its passengers Frank Borman, Jim Lovell, and Bill Anders, became the first humans to come under the dominant influence of the gravitational field of a body other than the Earth, namely, the Moon.

This was the point in time when spaceflight (and indeed all human flight) "stopped going around in circles" and had a destination other than somewhere on or around the Earth. This paradigm shift for humanity's place in the universe began on December 21, 1968, when the command was given by NASA Mission Control for the "trans-lunar injection" rockets on *Apollo 8* to be fired to boost the speed of the spacecraft to a point where it exceeded the Earth's escape velocity of 11.2 km/s (40,000 km/h) and could move out of the Earth's gravitational field.

At this moment in the flight, the "CapCom" for the mission, Mike Collins,[2] who was about to become the pilot of the Command and Service Model for *Apollo 11* some seven months later, stated, "Mankind, the time has come to leave your first home." Later, he expanded on this comment, stating:

> *For the first time in history, man was going to propel himself beyond escape velocity, breaking the clutch of our earth's gravitational field and coasting into outer space as he had never done before. After TLI [trans-lunar injection] there would be three men in the solar system who would have to be counted apart from all the other billions, three who were in a different place, whose motion obeyed different rules, and whose habitat had to be considered a separate planet . . . leaving us stranded behind on this planet, awed by the fact that we humans had finally had an option to stay or to leave—and had decided to leave. (Collins 2019)*

Notwithstanding all the undoubted achievements of these spectacular and groundbreaking flights, including six Moon walks by 12 individuals and lengthy space station missions, they pale into insignificance when considering the challenges of the establishment of a permanent settlement in space, whether this be in Earth orbit, in the near-Earth environment farther afield (the Moon and Lagrangian points), or on or around another Solar System body.

The closest humans have come to semipermanent occupation of another body are the Apollo Moon landings, but these involved a combined stay of "just" 12 days, 11 hours, and 35 minutes, including a total time of 80.6 hours during surface activities on foot and/or on a rover vehicle.

The combined length of stays of humans on the various space stations in Earth orbit can be counted in several decades, but the record of time in space by any individual is 2.4 years by cosmonaut Gennady Padalka. In all these examples, the humans' time in space has relied either on the very limited supplies of food, oxygen, and fuel that they took with them at launch or on the continuous resupply of these life-sustaining resources by regular cargo missions (to the corresponding space stations only).

Moreover, even the longest *continuous* period in space by any individual so far (438 days, or 14.6 months, by Valeri Polyakov in 1994–1995) is significantly shorter than the 22 months required to get to and from Mars (using a Hohmann transfer orbit,[3] or HTO, in both directions). During the Mars flight and at the destination there will be no opportunity to resupply essential resources like food, air, and water, and there will be no protection from cosmic radiation that is provided to the ISS by Earth's magnetic field for at least part of the time in its low Earth orbit.

Discussion of these and many other very difficult challenges of journeys beyond the Moon is provided in chapters 8 and 9.

CHAPTER FIVE

Options for Establishing a Permanent Settlement in Earth's Immediate Environment

Many, and some of the most pressing, of our terrestrial problems can be solved only by going into space. Long before it was a vanishing commodity, the wilderness as the preservation of the world was proclaimed by Thoreau. In the new wilderness of the Solar System may lie the future preservation of mankind.
—Arthur C. Clarke, in the essay "What Is to Be Done," 1992

With all the spaceflights before and since the paradigm-shifting *Apollo 8* mission to the Moon in December 1968, the scene was being set for human settlement of life beyond Earth, whether that be in near-Earth environments, in more remote locations in the Solar System, or places outside our Sun's direct influence.

Some of the potential platforms for settlement of space in a relatively near-Earth environment (including low Earth orbit,[1] the Moon, and Lagrange points[2]) are summarized in table 5.1. Their challenges and the ways in which they might be overcome are discussed below.

Table 5.1. Location options for near-Earth human settlements, including some of their relevant characteristics and major issues.

Location	Distance from Earth	Characteristics	Major Issues
Space station in low Earth orbit	Up to 2,000 km	Technically feasible (viz, Mir, ISS, etc.). Easy contact with Earth. Weightlessness (i.e., "zero gravity"). Some protection from cosmic rays due to the Earth's magnetic field. The ISS is built for long-term accommodation of just seven people.	Possibility of collisions with other satellites. Resources must be produced sustainably on board in case resupply from Earth is not possible. The orbit gradually decays due to atmospheric friction and must be reestablished periodically. Subject to potential attack from Earth. Expansion is limited.
The Moon	384,000 km	Technically feasible (viz., the Apollo missions). One-sixth of Earth's gravity. Four days' travel from Earth. Local resources are available (e.g., water at the poles). Space to expand. Effective launch pad to more-distant settlement options.	Exposure to cosmic rays. No atmosphere. Extremes of temperature (−173°C to +127°).
Lagrange point	L1 and L2 (1.5M km) L4 and L5 (150M km) L3 (300M km)	Technically feasible (viz., the James Webb Space Telescope at L2). Days to weeks travel time to Earth. Weightlessness ("zero gravity").	Resources must be resupplied from Earth. Exposure to cosmic rays and micrometeorites. Possible collisions with other objects at the location.

A Large Space Station in Earth Orbit

The International Space Station (ISS) is a habitable artificial satellite in low Earth orbit (figure 5.1), with its first component launched in 1998 and an operational lifetime recently extended to 2030. Due to a long history of continual development, it is now the largest artificial body in orbit and, at a cost of $100 billion and rising, is arguably the most expensive single item of infrastructure ever constructed by humanity. It is a multinational collaborative project involving five participating space agencies: NASA (United States), Roscosmos (Russia), Japan Aerospace Exploration Agency (JAXA), European Space Agency (ESA), and Canadian Space Agency (CSA). The ownership and use of the space station is established by intergovernmental treaties and agreements (International Space Station legal framework, ESA, November 19, 2013).

The ISS is the ninth space station to be inhabited by crews. These include the four Soviet Union– and later Russian-operated Salyut scientific research space stations from 1971 to 1986, two military recon-

Figure 5.1. The International Space Station photographed from the SpaceX *Crew Dragon Endeavour* during a fly-around of the orbiting space station that took place following its undocking on November 8, 2021. For scale, the width of the space station from left to right is about 75 meters. (*Public domain from NASA.*)

naissance stations Almaz launched between 1973 and 1976, and the first modular space station, Mir, assembled in orbit from 1986 to 1996. The American Skylab station, launched on May 14, 1973, by the last flight of the complement of Saturn V rockets used for the Moon landings, was occupied by three crewed missions for a total of 24 weeks between May 1973 and February 1974. Unable to be serviced by the Space Shuttle program (which did not launch until 1981), it was destined to remain in a decaying orbit until it broke up and strewed several large and many small pieces near the Australian coastal town of Esperance on July 11, 1979. No one was injured, but the Shire of Esperance lightheartedly issued NASA a fine of AUS$400 "for littering."

The ISS circles the Earth every 93 minutes and is visible routinely to the naked eye from the surface. To counteract the natural decay of its orbit, it is maintained at an average altitude of 400 kilometers using the engines of the *Zvezda* service module on board or those of docked visiting spacecraft. The first modular component was launched in 1998, and the first long-term residents arrived on November 2, 2000, from the Baikonur Cosmodrome. It is now 73 meters long and 109 meters wide, with a mass of 445 tons. The pressurized/livable space is 915 cubic meters, about the same size as a five-bedroom house or the cabin of a Boeing 747 aircraft. At the time of this writing, the ISS had been continuously occupied for more than 21 years, surpassing the previous record of nearly 10 years held by the Mir space station.

The ISS consists of a massive and complex series of interconnected pressurized habitation modules, structural trusses, photovoltaic solar arrays, thermal radiators, docking ports, experiment bays, and robotic arms (figure 5.2). Major ISS modules have been carried to the space station by Russian Proton and Soyuz rockets and US Space Shuttles. The station is serviced by the Russian *Soyuz* and *Progress*, the SpaceX *Dragon 2*, and the Northrop Grumman Space Systems *Cygnus*, and previously by the European Automated Transfer Vehicle, the Japanese H-II Transfer Vehicle, and SpaceX *Dragon 1*.

The ISS has been constructed, with modules continuing to be added, over a period of more than 25 years, yet it can still accommodate only around seven residents. The time taken to build the ISS, its huge cost, and

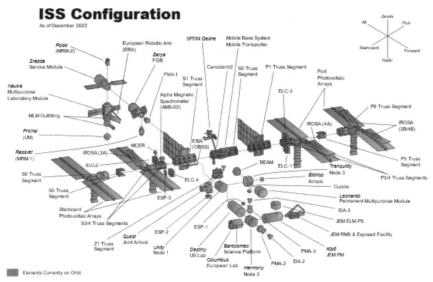

Figure 5.2. Exploded schematic of the International Space Station, showing all major components as of December 20, 2022. (*NASA and various Wikimedia editors including Colds7ream, Fritzbox, Johndrinkwater, Ras67, and Chepry. Derivative work: Trex2001 [talk]—ISS Configuration as of July 2021. Public domain.*)

the fact that it is critically dependent on regular support/supply missions from Earth point to the difficulty of constructing a sustainable platform for a much larger complement of humans in Earth orbit (or anywhere else, for that matter). This point is discussed at greater length in later chapters.

The ISS is designed to undertake a wide range of physical, chemical, and biological experiments, including extensive and comprehensive studies of human and plant responses to zero gravity and the testing of spacecraft systems and equipment that might be required for extended missions to the Moon and Mars. While this has produced a remarkable database of extremely useful information for future spaceflight, there have been calls to privatize/commercialize ISS operations beyond its current funding period to 2030, to free public expenditure to focus on the Moon and Mars missions themselves.

The ISS has demonstrated that long-term occupancy by humans of spacecraft in Earth orbit (viz., hundreds of days by individuals and decades collectively, so far) is feasible, albeit requiring continuous changeover of personnel and regular service visits from support craft. However, for a space station to qualify as a viable platform for the protection/preservation of a subset of humankind while waiting for a resolution of the chaos of one sort or another that might be playing out on the surface of the Earth, it would need to be far larger, significantly more complex, and completely self-sufficient. There would also need to be multiple space stations of this type to house the very large number of humans required for this "preservation" experiment.

Indeed, a platform or multiple platforms like the ISS in low Earth orbit is not ideal for preserving humanity's future, from a variety of critical perspectives (viz., the right-hand column of table 5.1 and details below):

1. Energy, fuel, water, food, and oxygen would need to be produced independently on the space station to maintain the large population (up to several thousand individuals) since supplies from Earth may not be able to be guaranteed, if indeed they could be delivered at all.

2. To maintain a stable orbit for the platform in space, the natural decay in its orbit due to friction with the (however tenuous) atmosphere around the Earth needs to be continually counteracted by the expenditure of significant quantities of energy that (1) has to be supplied from Earth or generated on board (very unlikely) and (2) could otherwise be put to alternative use.

3. The ever-present visibility of the Earth below and the relentless reminder of the suffering of fellow humans—whether it be from climate change, pollution, food and water shortages, asteroid impact, radiation and chemical contamination, etc.—would play heavily on the mental well-being (including "survivors' guilt") of the colonists.

4. Limited prospects for expansion of the physical dimensions of the platforms would exert significant pressure on the need for population control and the maintenance of genetic diversity.

5. The long-term effects of zero gravity on humans are still not well understood, but none of them are good. These include, inter alia, the

potential for reduced cognitive ability, eyesight damage, and bone and muscle wastage (see chapter 9 for a detailed discussion of these issues).

6. Without Earth's protective atmosphere and magnetic field, cosmic radiation doses can be more than 200 times higher than levels on Earth, with consequential effects on cancer incidence and damage to human genetic material. Measurements by the *Voyager 2* spacecraft showed a gradual increase in cosmic radiation up to five times the level experienced on Earth upon the craft reaching the heliosphere in 2018, some 119 AU away.

7. The platform would be extremely vulnerable to physical and electronic attack from some/most of the humanity that has not been included in such a settlement and has been left "in trouble" on the nearby ground below.

The parallel and related major challenges of who would be selected to occupy the platform and how it would be governed/managed from the political, psychological, and physical perspectives are very significant and are indeed common to all of the proposed settlement platform systems, whatever their nature or location. These "social" issues are discussed in detail in chapter 12.

THE MOON

The Moon has been the target for intense and consistent spaceflight exploration since January 2, 1959, when the Soviet Union's *Luna 1* became the first man-made object to escape geocentric orbit. Unfortunately, it missed its target (the Moon) but became the first craft to escape the Earth-Moon gravitational system. Other notable achievements of spaceflights to the Moon have been detailed in chapter 4.

At the time of the launch of *Sputnik 1* in October 1957, Soviet scientist U. Khiebtzevich was quoted in newspapers as saying that he "expects men to land on the moon within five or ten years of the launching of the first man-made satellite." This was nearly four years before the United States committed to the same goal! Whether President John F. Kennedy knew of the Soviet statement of intent is not known, but the early resounding success of the Soviet space program and its undoubted leadership in the so-called space race at that time certainly prompted his dramatic and famous announcement before a special joint session of Congress on May 25, 1961, that the United States "should commit itself to achieving the goal, before this decade is out, of landing a man on the Moon and returning him safely to the Earth." In any event, from that point on, the US program for exploration of the Moon was clearly focused on crewed missions to our nearest celestial neighbor as a means of restoring national pride.

After the success of the Apollo program, the Soviet Union switched its space exploration focus to uncrewed lunar missions that deployed rovers and returned samples to Earth. Three rover missions were launched, of which two were successful (*Luna 17, 21*), and 11 sample return flights were attempted with three successes (*Luna 16, 20, 24*). In total, the Soviet Union produced 18 successful missions to the Moon over the period 1959 to 1976, including 1 impactor (*Luna 2*), 1 flyby (*Luna 3*), 6 orbiters (*Luna 10, 11, 12, 14, 19, 22*), 2 soft landings (*Luna 9, 13*), 2 rovers (*Luna 17, 21*), 3 flyby and returns to Earth (*Zond 5, 7, 8*), and 3 sample returns (*Luna 16*, 101 g; *Luna 20*, 30 g; *Luna 24*, 170 g). A total of 0.3 kilogram of Moon rocks were brought back to Earth from the three sample-return missions.

The US lunar exploration program began on March 3, 1959, with the *Pioneer 4* flyby mission. Up to the present day, the United States has completed 28 successful Moon missions, with some of them continuing as orbiters. The total includes 2 flybys (*Pioneer 4* and *Mariner 10*); 3 impactors (*Ranger 7, 8, 9*); 12 orbiters (not including 8 in the Apollo program), with 3 of these combined with impactor probes and one with a lander probe); and 11 landers (*Surveyor 1–5*; *Apollo 11, 12, 14–17*). Interestingly, *Surveyor 6* executed a brief lift-off from the surface, landing 2.4 meters away from its landing spot.

Between 1968 and 1972, nine crewed missions to the Moon were conducted as part of the Apollo program. *Apollo 8* was the first crewed mission to enter orbit in December 1968, and it was followed by *Apollo 10* in May 1969, during which the landing module came within 14.3 kilometers of the surface before returning to dock with the command module in orbit. Six missions landed humans on the Moon, beginning with *Apollo 11* in July 1969, when Edwin "Buzz" Aldrin and Neil Armstrong became the first two people to walk on the Moon. *Apollo 13* was intended to land, but it was restricted to a flyby and return trajectory due to a malfunction aboard the spacecraft. All nine crewed Apollo missions returned safely to Earth.

In addition, Japan (from 1990), the ESA (from 2003), and India (from 2008) have completed orbiter and flyby missions and a soft landing near the lunar south pole, while the People's Republic of China (from 2007) has completed two orbital missions (*Chang'e 1, 2*) and three soft lander/rover missions (from 2013, *Chang'e 3, 4, 5*), the last of which was the first such landing on the far side of the Moon.

From 1959 to the time of this writing, a total of 146 missions to the Moon have been made, including failures, flybys, impactors, orbiters, landers, rovers, and crewed missions. Of these missions, 22 have successfully soft-landed (figure 5.3) to take pictures of the surroundings, execute scientific experiments, analyze samples of the regolith (soil), and, in 10 cases, return samples to Earth.

Over six Apollo missions, 12 humans explored the surface on foot and on rovers and brought back carefully selected samples (informed by earlier remote scientific assessments of the lunar geology and mineralogy by orbiters and landers) that have provided answers to many of the questions about the detailed geology and origin of the Moon. The preponderance of these crewed and uncrewed missions landed in the vicinity of the lunar equator, for reasons of trajectory and communication convenience (see further discussion below). More recent soft landings by a variety of nations (not all successful) have concentrated on the south polar regions because of the higher likelihood of water remaining in these locations and the importance of this for the sustainable operation of a future lunar base.

Human Occupation of the Moon

Humans have lived and worked on the Moon for a total of 12 days, 11 hours, and 35 minutes. The six visits by Apollo crews remain as the only time that humans have occupied a celestial body anywhere, aside from the Earth, although many semi-autonomous uncrewed spacecraft have spent, and continue to spend, considerable time on the Moon and several other bodies in the Solar System. The Apollo astronauts ventured out of their

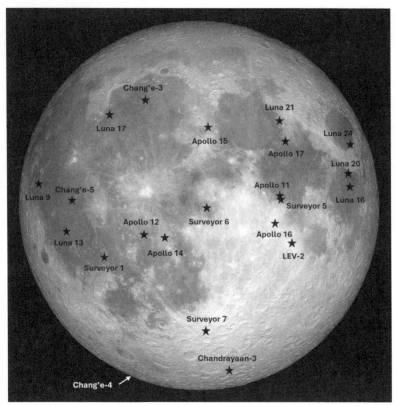

Figure 5.3. Location of the five Surveyor (USA), six Apollo (USA), seven Luna (Soviet), three Chang'e (China), one Chandrayaan (India), and one LEV-2 (Japan) landing sites on the Moon. The *Surveyor 3* site was within walking distance of *Apollo 12* and so lies under that symbol, and the *Chang'e-4* landing site is on the far side. (Compilation of landing site data from several sources, including https://en.wikipedia.org/wiki/Moon_landing.)

landing vehicle onto the surface 14 times for a total of 80 hours and 48 minutes (figures 5.4 and 5.5), during which they set up a wide range of scientific experiments, examined the geology at close quarters, and collected 2,415 samples of Moon rocks and soil, weighing 382 kilograms, for return to Earth.

All six Apollo landings were carefully planned to take place near lunar dawn. This was prescribed to ensure that the landing site was in full

Figure 5.4. Pete Conrad, commander of *Apollo 12*, standing on the Moon during his extravehicular excursion on November 20, 1969, next to the *Surveyor 3* lander that had arrived there 17 months earlier, on April 19, 1967. About 200 meters away on the horizon is the *Apollo 12* lunar module lander, *Intrepid*, in which Conrad and his pilot, Alan Bean, had descended to the surface on November 19. The television camera and other components were taken from *Surveyor 3* and brought back to Earth for scientific analysis. (*Alan L. Bean, lunar module pilot. Public domain from NASA.*)

sunlight for the duration of their stay and the temperature of the surface was not too hot or cold. Since none of the missions stayed on the surface more than 3 Earth days and the lunar day is 14 Earth days long, this ensured the surface did have time to heat up fully to its maximum daytime temperature of around 127°C. Notwithstanding this planning, the astronauts' spacesuits were designed to reflect almost 90 percent of the direct radiation from the Sun and its reflections from the surface. Also, the astronauts' boots were heavily insulated, slowing down the already

SPACE TO LIVE

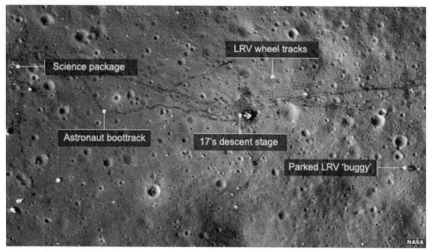

Figure 5.5. High-resolution image of the *Apollo 17* landing site acquired on September 15, 2009, by NASA's robotic *Lunar Reconnaissance Orbiter*, some 37 years after the mission was completed. The tracks laid down by the lunar rover vehicle (LRV) during this, the 11th and final mission of NASA's Apollo program, from December 7 to 19, 1972, are clearly visible, along with the last human foot/boot trails left on the Moon. The annotations also show where the astronauts placed some of the scientific instruments. (*Public domain from NASA, found at https://ichef.bbci.co.uk/news/976/mcs/media/images/55193000/jpg/_55193708_nasa.jpg.*)

relatively poor conduction from the lunar regolith to provide thermal protection for their feet.

Landing at night would have severely restricted the astronauts' visibility of the dangerous and unknown lunar surface, and would also have exposed them to temperatures as low as −173°C. In other words, even these landings on a celestial body very close to "home" had to be carefully planned to operate in the "Goldilocks zone" of the Moon's surface, where it is not too hot and not too cold for humans, even with their advanced technology to cope with the life-threatening local conditions!

Landing sites near the lunar poles, where the angle of the sunlight is more uniform, would have allowed the astronauts to stay in one location for an extended period without being concerned with large temperature deviations. However, polar sites were not favored by the Apollo missions because:

- They are on the edge of the line of sight to Earth, and thus communication would be compromised (the same argument resulted in the early elimination of sites on the far side of the Moon).
- The first three missions needed a free-return trajectory (which restricts the inclination of the initial lunar orbital insertion to be close to the Earth-Moon plane), limiting the latitudes to ±5° (*Apollo 11*, *12*, and *14* in figure 5.3); later missions landed at higher latitudes of +26° for *Apollo 15* and +20° for *Apollo 17*.
- Landing farther away from the lunar equator results in progressive loss of the slight surface rotation velocity advantage both on landing and liftoff, and the need to change the command module's orbit nearer to polar inclination for docking with the landing module moving up from the surface.
- A polar site would have compromised lighting due to the low inclination of the Sun and would admit the possibility of the landing site being in shadow and/or subjecting the astronauts to solar glare.
- Terrain nearer the poles is, on average, more rugged than near the equator, making landing trickier.

It goes without saying that all of these issues (along with many others to be discussed below) will have to be dealt with as part of any future mission to occupy the surface of another celestial body, wherever that body may be. It is certainly not going to be easy!

In more recent times, interest in exploring the polar regions of the Moon (especially the south pole) has increased significantly. Spectroscopic analyses from orbiting spacecraft have shown that the polar regions offer the potential advantage of the presence of large volumes of water ice in permanently shadowed regions at the bottom of deep craters (Speyerer and Robinson 2013; figure 5.6). This water is likely to have been deposited by the impact of asteroids and comets (especially during the Late Heavy Bombardment period of lunar and Solar System history) that contained significant quantities of volatiles, including water. The recent *Chandrayaan-3* soft landing near the lunar south pole (figure 5.3) was the first successful mission to target on-ground exploration in the polar regions,

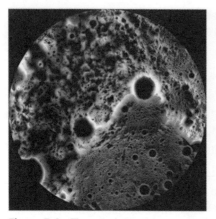 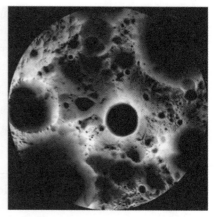

Figure 5.6. Time-weighted lunar north pole (left) and lunar south pole (right) illumination maps from 88 to 90 degrees, showing how much sunlight specific locations receive over the course of a lunar year. Areas that are nearly white are almost always in the sunlight, while black areas are permanently shadowed and are likely to contain vast amounts of water ice. (*Lunar Reconnaissance Orbiter Camera [LROC] and released by the Science Operations Center team at Arizona State University [Speyerer and Robinson 2013]. Reproduced with permission.*)

in advance of future missions that may be able to extract the water for human consumption and the generation of electricity and rocket fuel for a semipermanent settlement.

But temperature control was not the only life-support issue that needed to be addressed by the Apollo astronauts. Other challenges were:

- The need to maintain Earth-like pressure in the spacesuits in the face of the essentially pure vacuum of the Moon's surface.
- The continuous need to cool and dehumidify the spacesuits to deal with sweat and, of course, to cater to other natural bodily functions.
- In the absence of an atmosphere, the need for helmet visors to have ultraviolet radiation filters to avoid skin and eye damage from direct sunlight.
- The danger presented by lunar soil (regolith), which is composed of sharp, abrasive, and dangerous shards of silica that scratched visors,

clung to the spacesuits, and was inevitably transported into the landers, where some was no doubt inhaled.
- Again, without the presence of a protective atmosphere, there was an ever-present threat of the spacesuits and/or the landing vehicle being breached by micrometeorite (or larger) impacts.

The prospect of humans residing on the Moon for long periods has grown in recent years, with NASA's Artemis program targeting a return there by the late 2020s using its new Space Launch System and *Orion* spacecraft (possibly supported by the construction of a space station in lunar orbit). China is also planning its first crewed trip to the lunar surface the following decade, and SpaceX, collaborating with NASA on the *Starship* landing vehicle, has a long-term goal of settling Mars. The challenges discussed above will be at the front of mind in the planning for these missions, along with other equally challenging concerns such as:

- Sleep deprivation due to the difficulty of human circadian rhythms to adjust to a 28-day rather than the usual 24-hour cycle on Earth.
- The significantly reduced force of gravity—one-sixth of that on Earth—which, while not zero, will still affect humans in a range of undesirable ways (see chapter 9).
- Without Earth's protective atmosphere and magnetic field, cosmic radiation doses will be much higher than levels on Earth, higher than those experienced by a putative settlement on a spacecraft in low Earth orbit, and similar to those experienced by anything located at Lagrangian points. Protection from this radiation will require the construction of dwellings specifically designed to keep this radiation out or eliminated by their location underground.

The Advantages of a Settlement on the Moon
Despite the discussion above, not all aspects of the human settlement of the Moon are negative. The positive characteristics of such a location are provided, in no particular order, as follows:

1. The proximity of the Moon to the Earth (384,000 kilometers) provides several benefits:

   - The energy required to send objects from Earth to the Moon is lower than for more remote locations in the Solar System, say, at a Lagrange point or Mars, the asteroid belt, or other moons. This permits the construction of a much larger settlement than would otherwise be possible.

   - Transit time is short (around three to four days), which would allow supplies to quickly reach a Moon settlement from Earth during the set-up and operational phases (if the situation on Earth had not deteriorated too much by then).

   - The round-trip electronic communication delay to Earth is less than 3 seconds, allowing near-normal voice and video conversation, and allowing close to real-time control of machines that is not possible for more remote celestial bodies. For example, it takes a signal between 3 and 22 minutes to travel between Earth and Mars, depending on their relative orbital locations.

2. The Moon has been demonstrated to contain many (although not all) of the minerals and elements suitable for the construction of shelters and the manufacture of equipment. It may also have massive recoverable quantities of water ice present in deep craters near the lunar poles and perhaps in underground caverns elsewhere. These resources could eventually provide critical aspects of self-sufficiency for the settlement by delivering water for consumption by humans and for the growth of food and generation of fuel (through its solar electrolysis into the constituent gases hydrogen and oxygen). However, the Moon is highly depleted in carbon and nitrogen, and these deficiencies would severely compromise attempts to grow plants for food.

3. ^{3}He, a light, stable isotope of helium that has been deposited in the lunar soil from the solar wind over billions of years at a concentration of ~4.2 ppb or 0.007 gm^{-3} (Bruhaug and Phillips 2021), has been proposed as a feedstock for nuclear fusion energy on the Moon. How-

ever, this is unlikely to be realized anytime soon, given that fusion has yet to be demonstrated at useful/practical levels in even the simpler fusion experimental paths currently being explored on Earth (Simko and Gray 2014).

4. Lacking a significant atmosphere, the Moon would allow:
   - The production of solar electricity using a wider portion of the Sun's spectrum than is available on Earth.
   - The ability to construct lighter-weight and lower-strength buildings due to the absence of wind and rain, although they will have to be strong enough to resist damage and/or breaching by micro (and larger) meteorites unless the habitats are located underground in lava tubes or caves.

5. The lower lunar gravity means that less energy is required for the movement (viz., mining, transport, and pumping) of materials and for the launch of spacecraft from the surface back to Earth, or to other destinations. In contrast, some of the traditional mineral separation and processing operations (such as grinding mills that rely on gravity and mineral separation processes based on density contrast) will become more difficult.

6. Notwithstanding the prospect of feelings of "survivor guilt," being able to see the Earth in the sky continuously (at least on the hemisphere permanently facing Earth) may enable humans in a lunar settlement to feel less remote by providing every day the tangible prospect of a return to Earth when and if the need for their separation has passed.

On the other hand:

1. During the lunar night, temperatures quickly drop to around −173°C and stay there until dawn breaks two weeks later (except near the poles, as discussed above). This makes chemical and physical processing even more difficult and energy intensive than on Earth and will likely lead to embrittlement of important materials of construction.

2. Solar radiation, extreme temperatures, and micrometeorite strikes acting over billions of years create a lunar soil (regolith) that is exceptionally fine and dusty, and, without weathering, the particles remain very sharp and are "sticky" because they are electrostatically charged. Rovers and other equipment will need to operate under these harsh conditions for long periods, and humans will need to be very aware of the potential for lung damage if their breathing air becomes contaminated.

3. Human exposure to weightlessness over long periods on Mir and the ISS has been demonstrated to cause (mostly, but not always, temporary) deterioration of physiological systems, such as loss of bone and muscle mass and a depressed immune system (see chapter 9 for a detailed discussion). Similar effects are likely to occur in a low-gravity environment such as that on the Moon and Mars, although virtually all research so far has been limited to the health effects of zero gravity in low Earth orbit.

4. The spin axis of the Moon is nearly perpendicular to the ecliptic plane; that is, the plane of the Earth's orbit around the Sun. This results in the incidence of extreme lighting conditions at the lunar poles (Bussey et al. 2004).

    - The floors of impact craters near the poles may be ideal locations as future landing sites since some of the interiors of these craters never see the Sun, while topographically high regions there could be in near constant sunlight (>80%). These locations provide the potential for habitation in a much more benign thermal environment in which the temperature is around −50°C ±10°C (Bussey et al. 1999), rather than the extremes of +121°C and −133°C near the equator.

    - Despite arriving from low on the lunar horizon, the sunlight received at sites with higher elevation than the surrounding landscape would not be reduced by the usual atmosphere absorption, and its mostly continuous incidence throughout the lunar "day" and "night" would permit solar power generation for most of the time.

- Without the darkness of 14-day lunar nights, a community located near a pole could grow food crops essentially all the time, whereas elsewhere the plants could not survive the 14-day night without the provision of artificial light. In any event, plant growth on the Moon faces many difficult challenges due to continuous exposure to solar flares, lack of carbon and nitrogen in the soil and atmosphere, and lack of insects for pollination. Plants would need to be grown in sealed, transparent chambers with imported soil consisting of heavily modified lunar regolith and carefully controlled atmospheres to provide CO_2, but with the prospect of harvesting O_2 generated by the same plants.

5. When the Moon passes through the magnetotail[3] of the Earth, the plasma sheet passes across its surface, releasing electrons on the day side and causing a buildup of voltages on the dark side, which could have a significant negative impact upon a wide range of electronic equipment.

6. Issues regarding the rights of nations and their settlements to exploit resources on the lunar surface and to stake territorial claims and other issues of sovereignty would have to be agreed upon before a permanent presence on the Moon was established (see the next section for a discussion on this matter).

On balance, while it has severe challenges, the Moon is a very promising and perhaps the best candidate for the construction of, at the very least, a useful "demonstrator" platform/refuge and staging post for ex-Earth human settlement. It could also provide an intermediate platform for the assembly of residential and transport infrastructure, including the generation of fuel stock reserves intended for transport to settlements farther afield.

Claims on Extraterrestrial Real Estate

Each of the six manned landings on the Moon "planted" the US national flag on the surface, and numerous unmanned spacecraft have now landed on various other Solar System bodies. With interest in a return of humans to the Moon and potential exploitation of its resources rising fast, the issue of ownership of these territories and resources comes into sharp focus.

Agreement on the Outer Space Treaty was reached in the United Nations General Assembly in 1967 (Resolution 2222 [XXI]). The treaty was signed by the three depository governments (the Russian Federation, the United Kingdom, and the United States) in January 1967, and it entered into force in October 1967. It provides the basic framework on international space law, including the following principles:

- the exploration and use of outer space shall be carried out for the benefit and in the interests of all countries and shall be the province of all mankind;
- outer space shall be free for exploration and use by all States;
- outer space is not subject to national appropriation by claim of sovereignty, by means of use or occupation, or by any other means;
- States shall not place nuclear weapons or other weapons of mass destruction in orbit or on celestial bodies or station them in outer space in any other manner;
- the Moon and other celestial bodies shall be used exclusively for peaceful purposes;
- astronauts shall be regarded as the envoys of mankind;
- States shall be responsible for national space activities whether carried out by governmental or non-governmental entities;
- States shall be liable for damage caused by their space objects; and
- States shall avoid harmful contamination of space and celestial bodies. (United Nations Office for Outer Space Affairs, n.d.)

By 2023 the treaty had been ratified by 113 countries, including all the major spacefaring nations. It has also been signed but not yet ratified by 23 other nations.

Article 1 of the treaty (in full) states, "The exploration and use of outer space, including the Moon and other celestial bodies, shall be carried out for the benefit and in the interests of all countries, irrespective of their degree of economic or scientific development, and shall be the province of all mankind."

Article 2 (in full) states, "Outer space, including the Moon and other celestial bodies, is not subject to national appropriation by claim of sovereignty, by means of use or occupation, or by any other means."

A subsequent Agreement Governing the Activities of States on the Moon and Other Celestial Bodies (otherwise known as the International Moon Treaty) was finalized in 1979 and signed by five nations. It forbade private ownership of extraterrestrial real estate and, as of 2022, had been ratified by only 18 countries, none of which are involved in self-launched human spaceflight, and thus it has little to no relevancy in international law.

The implications of these treaties are clear—any nation or group of nations that land on another celestial body cannot make a successful ownership claim on that real estate, although resource extraction is not prohibited. If permanent occupation (i.e., settlement) of the Moon and other celestial bodies does take place in the future, it will be interesting to see if the treaties carry any weight.

Weinersmith and Weinersmith (2023) somewhat ominously pointed out that the technology of space exploration is difficult enough but it also needs to be accompanied by harmonious relationships between nations, and that bit isn't looking too promising at the moment. However, there is some hope in that humans have, so far, managed peacefully to abide by the regulations covering Antarctica and the ocean floor, despite the very likely possibility of massive mineral and fossil fuel resources in those regions.

Now, back to options for missions farther afield in the Solar System.

Artificial Satellite at a Lagrange Point in the Sun-Earth-Moon System

Another opportunity for the establishment of bases and/or settlements in the near-Earth environment, although considerably farther away than the Moon, are the so-called Lagrangian points L1–L5. These are five virtual points in the plane of any two co-orbiting large bodies (for example, the Sun and a planet, or a planet and its moon) where a much smaller object (such as an asteroid, meteoroid, interplanetary dust, or man-made satellite) can maintain a relatively stable position with respect to the two large bodies.

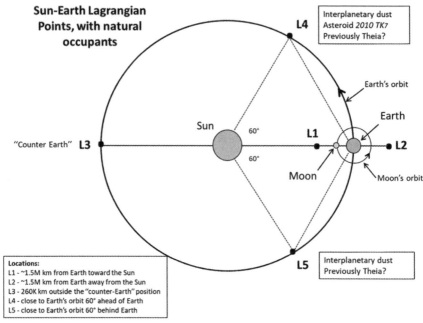

Figure 5.7. A schematic representation (not to scale) of the five Lagrangian points in the plane of the Sun-Earth system. Three of the points are colinear with the Earth and Sun (L1, L2, and L3), while the other two occupy positions (L4 and L5) 60° behind and 60° ahead of the Earth in its orbit, respectively. Only L4 and L5 are relatively stable orbital positions where interplanetary dust (Kordylewski clouds), asteroids ("trojans"), and even the protoplanet Theia (perhaps) are/were able to reside for long periods. The relative locations of each of the five points are provided in the key at the lower left of the figure.

The gravitational fields of the two massive bodies combined with the third body's acceleration are in balance at these Lagrangian points (figure 5.7).

Objects located at or close to L1, L2, and L3 have dynamically unstable orbits, and therefore cannot remain there for long periods without external intervention. Thus, a spacecraft can be "parked" in an orbit around these points if it is maintained there by relatively small amounts of fuel used for "station-keeping" adjustments. The L4 and L5 points represent stable equilibrium positions in the three-body system, provided that the ratio of the masses of the two large bodies is greater than 25. Indeed, it has been suggested that the hypothetical protoplanet Theia was located at either L4 or L5 in the Sun-Earth system (figure 5.7) prior to the destabilization of its orbit there, which resulted in its collision with the proto-Earth to create the Moon.

The Sun-Earth L1 point, some 1.5 million kilometers from Earth toward the Sun, lies outside the Moon's orbit and so is never shadowed by the Earth or the Moon, always sees the illuminated hemisphere of Earth, and has an uninterrupted view of the Sun. It has provided an ideal location for a host of past and current spacecraft missions and several future ones, including the Solar and Heliospheric Observatory (SOHO) to provide early warning of solar activity, Genesis to collect samples of the solar wind, the Advanced Composition Explorer (ACE) to study energetic particles from the solar wind and the interplanetary medium, and the Deep Space Climate Observatory (DSCOVR) to observe the sunlit side of the Earth.

The Sun-Earth L2 point, being 1.5 million kilometers on the outer side of the Sun-Earth pair (figure 5.7), sees only the "new Earth" (i.e., the unlit side) and has a clear view of deep space since it lies outside the Moon's orbit. Radio telescopes placed there would receive much less interference than existing telescopes on Earth or in low Earth orbit. Also, because it retains a more or less constant relationship to the position of the Sun and Earth, critical radiation shielding and temperature control of spacecraft located there are much simpler than in other positions. It is unsurprising, therefore, that the much-awaited James Webb Space Telescope was positioned there in January 2022.

Past missions to Sun-Earth L2 include the 3.5-meter Herschel Infrared Telescope and the Planck Observatory, which studied the cosmic microwave background.[4] Among current missions to orbital positions around L2, the Gaia Observatory is used to determine the precise locations and velocities of up to a billion stars, thereby creating a precise 3D catalogue of the Milky Way galaxy.

No spacecraft has been sent to Sun-Earth L3 (the "counter Earth" position) yet, despite the fact that it would be able to closely monitor the evolution of active sunspot regions on the Sun before they rotated (after seven days) into a position on the side facing Earth, where the particles emitted from these regions might endanger communications and life on Earth. An early warning of such an event could then be issued by the National Oceanic Atmospheric Administration (NOAA) Space Weather Prediction Center.

L3 has been described in science fiction as the site of an invisible "counter Earth" body harboring unfriendly aliens. Fortunately (in this case), the Sun-Earth L3 is unstable and could not contain an object, large or small, friendly or unfriendly, for very long because the gravitational forces of the other planets are stronger than that of Earth (Venus, for example, comes within 0.3 AU of this L3 every 20 months).

The L4 and L5 points are stable in any Lagrangian system, and therefore asteroids, dust, and other objects have been observed in orbits around these positions on several planets. Many thousands of asteroids (figure 5.8) have been observed around Jupiter at L4 ("Trojans," with individual occupants named after characters in Homer's *Iliad*) and L5 ("Greeks"), and the Neptune points are even more thickly populated, perhaps up to 10 times more so than for Jupiter. The same nomenclature is applied to the L4 and L5 positions on planets other than Jupiter, but these are distinguished by use of the lowercase "trojans." A further set of asteroids, named after asteroid 153 Hilda, is a dynamical group of more than 5,000 bodies located roughly inside the L3, L4, and L5 positions in Jupiter's orbit, kept in place by a 3:2 orbital resonance with that planet.

The same generic Lagrangian geometry applies in the Earth-Moon system (figure 5.9). Only two asteroids, 2010 KT7 (400 meters in diameter) and 2020 XL5 (1.2 kilometers), have been located at L4 so far, and none at L5, but large concentrations of dust have been discovered at both points, collectively known as the Kordylewski clouds.

Earth-Moon L1 and L2 have been, and are, of great interest for lunar exploration, including for the provision of a staging location for assembly of telescopes and human exploration of planets and asteroids, and as communication relay locations for a future lunar outpost. L1 allows easy access to lunar and Earth orbits with little change in velocity and therefore has advantages for the transport of cargo and personnel to the Moon and back. NASA's *Artemis-P1* visited both L1 and L2 in 2010, and *Artemis-P2* visited L2 in the same year. Both craft were on missions to determine how mass and energy move through near-Earth space to produce auroras. The service module of China's *Chang'e-5-T1* sample return mission to the

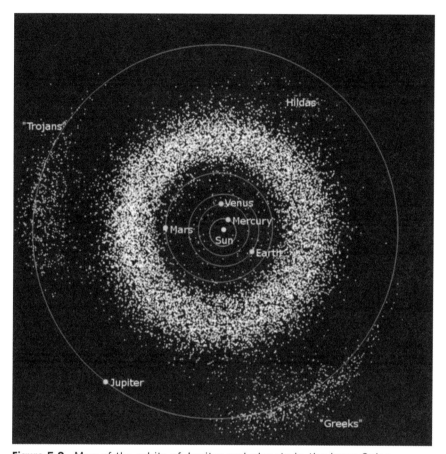

Figure 5.8. Map of the orbits of Jupiter and planets in the inner Solar System, together with clusters of specific "families" of asteroids. Asteroids linked gravitationally to Jupiter are shown as broad patches of gray dots 60° in front of (L4 "Trojans") and 60° behind (L5 "Greeks") the planet on both sides of its orbital path. Also shown is a more or less continuous broad band of white dots, representing the main asteroid belt between the orbits of Mars and Jupiter, and the "Hilda" family of asteroids located in broad patches inside the L3, L4, and L5 positions. The image is based on data in the en:-JPL DE-405 ephemeris and the en:Minor Planet Center database of asteroids (etc.) published July 6, 2006. (*Wikiedia Commons.*)

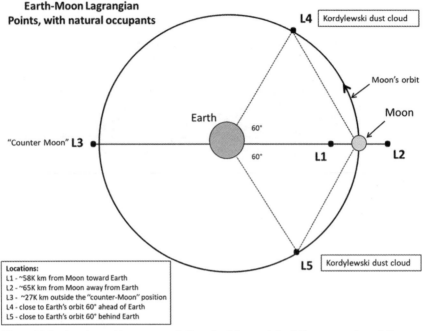

Figure 5.9. A schematic representation (not to scale) of the geometry of the five Lagrangian points for the Earth-Moon system, with the relative positions of the points provided in the key at the lower left. Unlike the Sun-Earth Lagrangian system (figure 5.7), the only "natural" occupants of the relatively stable L4 and L5 positions found to date are asteroids 2010 K7 and 2020 XL5 and interplanetary dust particles comprising the corresponding Kordylewski clouds.

Moon in 2014 was relocated temporarily to L2. The *Chang'e-4* mission to the far side of the moon in 2018 included a lander and rover (*Yuto-2*), with communications supplied by the *Queqiao* relay satellite that had been positioned earlier at L2.

No spacecraft have been parked at L4 or L5 in the Earth-Moon system, but Japan's *Hiten* spacecraft, following studies of the Moon, was put into a looping orbit that passed through both of these points to look for trapped dust particles in the so-called Kordylewski clouds (none were found). In 2006 NASA's Solar Terrestrial Relations Observatory (STEREO) was launched in an orbit around the Sun, consisting of two identical craft that passed by the L4 and L5 positions during their mission to provide stereoscopic imaging of the Sun and coronal mass ejections. Several other spacecraft (*OSIRIS-REx*, the Spitzer Space Telescope, and *Hyabusa 2*) have also passed close to L4 and L5 at one stage of their missions to other targets.

Implications for the Location of Human Settlements and/or Supply Depots
Sun-Earth Lagrange Points

The Sun-Earth L1 and L2 positions would be useful for settlements, but they are not stable and spacecraft located there would need to be maintained in their positions using station-keeping procedures. The L1 position in the Sun-Earth Lagrange system is relatively close to the Earth and is bathed in constant sunlight (figure 5.7). It would provide the opportunity for the continuous collection of solar power in support of a large settlement. On the other hand, the L2 point is 1.5 million kilometers from Earth, just outside the tip of the Earth's umbral shadow (1.4 million kilometers from Earth). This means that, depending on the specific orbit of the settlement around L2, an observer looking back from L2 in the direction of the Sun would see an occasional partial or full annular eclipse of the Sun by the Earth. A spacecraft/settlement located exactly at or close to L2 would have the advantage of being shielded from most of the Sun's heat and solar wind.

Both L1 and L2 can be used to deliver communications covering the Moon's far side (depending on the position of the Moon in its orbit around Earth) and is a potential location for a propellant depot for the settlement and/or lunar and planetary missions.

The L4 and L5 positions are relatively stable, requiring only minimal fuel for station-keeping procedures, and so could be used as settlement platforms and/or for deep space mission support, including waypoints in manned mission to Mars and near-Earth asteroids (for commercial extraction of valuable metals and gases) and to expand the practical launch window for travel to and from the Earth and other planets. They are, however, 150 million kilometers from Earth, meaning that routine and emergency resupply missions would be lengthy.

Earth-Moon Lagrange Points

A satellite at the L1 Lagrangian point in the Earth-Moon system is in a good location for monitoring and coordinating communications in support of operations based on the nearside of the Moon. The L1 position allows comparatively easy access to lunar and Earth orbits with minimal change in velocity and can therefore provide advantages as a position for a space station settlement and/or to transition cargo and personnel to and from settlements on the Moon.

A vessel launched from L1 could reach any place on the Moon within a few hours to a day, making it ideal for crisis management if an emergency occurred on the Moon, and would allow it to serve as a repair center for ships moving throughout the Solar System. However, both L1 and L2 require active station-keeping since neither is fully stable.

On the other hand, settlements at L4 and L5 would have the advantage of being stable without any need for substantial station-keeping corrections and could also be used as waypoints for travel to and from locations between the Earth's upper atmosphere and the Moon's orbit, including other Lagrangian points. In addition, they would significantly reduce the velocity change needed to move from one body to another, or to enter or leave Earth orbit, an important drawback of any lunar surface station, which demands high energy expenditure to escape and a comparable or greater amount to soft-land.

Disadvantages of the Lagrange Points

The main risks to any extended human presence on a Lagrangian point are significant exposure to particles in the solar wind and the health threat from cosmic rays. In the Earth-Moon system, the L1 to L5 points all revolve with the Moon around the Earth and this takes them outside the protection of the Earth's magnetosphere for approximately two-thirds of the time (as it does for settlements on the Moon itself). Settlements at L1 (located between the Earth and Moon) will experience this to a lesser degree, while L2 (located beyond the Moon) will experience this to a greater degree, and all of them will be exposed to the little-understood plasma sheet of the magnetotail.

In the Sun-Earth system, L1 and L3–L5 all lie on the Sun side of the Earth and so are all outside the protection of the Earth's magnetosphere. L2 periodically transfers from within the magnetotail, plasma sheet, and solar wind depending on the intensity and direction of the solar wind.

Similar health issues for humans apply when they are located on space stations with elongated elliptical orbits around Earth.

On balance, while Lagrangian points offer some advantages as way stations for other potential settlement platforms (especially those on the Moon) and for the location of specialist astronomical experiments, they are not ideal locations for a permanent human settlement.

Near Earth—Conclusions

Several reasonable options exist for the establishment of a human settlement off the surface of, but relatively close to, the Earth, each with considerable variations in technical difficulty and cost. Near-Earth options like large(ish) space stations in low Earth orbit or at certain Sun-Earth and Earth-Moon Lagrange points and settlements on the Moon offer the simplest options. Several missions have already provided humans with some degree of experience in remote operation of uncrewed spacecraft occupying these locations and experience from a small number of crewed lunar missions, albeit in the latter case for periods of less than 3 days of continuous occupation and only 12 days in total.

The crewed mission experiences have not only provided a comprehensive indication of the technical difficulties of establishing and maintaining a human presence at these sites, but have also permitted knowledge to be gained about the social and political aspects of occupation by small isolated and confined groups (e.g., up to 13 people at any one time on the ISS, so far) of multinational humans.

In the case of uncrewed craft located in Earth orbit and at Lagrange points, the missions have provided an increased understanding of the challenges in providing supplies to and operating these craft remotely, and the ability to implement any required fix-up actions locally through "space walks."

Of note in this respect are the repairs that were needed to be undertaken to operationalize the Hubble Space Telescope, which was launched with a fundamental flaw in its optics (namely, spherical aberration in the incorrectly ground main mirror). A rescue mission by the Space Shuttle in 1993 provided a correction to the manufacturing flaw, and another four servicing missions between 1997 and 2009 (when the shuttle went out of service) allowed other equipment repairs, replacements, and upgrades. These missions were also able to progressively extend the life of the telescope by using the Space Shuttle's thrusters to move Hubble into a higher orbit, thereby delaying the natural deorbiting process due to atmospheric drag. The aborted *Apollo 13* mission to the Moon in 1970 is another example of the need for intense engagement from home base on Earth when trouble arose, in that case to affect a successful return of the three-person crew when an oxygen tank explosion on the way to the Moon rendered a lunar landing impossible.

The hugely successful International Space Station's days are now numbered. Originally commissioned in 1998 in anticipation of a 15-year lifespan, the ISS has already been in operation for 26 years, but plans are afoot to decommission it around 2030. The decommissioning operation will most likely culminate in arranging for it to plunge safely into the middle of the southern Pacific Ocean at a location called "Point Nemo," also known as the "spacecraft graveyard" or "pole of inaccessibility," the farthest point from any coastline on Earth.

The only other current "permanently" crewed space station is China's Tiangong, the first module of which was launched in 2021 in a low Earth orbit between 340 and 450 kilometers above the surface. After the addition of two further modules, it has a pressurized volume of 340 cubic meters (the habitable space is 122 cubic meters), slightly over one-third the size of the ISS, with plans to expand this through a fourth module in 2024. It has an expected operational life of 15 years, including accommodation for up to three crew.

The promise of private sector "space tourism" is emerging through companies such as Virgin Galactic, following in the footsteps of paying customers being able to visit the ISS on *Soyuz* supply vessels. So far, these initiatives have been limited to orbital and suborbital ventures, but it is perhaps a sign of the future that the next major developments in human occupancy of space will be driven by private-sector interests.

At the time of this writing, there have been no human excursions beyond low Earth orbit since the Apollo program over 50 years ago. The prospects for expansion of the human experience in near-Earth space are uncertain in the near term due to the need to reinvent the launch vehicle, a redesign of the human spacesuits and cabin, and the involvement of multiple national and private-sector parties, some of which have quite different visions for the end game, namely, tourism, Moon orbital and/or surface bases, and a settlement on Mars, and various combinations of these.

The Lunar Gateway space station proposed by NASA, the European Space Agency, Canada, and Japan is currently set for launch in late 2028 and will serve as a science platform and staging area for the anticipated lunar landings of NASA's Artemis program and subsequent human mission(s) to Mars. China and India are also ramping up their space programs and already have documented plans to establish bases on the Moon.

There is talk of expeditions being mounted to Mars, not just by NASA but also by several private entrepreneurs, including Elon Musk, who has stated publicly that he intends to launch a crewed Mars mission by 2029 (Torchinsky 2022).

But more about Mars in chapter 6.

CHAPTER SIX

Options for Settlement Beyond the Earth-Moon Setting but Within the Solar System

> *There will, from desire or necessity, come the idea of building a permanent home for men in space. . . . At first space navigators, and then scientists whose observations would be best conducted outside the earth, and then finally those who for any reason were dissatisfied with earthly conditions would come to inhabit these bases and found permanent spacial colonies.*
> —J. D. BERNAL, THE WORLD, THE FLESH AND THE DEVIL, 1929

Robotic, uncrewed rovers and other static equipment have survived and operated successfully for long periods (decades) and distances (billions of kilometers) and in extreme environments, including close proximity to the Sun; the surface of Venus (for short periods of time); the Moon and Mars (for several years in some past and ongoing examples); Saturn's moon Titan; the asteroids Ryugu, Eros, and Itokawa and comets Churyumov-Gerasimenko and Tempel; and in deep space (*Voyager*, *Pioneer*, and *New Horizons*).

The complexity and duration of these missions provides some confidence in humans' ability to engineer, deploy, and operate man-made structures on and around other bodies in the near and far reaches of the Solar System. However, in only one case—the Moon—have the missions involved human passengers with the associated need to deal with the multidimensional and very complex issues of life support for a biological entity.

Many nations, but more particularly the United States and the Soviet Union, have undertaken a bewildering array of exploratory missions to the

Sun, all the planets in the Solar System, most of the larger moons of the planets, and several asteroids and comets. One craft, *New Horizons*, has explored (in flyby) the dwarf planet Pluto and the 35-by-20-by-10-kilometer object named Arrokoth in the Kuiper Belt beyond the orbit of Pluto and is now on its way into deep space, like the earlier *Pioneer* and *Voyager* craft, to undertake scientific observations of relevance to the origins and nature of the Solar System.

RECORD OF PREVIOUS SPACECRAFT MISSIONS BEYOND THE MOON

The first interplanetary mission to fly past another planet was the Soviet Union's *Venera 1* in 1961. The robotic spacecraft was intended to enter orbit around Venus, but suffered a malfunction, contact was lost, and it passed within 100,000 kilometers of Venus before assuming an orbit around the Sun. The United States' *Mariner 2* became the first successful interplanetary mission in December 1962 when it collected photographic and other physical data from 35,000 kilometers above Venus.

Since these pioneering missions, spacecraft, or small probes launched from them, have variously orbited, impacted, flown by, soft-landed stationary or mobile rovers, and, in some cases, returned samples of these bodies to Earth. Highlights of the achievements from these missions have been provided in chapter 4. As mentioned above, five of these spacecraft—*Voyager 1*, *Voyager 2*, *Pioneer 10*, *Pioneer 11*, and *New Horizons*—have speeds that are greater than the escape velocity of the Solar System and so are on paths that either will take or have already taken them out of the Solar System. Their trajectories are shown in figure 1.7. From 1973 *Pioneer 10* was the most distant human-made object from the Sun, but in 1998 *Voyager 1* caught and passed it. *Voyager 1* is now about 160 AU away and is, and will remain, the farthest human-made object from Earth.

Voyager 1 is a 722-kilogram space probe launched by NASA on September 5, 1977, to study the outer Solar System. It has operated for more than 47 years to the time of this writing and is expected to continue its mission until 2025, when the (decaying) electrical output from its thermonuclear power unit will not be sufficient to drive any of its instruments. None of the other four interstellar spacecraft will ever catch it, as they all have lower velocities than *Voyager 1*'s 16.9 km/s (3.57

AU/yr). After about 2036, both *Voyager* probes will be out of range of the Deep Space Network and will no longer be able to be tracked.

Voyager 1 completed its primary mission on November 20, 1980, after engaging with the Jupiter and Saturn systems in 1979 and 1980, respectively. Like *Voyager 2*, it is now in an extended mission to locate and study the regions and boundaries of the outer heliosphere, and then to explore the interstellar medium. It was the first human-made object to cross the heliopause and enter interstellar space on August 25, 2012.

Pioneers 10 and *11* both carry small metal plaques identifying their time and place of origin for any intelligent life that might find them in the distant future. Both *Voyagers* took this messaging a step further by carrying a kind of "time capsule," intended to communicate a more descriptive story of our world. This includes a phonograph record, a 12-inch gold-plated copper disk containing sounds and images selected to describe humans and to portray the diversity of life and culture on Earth. It indicates where Earth is located in the Milky Way galaxy in case the probes are intercepted by another intelligent life-form.[1] Some worry that it might have been a mistake to have provided this location information, but others have claimed that the instructions, based on the locations and rotation periods of nearby pulsars (highly magnetized rotating neutron stars), although clever at the time, are flawed because of inaccuracies in the rotation timings and will be of no help (Siegel 2017).

Opportunities for Settlement

Without wishing to diminish the extreme additional complexity of transporting humans (as opposed to inorganic structures) on any of the mission trajectories described in chapter 4 and above, they demonstrate that it is possible, with current technology, to send craft to just about all corners of the Solar System and beyond. Several of the Solar System destinations have been identified as possible locations of past and/or present life-forms, albeit probably only in bacterial or archaean form. Indeed, several of the missions have been specifically designed to test for evidence of life on the target body. More recently, many of these celestial bodies have been suggested as likely target platforms for settlement, or at least visits, by humans (table 6.1).

Table 6.1. Options for Solar System environments that might be suitable for human settlement, including some of their relevant characteristics and major issues.

Destination	Distance from Earth / Diameter	Characteristics	Major Issues and Positives
Mars	55–401M km 6,779 km	Solid surface. 38% of Earth's gravity. Thin, unbreathable atmosphere (95% CO_2, 3% nitrogen, 2% argon). Low atmospheric pressure (0.6% of Earth's). Temperature −153°C (poles) to +20°C (equator).	Technologically feasible—many spacecraft have successfully orbited and soft-landed. Local resources available for production of structures and generation of fuel, energy, and water. Plenty of room to expand, subject to the availability of local resources. Unknown long-term effect on humans of reduced gravity.
Mercury	79–228M km 4,879 km	Solid surface. 38% of Earth's gravity. No atmosphere. Temperature −180°C (side opposite the Sun) to +427°C (side facing the Sun).	Practical long-term habitation only near the poles. Local resources available for utilization. Plenty of room to expand, subject to the availability of local resources.

(*continued*)

Table 6.1. *Continued*

Destination	Distance from Earth / Diameter	Characteristics	Major Issues and Positives
Venus	38–261M km 12,104 km	Solid or partially molten surface. 91% of Earth's gravity. Atmosphere is mainly 96.5% CO_2 and 3.5% N_2. Surface pressure is 93 bar (equivalent to 900 m underwater on Earth). Temperature 438 to 482°C.	Far too hot for life to exist on the surface, but it may possibly survive in the upper atmosphere.
Titan (the largest moon of Saturn)	1.27B km 5,148 km	Only body other than Earth in the Solar System known to have liquid lakes of ethane (C_2H_6) and methane (CH_4) on its surface. 13.8% of Earth's gravity. Atmosphere is 98.4% nitrogen, 1.4% methane. Surface atmospheric pressure is 1.45 times Earth's. Temperature –179.5°C.	Very cold and toxic environment, but practical scenarios exist for long-term settlement (Wohlforth and Hendrix 2016). Atmosphere and surface could provide abundant water and ingredients for construction materials and fuel. Expansion possible using plentiful local resources.

Destination	Distance from Earth / Diameter	Characteristics	Major Issues and Positives
Dwarf planet Ceres (in the main asteroid belt between Mars and Jupiter)	414M km (at closest approach) 974 km	Solid surface of water ice and clay minerals. 2.9% of Earth's gravity. Very tenuous atmosphere. Temperature −168°C.	Most likely option is as a potential way station and/or launch platform for more-remote settlement destinations.
Enceladus (one of the small moons of Saturn)	1.27B km 252 km	10 km thick reservoirs of liquid water lie beneath the frozen exterior. 1.2% of Earth's gravity. Atmosphere is 91% water vapor, 4% N_2, 3.2% CO_2, 1.7% CH_4. Temperature −198°C.	Potential launch platform for more-remote destinations.
Europa, Callisto, and Ganymede (moons of Jupiter)	645M km 3,121 km, 4,820 km, and 5,268 km, respectively	Europa may have more liquid water than in all of Earth's oceans, but it is buried below kilometers of ice. Callisto and Ganymede may also have buried, salty oceans located 100 km below the surface.	Any life forms existing there might be able to use the energy from the moons' internal heat, generated by gravitational flexing.

Of particular note is the surface of Mars, where conditions seem to be favorable for the presence of bacterial or archaean life and, perhaps, more complex organisms (very unlike Venus). As a result, the focus and intent of most of the scientific experiments that have been installed on the landers and rovers that have been sent to Mars have been to seek evidence for past or present life-forms (mainly by "chasing the water"). These missions have covered a broad area of Mars's surface from the poles to the equator but have, so far, failed to provide direct evidence of life, past or present (see the detailed discussion in chapter 1).

The subsurface, surface, and atmosphere of several of the moons of Jupiter and Saturn are also considered prospective for the discovery of life, albeit primitive, in the Solar System (table 6.1). This is because large "oceans" of liquid organic molecules, including compounds of water, are inferred to be present from orbital spacecraft radar and optical imagery of surface structure and emissions from geysers.

On balance, these planetary moon locations have very severe and challenging environments (very low gravity, very low surface temperatures, and mostly toxic or nonexistent atmospheres) and are hardly representative of warm and inviting "Goldilocks" platforms for human settlement. They are also very long shots for the discovery/existence of anything beyond the simplest and hardiest of bacterial and archaean life-forms.

Terraforming

A different and much longer-term settlement scenario for any of these platforms, especially ones similar to Earth (like Mars), emerges if the prospect is considered of transforming, or "terraforming," the entire environment of the body to make it more compatible with the life-supporting conditions on Earth (Fogg 1995). This would allow much longer, potentially indefinite periods of occupancy on the planet than would be possible through the construction of localized "bubbles" of human-compatible environments on the surface (or underground), as occurred temporarily on the Moon during each of the Apollo lunar landing missions when the astronauts retreated to their lunar (descent and ascent) module after each extravehicular excursion.

However, the time required to modify the entire atmosphere of a large celestial body enough to provide a breathable composition, to increase or decrease the temperature to permit liquid water (and perhaps rain) on the surface, and to enable the widespread growth of food would be of the order of centuries, if not millennia. Furthermore, the requisite engineering effort (and physical resource and energy costs) would be gargantuan. Nevertheless, were the body to be successfully transformed in this manner, the path would then be clear for significant numbers of humans to settle it for the purposes of escaping a cataclysmic end to life on Earth, or at least to provide a "pressure release valve" for population expansion on Earth.

If the driving force for such terraformation of anther celestial body is the need to evacuate Earth due to one or more of the anthropogenic extinction scenarios confronting Earth (as described in chapter 2), then one caveat raised by Sagan (1994) was his prescient question of whether humans can be trusted to occupy and modify another celestial body when we have performed so poorly in the sustainable management of our current home. It seems hard to argue sensibly against this proposition unless the driver for settlement is one of the extinction scenarios that is beyond our control.

If humans can "get their act together" on Earth regarding all those scenarios over which we *do* have control, it has been suggested (Dyson 1979) that the proposed settlement platforms may not be needed to accommodate large populations of humans fleeing Earth. Rather, they might provide a place off-Earth where the "dirty" and energy-intensive manufacturing and resource extraction and processing operations might be relocated and where unwanted, non-recyclable waste (including CO_2) might be deposited, leaving Earth to remain in a more pristine and sustainable state for humanity to live.

If this nirvana is deemed impossible to achieve, and the primary driver is the search for an alternative home for all of humanity rather than just the "industrial" activities, then in order to terraform, say, Mars, three initial massive, interconnected transformations of the Martian environment would need to be made (Fogg 1995): (1) strengthening the magnetosphere to provide radiation protection and to prevent atmospheric stripping,

(2) thickening the atmosphere and changing its composition to a nontoxic mix of breathable gases, and (3) raising the surface temperature to a level compatible with the growing of food and well-being of humans.

A thicker atmosphere that is protected from stripping and composed of greenhouse gases like CO_2 and water vapor and/or the addition of more powerful types, like methane, sulphur hexafluoride, chlorofluorocarbons (CFCs), or perfluorocarbons (PFCs), would trap even more incoming solar radiation, and the resultant raised temperature would add further greenhouse gases to the atmosphere by subliming or melting of polar CO_2 ice. All three terraforming processes would augment each other until enough warming had been created for the system to then be stabilized (by a yet-to-be-determined process).

The source of the required gaseous CO_2 could be through initial localized warming of existing copious quantities of frozen CO_2 at the poles, but much more challenging proposals for warming the planet could be to (1) transport ammonia (NH_3) from one of the asteroids in the main belt of these objects just outside Mars's orbit, whereupon its conversion to nitrous oxide via bacteria and archaea added to the Martian soil (Prosser et al. 2020) would provide the requisite greenhouse effect; (2) use gigantic mirrors in orbit around Mars to provide direct local heating of the surface; and/or (3) lower the albedo of the Martian surface through application of a thin layer of darker material, especially on substantial parts of the polar ice caps, to accelerate evaporation of CO_2.

Creating and/or strengthening the magnetosphere around Mars would be an even more complex and costly exercise. Various "out there" options to achieve this have been proposed, including (1) increasing the temperature of the core of Mars by passing a *very* strong electrical current down a series of *very* deep holes, (2) increasing the rotation speed (and simultaneously, the overall temperature) of the planet by arranging a collision with another large body, and (3) constructing a swarm of solar-powered satellites or superconducting rings that generate a gigantic artificial magnetic field.

Clearly, these latter proposals lean into science fiction and would require a gigantic and very costly engineering and materials effort at a scale that is currently beyond our capability.

Paraterraforming

An alternative but related concept is known as "paraterraforming," or the "worldhouse" concept (Fogg 1995). This involves the construction of a habitable enclosure (viz., a contained biosphere) on a planet or moon that starts small but may eventually grow "organically" to encompass a large part of the planet's potentially usable area. One of the more common proposals involves the construction of a scalable enclosure consisting of a transparent roof extending one or more kilometers above the surface, pressurized with a breathable atmosphere and anchored with tension towers and cables.

Paraterraforming has several advantages over terraforming an entire celestial object (Fogg 1995). For example, proponents claim worldhouses can be constructed with technology known since the 1960s and that they provide a shorter-term payback to investors (assuming a capitalistic financing model). The model can be applied readily to a small area (perhaps a small domed community), where it quickly provides habitable space for the initial colonists and allows for expansion through a modular approach that can be tailored to the needs and scale of the occupant population, growing only as fast as necessary and only in those areas where it is required. Importantly, it can be designed to utilize the indigenous resources of air and solid materials following the initial establishment of shelter and manufacturing capability.

The solid envelope of the habitat dramatically reduces the volume of atmosphere that is needed to provide an Earth-like (breathable) composition and pressure. The atmosphere would not need to be replenished continuously to compensate for the usual loss of gases to space and so could be used on bodies that would otherwise be unable to retain an atmosphere at all (such as small moons and asteroids). Furthermore, the composition of the atmosphere (and the materials used for the transparent window of the enclosure's roof) could be manipulated to provide positive and negative greenhouse effects as needed, thereby allowing control of the habitat temperature.

Paraterraforming is also less likely to cause harm to any native lifeforms (simple or complex) that may already inhabit the planet, as the areas outside the enclosure will not normally be affected. Finally, it would be possible (if desired by the inhabitants) to embark upon a controlled, long-term process of terraforming the entire planet without the need to evacuate it during this development phase.

Disadvantages of worldhouses include the susceptibility of the containment structure to catastrophic failure if a major breach occurred, though this risk might be reduced by compartmentalization and other active safety precautions. Meteor strikes on a body that has no or very little atmosphere are a particular concern because without this usual protective shield, they would not burn up or slow down and would reach the surface with their full force. Also, large habitats would require massive amounts of construction materials, energy, labor, and maintenance activity, although some of this responsibility might be delegated to automated/robotic manufacturing and repair processes prior to human occupation.

An alternative habitat proposal is that large underground lava tubes, formed during the eruption of basaltic lava flows and known to be quite common on the Moon, might provide a viable environment to accommodate settlers (Sauro et al. 2020). This habitat would not require the construction of a full-blown enclosure from scratch, since it could utilize the strength and structure of the walls and roof of the tube itself to provide the enclosure. The outer margins of the tube would need to be sealed/coated to permit the creation of a pressurized atmosphere and temperature control, and artificial lighting would be required all the time, rather than just during the long lunar night (as for a habitat on the lunar surface). However, the tubes being underground, they are insulated from the extreme temperature variations on the surface, and there would be no need to provide the protection from meteorite impacts and cosmic rays that is required of a structure aboveground.

LIFE-SUPPORT SYSTEMS IN CONFINED ENVIRONMENTS

Evidence from experiments conducted on confined small human populations on Earth (e.g., Antarctic expeditions and various artificial constructs discussed below) suggest that full recycling of water and waste products and the viability of agriculture and animal husbandry for sustainable food production is likely to be problematic.

To date, the only "experiments" involving humans in real-world space environments are the more than 400 low Earth orbit, mostly short-term crewed missions in the Vostok (6 flights), Mercury (4), Voskhod (2), Gemini (10), Soyuz (141), Apollo (11), Salyut (6), Skylab (3), Apollo-Soyuz (1), Space Shuttle (135), Mir (28 long-duration "crews"), International Space Station (69 expeditions), Shenzhou (6), and Crew Dragon (11) programs (see chapter 4).

The longest single human spaceflight was executed by Valeri Polyakov, who spent 437 days and 18 hours on the Mir station between 1994 and 1995, and the most time accumulated in space by any human is 878 days, set by Gennady Padalka on Mir and the ISS over five spaceflights. Many other astronauts and cosmonauts have spent extended periods of six months or more on Mir and the ISS. Only 27 individuals have spent time out of low Earth orbit during the nine Apollo missions that left low Earth orbit, with the longest time in this situation by any individual being around 10 days. While these missions are very significant and many were groundbreaking, they were all within "easy" reach of Earth and have provided nowhere near the length of exposure that settlers will need to endure in getting to and living on, say, Mars or even the Moon.

Moreover, virtually all these expeditions were completely dependent on resources sourced from Earth. The only exception was/is the production of electricity entirely on the ISS, generated from a vast array of onboard solar panels (figure 5.1). The air and water on the ISS all originally came from Earth with the astronauts and cosmonauts when they arrived, or via uncrewed resupply spaceships. Even the initial supplies from Earth needed to be conserved carefully and replenished by complex, onboard recycling systems in which pure drinking water is regenerated from wastewater, sweat, and urine, and oxygen is replaced by electrolysis of water. The hydrogen produced as a by-product of the electrolysis is turned back into water (and methane) by reacting it with CO_2 separated from the exhaled breath of the occupants. The methane is then vented into space, so the ISS is not a closed system.

Aside from these off-Earth experiments, there have been several programs conducted on Earth that have explored the creation and habitation of spatially and time-limited, carefully controlled environments, many as models for long-term settlements being contemplated for the Moon. The best known of these are the isolated, very small communities that have been established in Antarctic bases, but these have characteristics that make them imperfect models of the true isolation and containment typical of a long-term spaceflight and/or settlement on a remote celestial body. Specifically, the participants in these Earth-bound communities know that fellow humans are nearby and that help is at hand immediately if issues arise.

The Biosphere 2 project (figure 6.1) is the largest closed-system research facility ever created. It had a mission to "demonstrate the viability of closed ecological systems to support and maintain human life in outer space . . . and explored the interactions between humans, farming and technology and the rest of nature" (Wikipedia 2024c). Located in Arizona, the 3.14-acre facility was constructed from 1987 to 1991 using funding of US$150 million from businessman and billionaire philanthropist Ed Bass, in collaboration with systems ecologist John P. Allen (Broad 1991).

Figure 6.1. The Biosphere 2 project compound. (*Wikimedia Commons.*)

The initial project involved the isolation of eight people selected for their complementary skills and experience. It encountered many operational difficulties, including inter alia unanticipated population explosions of some plants and animals; a lower-than-expected production of oxygen by plants, requiring supplementation of oxygen from outside the system; and the emergence of factions among participants inside the bubble due to rifts arising from power struggles between the joint venture partners on how the science should proceed (Wikipedia 2024c).

The small size of the system (relative to natural ecosystems) meant that the associated buffers and higher concentration of organisms produced greater fluctuations and more rapid biogeochemical cycles than are found in Earth's biosphere. Also, the experiments were terminated earlier than expected, meaning that they were too short to allow any substantial agriculture or animal husbandry conclusions to be drawn, and no data were gathered that might have been useful in estimating whether the Biosphere itself could sustain eight people for the planned period of two years.

In 1995 management of the facility transferred to Columbia University, which ran it as a research site and campus until 2003. In 1996 Columbia halted closed-system research and moved to a "flow-through" system, for global warming research, and in 2007 the University of Arizona took over the research, focusing on the water cycle and how it relates to ecology, atmospheric science, soil geochemistry, and climate change.

The troubled history of Biosphere 2 and the combination of short isolation periods and the need to breech the containment to deal with life-support issues means that while much has been learned from the experiment, the feasibility of creating a planetary biosphere that mimics Earth on another planet has yet to be tested at meaningful scales of time and size.

Despite this, Mars and the Moon continue to be considered (largely, it must be said, by enthusiastic advocates for space travel, namely, Aldrin [2009], Zubrin [2011 and 2019], and Kaku [2018], among many others) the most likely candidates for the establishment of settlements of humans, usually under a paraterraforming scenario of one sort or another. While occupation of the Moon by modest numbers of humans for relatively short periods seems achievable in the short term (the Apollo program has already demonstrated this for groups of two individuals for up to three days on the lunar surface), the establishment and even short-term viability of settlements on Mars are not so assured (see the next section).

As proposed in table 6.1, other possible candidates for terraforming (most likely only paraterraforming) outside the Earth-Moon system include Mercury; Venus; Jupiter's moons Europa, Callisto, and Ganymede; Saturn's moons Titan and Enceladus; and the dwarf planet Ceres, the largest body in the asteroid belt between Mars and Jupiter. Most, however, have too little mass and gravity to hold an atmosphere indefinitely, although it may be possible that, once generated, an atmosphere could remain for tens of thousands of years and only need to be replenished periodically. Also, aside from Mercury and Venus, most of these worlds are so far from the Sun that generating sufficient heat and electricity from solar insolation to render the temperature more amicable for human life would be much more difficult than it would be even for Mars.

The next section examines, in turn, each of these options outlined in table 6.1 as potential platforms for settlement in the Solar System beyond Earth and its Moon.

Potential Platforms for Settlement
Mars

For centuries people have speculated about the possibility of past and/or present life on Mars due to the planet's relative proximity to Earth and its similarity to what is known or assumed about the conditions prevailing on Earth throughout its history. At least two-thirds of Mars's surface is more than 3.5 billion years old, and it may provide a record of the conditions leading to the generation and maintenance of life there, even if it never actually did emerge or was snuffed out at some time in the distant past.

As discussed in chapter 1, equipment and experiments on Mars orbiters, landers, and rovers have mostly been directed to the determination of the chemical, geological, and physical properties of the planet and its surface and subsurface to provide data on the origins of the planet and its history. However, most of these missions have also sought to determine whether life exists, or once existed, on the planet. The search for evidence of life has concentrated on detecting (1) the presence of water and/or signs in the landscape that liquid water has been present in the past, most likely in the form of erosion patterns characteristic of flowing water or minerals that can only form in aqueous environments; (2) chemical biosignatures in the soil and rocks on the planet's surface, including fossils of life-forms, and the presence of organic carbon and certain types of nitrogen-containing compounds that could only have been formed by living organisms; and (3) biomarker gases in the atmosphere that represent the by-products of living organisms.

The book *The Rock from Mars* (Sawyer 2006) documents the scientific excitement, political interest, and personal/professional intrigue that can be associated with the search for evidence of life ex-Earth. This nonfiction book follows the announcement in 1993 (by no less than the then-president of the United States, Bill Clinton) that a meteorite from Mars recovered from the Antarctic ice sheet in 1984 contained evidence of microbial life. The claims and counterclaims divided the scientific community and raged for more than a decade until a consensus emerged that the evidence did not uniquely support a biological origin for the signatures. In the process, valid questions were raised about the very nature and recognition of biosignatures in the geochemical record and, indeed, about the definition of life itself in this context.

At the time of this writing, Mars is host to 10 functioning spacecraft (Wikipedia 2024d), with 7 in orbit—*2001 Mars Odyssey* (NASA), *Mars Express* (ESA), *Mars Reconnaissance Orbiter* (NASA), *MAVEN* (NASA), *ExoMars Trace Gas Orbiter* (Roscosmos/ESA), *Emirates Mars Mission Hope* (UAES), and *Tianwen 1* (China)—and 3 on the surface: *Perseverance* (NASA), its helicopter *Ingenuity*, and *Curiosity* (NASA). There are an additional 8 once operational but now defunct spacecraft on the surface (figure 6.2) along with the remains of many others (not marked in the figure) that did not soft-land when they should have.

In many respects, Mars is the most Earth-like of all the other bodies in the Solar System, although Venus is closer in size and Saturn's moon Titan has a thick atmosphere and many of the characteristics of Earth's climate system, with analogous chemical cycles. Mars is about half the diameter of Earth (0.53), with a solid surface, polar water (and CO_2) ice caps, and an axis tilt of 26.2 degrees (like Earth's tilt of 23.4 degrees), leading to the presence of seasons like those on Earth. However, due to its

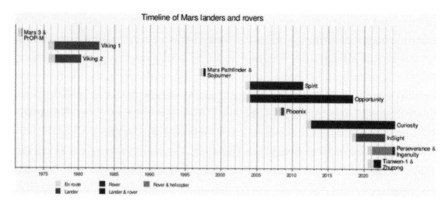

Figure 6.2. Timeline of Mars landers and rovers, including their operational periods. Only the three machines with their rectangles projecting into 2024 (*Curiosity*, *Perseverance*, and *Ingenuity*) are still operative. (*Wikimedia Commons*.)

much greater distance from the Sun (1.52 AU) and location at the outer edge of the "Goldilocks zone" where things are starting to get rather cold, its temperature range (−153°C to +20°C) is wider than Earth's (−88°C to +58°C) and skewed to lower values. Under these conditions, liquid water is present ephemerally, at best.

Mars's atmosphere is also quite different from Earth, being very thin (0.6 percent of Earth's),[2] and quite different in composition, with 96 percent CO_2, 2 percent nitrogen, and 1.6 percent argon (compared to 78 percent nitrogen, 21 percent oxygen, and 1 percent argon for Earth), along with only traces of oxygen and water.

Like Earth, it is a "differentiated" planet, meaning that it was hot enough during its formation period to allow heavier elements (iron, nickel, and sulphur) to melt and separate out from the lighter materials to form a partially liquid core, with the lighter silicate and other minerals forming a mantle and crust. This means that Mars once sported a strong magnetic field, arising from convection in the fluid metallic core. However, the global dynamo ceased to operate about 3.8 billion years ago, most likely due to relatively (compared to Earth) rapid cooling and partial solidification of the core due to Mars's small size. A patchy magnetic field continues to exist owing to remnant surface magnetization in the crust.

Because of these similarities to Earth, Mars remains the most likely candidate for settlement by humans. It is the subject of significant past and ongoing mission development by NASA, the ESA, and privately owned companies such as SpaceX (see discussion below for further details) and has been strongly recommended as a viable platform by Zubrin (2011 and 2019).

Spectroscopic, geological/mineralogical, and geomorphological evidence gathered by orbiting spacecraft and various landers and rovers suggest that Mars once had large-scale water coverage on its surface and that the flow of water during the warmest months left its unambiguous impact on landforms in many areas. Radar data has revealed the presence of large quantities of water ice at the poles and midlatitudes.

It is thought that Mars had a more Earth-like environment early in its history, with a thicker atmosphere and abundant water that was lost over the course of hundreds of millions of years. Three mechanisms for this loss of atmosphere have been proposed.

First, whenever surface water is present, carbon dioxide in the atmosphere reacts with rocks to form carbonates, thus drawing this gas off and binding it to the planetary surface. On Earth, this sequestration process is thwarted by volcanic eruptions associated with ongoing plate tectonics that decompose the carbonates and vent carbon dioxide back to the atmosphere. However, while there is no evidence that plate tectonics is currently active on Mars (thereby not permitting the recycling of gases locked up in sediments), magnetic anomalies observed from orbiters suggest that at least some parts of Mars may have undergone plate tectonics early in its geological past (Yin 2012).

Second, the lack of a uniform protective magnetosphere surrounding Mars allows the solar wind to gradually "erode" the atmosphere, especially since the gravity of Mars is only 38 percent of Earth's. And third, asteroid impacts during the Late Heavy Bombardment period approximately 4 billion years ago could also have ejected much of the Martian atmosphere into deep space.

All of these physical and environmental features conspire to make Mars a relatively harsh platform for human settlement but not necessarily an impossible one, if relevant life-support mechanisms can be put in place. In many ways it is more hospitable to humans than is the Moon in that (1) it has an atmosphere that contains CO_2 that can be used as an ingredient for the synthesis of fuel, (2) its gravity is about twice as strong, (3) it contains copious quantities of water as polar ice or underground, and (4) it has less dramatic variations in temperature.

On the negative side, in 1997 the first Mars rover *Sojourner* detected 0.5 percent of perchlorate compounds in Martian soil, a finding that was confirmed by later rovers. Unfortunately, these compounds are highly toxic to plants and humans. NASA plans to undertake deeper studies of these soils on Earth (including ways in which the toxicity might be counteracted) through the return of samples that have been gathered and stored by the *Perseverance* rover. Clearly, the spread of these perchlorates by regular global dust storms on Mars will also prove to be a major nuisance.

Similar to all potential habitation platforms located farther away than the Moon, Mars's main disadvantage is its much larger distance from Earth, requiring a travel time of six to eight months, rather than just three to four days to get to the Moon. This means that any resource that cannot be extracted in situ on the Martian surface (like fuel and food) during the early periods of occupation must be transported to Mars on the forward journey to be used on the equally long return journey. This is why some exploration strategies for the initial visitation to the planet involve one-way missions populated by people who agree that they will never return to Earth. Alternatively, the construction of habitation buildings and infrastructure for resource extraction and processing might be undertaken by robotic expeditions in advance of the arrival of human settlers.

Some of the potential methods of altering the atmosphere and climate of Mars on a small (paraterraforming) scale may fall within humanity's current technological capabilities. However, the huge financial and material resource requirements, human health and well-being risks (both in travel and sustained occupation), and the long timescales required to undertake even modest local paraterraforming on a platform so remote from Earth suggest that the prospects for practical implementation are very unlikely in the near to medium term.

Even if the technical and logistical questions can be answered, other equally if not more difficult challenges remain in respect to social, ethical, political, and governance issues in an extraterrestrial world (see chapters 9 and 12 for a detailed discussion of these issues).

Mercury

Mercury is only slightly larger than the Moon but is much closer to the Sun, so the intensity of solar radiation is much higher. It has no atmosphere and rotates (59 Earth days) only three times for every two revolutions (88 days) around the Sun, meaning that it is partially tidally locked. It has a large metallic core and a surface that is intensely pockmarked with craters, like the Moon.

Terraforming Mercury (figure 2.9; table 6.1) would present similar challenges to the Moon and Mars. Temperatures on Mercury have a wider range than the Moon, from the unbearable heat of day (430°C) to the cold of night (-180°C), but settlements could be established near the poles where the range would only be from around 0 to 50°C, and where water ice may be found in deep, permanently shaded craters (as on the Moon).

While Mercury is smaller than Mars and the large Solar System moons Ganymede and Titan, its higher density means that it has about the same surface gravity as Mars. Its magnetic field is only 1.1 percent that of Earth's, so that its surface has little protection from the more intense solar wind, but if a shield could be placed at Sun-Mercury Lagrangian point L1 and an artificial magnetic field generated by the mechanisms proposed above for Mars, Mercury's magnetic field would be increased in intensity to a point where it would be self-sustaining. With this enhanced magnetosphere, Mercury would be able to hold on to a nitrogen/oxygen atmosphere for millions of years.

Despite these positive statements, (para)terraforming of Mercury is a gargantuan task and human settlement is likely to be far less achievable than the opportunities provided by the Moon and Mars. Furthermore, Mercury is far removed from the Earth, in an orbit that is not well suited for it to be used as a "way station" for missions farther afield, whether in or outside the Solar System.

Venus

Terraforming Venus was first proposed by Carl Sagan in 1961 (Sagan 1994), when he proposed seeding the clouds with genetically engineered organisms that would convert CO_2, N_2, and H_2O from the atmosphere into other molecules with lower or absent greenhouse gas effects. At the time, the world was unaware of the real composition of Venus's atmosphere, or its crushing surface pressure (table 6.1), which ultimately rendered Sagan's proposal inappropriate.

Modifying the surface environment of Venus to support human life requires at least three major changes to the planet's fundamental characteristics: (1) significant reduction of the surface temperature from its average of 464°C by reversing the existing runaway greenhouse warming; (2) elimination of most of the planet's dense carbon dioxide and sulfur dioxide atmosphere, thereby reducing the surface pressure from its current crippling 93 atmospheres; (3) the addition of breathable oxygen.

Proposals for dealing with these challenges include, inter alia, using giant solar reflectors in the upper atmosphere or at the Sun-Venus L1 Lagrangian point to reduce insolation, increasing the albedo of the planet, converting and sequestering the CO_2 to carbonates, introducing vast amounts of hydrogen (to react with the CO_2) and water (sourced from asteroids or elsewhere), and thinning the atmosphere through the engineered impact of large bodies from the asteroid belt. Whatever method or combination of these methods is used, the task is truly massive, well beyond our current technical capability, and probably much less feasible than several of the other options in the Solar System.

Titan

Saturn's moon Titan offers several unique advantages for human settlement, including an atmospheric pressure like Earth's and an abundance of nitrogen and frozen water, while Jupiter's moons Europa, Ganymede, and Callisto also have an abundance of water ice (probably deep below the surface). Titan is the second-largest moon in the Solar System after Ganymede, being 50 percent larger (in diameter) than Earth's Moon (and 80 percent more massive) and 10 percent larger than the planet Mercury (but only 40 percent as massive).

Titan orbits Saturn at just 20 Saturn radii (compared to the Moon's distance of 60 Earth radii), so when viewed from Titan's surface, Saturn subtends an arc of 5.09 degrees, and if it were visible through the moon's thick atmosphere, it would appear 11.4 times larger than the Moon does from Earth, not counting Saturn's rings—a truly magnificent sight, despite the rings being edge-on.

Titan is primarily composed of ice and rocks, which is likely differentiated into a rocky core surrounded by a crust of ice and probably a subsurface layer of ammonia-rich liquid water. The dense opaque atmosphere prevented detailed characterization of Titan's surface from orbiting or flyby spacecraft until the *Cassini-Huygens* mission in 2004 provided new information based on *Cassini*'s radar images from orbit and data provided by a soft descent of the *Huygens* probe to the surface (figure 1.11). The geologically active surface is generally smooth, with few impact craters, although mountains and several possible cryovolcanoes have been found.

The atmosphere of Titan is 95 percent nitrogen and 5 percent methane, with minor other hydrocarbons, leading to the formation of methane and ethane clouds and a heavy organonitrogen haze. The climate is dominated by seasonal weather patterns as on Earth, including wind and rain, leading to the formation of surface features that bear close similarities to those of Earth, such as dunes, rivers, lakes, seas (probably of liquid methane and ethane), and deltas. With the presence of liquids (both on the surface and in the subsurface) and an atmosphere dominated by nitrogen, Titan's hydrocarbon (methane) cycle is closely similar to Earth's water cycle, although operating at the much lower temperature of around −180°C.

Due to these factors, Titan has emerged as a promising platform for the emergence of life based on abundant complex hydrocarbons in the atmosphere and on the surface and/or in the water oceans underground. Consequently, it has frequently been proposed as a viable settlement target (Wohlforth and Hendrix 2016; Lorenz 2020).

Specific advantages of such a settlement include the following:

- An essentially limitless abundance of fuel for energy production for light, heat, and transportation using methane and polycyclic aromatic hydrocarbons.

- The prospect of the production of oxygen for breathing and as an oxidant for burning the methane for lighting and in power plants, through electrolysis of the water ice in the oceans below the surface.
- The ability to manufacture lightweight plastics from hydrocarbons for the construction of dwellings, buildings for manufacturing processes, and greenhouses for growing food.
- Human settlers could walk around without the need for pressurized spacesuits and requiring only heated, insulating clothing and oxygen masks, and they could live in unpressurized buildings (Lorenz and Mitton 2010).
- The very thick atmosphere (one and a half times the pressure on Earth and four times denser) provides in-built protection from cosmic radiation and micrometeoroids and together with the low gravity (14 percent of Earth's) would allow flight by humans with only a little mechanical assistance.

Disadvantages include: the carcinogenic nature of many of the ubiquitous hydrocarbons; the need to mine for water (for drinking and its contained oxygen); issues arising from the long-term effects on the human body (see chapter 9) of low gravity (slightly less than that on the Moon); and the need to manage the negative impact on diet (specifically, lack of vitamin D) and mood (depression) arising from weak sunlight, the constant glow from Saturn's disc (to which Titan's rotation is locked), and the "days" and seasons lasting 16 days and about 7 years, respectively.

Furthermore, this settlement would be so distant from Earth that even with gravity-assisted trajectories, it would take between three (*Voyager 1* and *2*) and seven (*Pioneer 11* and *Cassini*) years for a one-way trip. With these lengthy travel times from Earth, the settlements would need to be totally self-sufficient from the get-go and, like all the other remote location options, would probably require lengthy robotic setup periods before humans could assume sustainable residence.

While not to be underestimated, the challenges of life on Titan would, in many ways, be less daunting than most of the other potential settlement platforms in the Solar System, apart from the Moon and Mars.

Ceres

Ceres is the largest body in the asteroid belt and the only dwarf planet in the inner Solar System. It is a roughly spherical rock-ice body 950 kilometers in diameter and is to date the smallest identified dwarf planet. It contains about one-third of the mass of the asteroid belt. Discovered on January 1, 1801, by Giuseppe Piazzi, it was the first asteroid to be identified, though it was classified as a planet at the time. It is named after Ceres, the Roman goddess of growing plants, the harvest, and motherly love. It was the subject of intense study by the *Dawn* spacecraft in 2015 (following its visit to 530-kilometer-diameter Vesta in 2011–2012).

The Cerian surface is likely to be a mixture of water ice, carbonates, and clays (table 6.1). It appears to be differentiated into a rocky core and a 100-kilometer-thick icy mantle that may harbor oceans of liquid water with a volume of 200 Mkm,³ more than the amount of freshwater on the Earth. It is not clear how these proposed oceans can stay in a liquid state, since Ceres does not have enough mass to sustain a long-term molten core that would provide significant tectonic activity, and it is not orbiting a tidally disruptive body like the moons of Jupiter and Saturn, where tidal frictional forces provide the necessary heat.[3]

Habitation of Ceres presents issues like those that would be encountered for Enceladus and Europa, both satellites of Jupiter, namely, very low gravity (3 percent of the Earth's, and unable to retain an atmosphere), extremely low temperature, and no magnetic field. In addition, being resident in the asteroid belt, there would be the ever-present danger of an impact by another large body, which would cause massive damage to any biosphere that was present. Limited paraterraforming would appear to be the best option for human occupation, although occupation solely by robotic machines might be more practical (as discussed more broadly in chapter 13).

Its potential advantages are (1) its use as a base and transport hub for future asteroid mining infrastructure, allowing mineral resources to be transported to Mars, the Moon, and Earth; (2) its role as an intermediate step in the path to settlement of the objects in the outer Solar System, such as Titan or the moons of Jupiter; and (3) because of its small escape velocity and the presence of large amounts of water ice on or near the surface, it could serve as a source of water, fuel, and oxygen for ships going through and beyond the asteroid belt. Indeed, transportation of payloads from Mars or the Moon to Ceres would be more energy-efficient than carriage from Earth to the Moon.

WITHIN THE SOLAR SYSTEM—CONCLUSIONS

Options for human settlement beyond the Earth-Moon pair and their Lagrangian points within the Solar System have only been investigated so far with uncrewed vehicles. While "soft," that is, controlled, landings have been made on Venus, Titan, and several comets and asteroids, these expeditions have been limited to very short periods on the surface of these bodies to undertake photography, physical experiments, and sample collection and analysis with, in several cases, samples being returned to Earth for closer examination.

The major exceptions to this experience are the wide range of past and current expeditions to Mars, with several "rovers" operating successfully on the surface for many months and sometimes years. These missions have demonstrated that it is possible to operate uncrewed, remotely controlled vehicles on and in orbit around another planet for long periods of time, using nuclear and/or solar power and onboard energy storage. Mission controllers have become expert at the repair and upgrading of onboard computer software remotely after decades of operation in deep space as systems fail, perhaps due to cosmic ray or dust particle impact damage, or are required to execute more-complex tasks.

To date, 10 batches of soil and rocks from the Moon have been collected and returned to Earth by robotic and crewed missions (see chapter 1), and small samples of material have been brought back from the comet Wild 2 and the asteroids 25143 Itokawa, 162173 Ryugu, and 101955 Bennu. Samples of the solar wind have also been returned by the robotic *Genesis* mission to Sun-Earth Lagrange point L1 and from the Earth-Moon Lagrange point L1 by *Apollo 16* on its way to the Moon.

The requirement to transport on the outward journey to Mars the required fuel for the return leg, especially the fuel cost of escaping the planet's relatively deep gravity well, has meant that no sample return trip has been made to date. Indeed, discussion of long-term crewed flights to Mars often involves the scenario that the journey will be one way (either by design or mishap), thereby opening up a major dialogue of the ethics of a "suicide" mission of this sort. Add to this the as yet unknown long-term impact of damage to human health (and perhaps software systems) from cosmic rays and low gravity (chapter 9), massive resource requirements, social acceptance of the costs of technology development and execution of the mission, and poorly understood sociopolitical issues in transit and in the settlement after arrival, and it is very doubtful that crewed missions to Mars or anywhere else outside the Earth-Moon system will be successfully completed within the next 25 years or so, at best.

In the case of the "space race" between the Soviet Union and the United States in the 1950s and '60s, public acceptance of the financial cost of this technology development was driven largely by national pride on both sides. In contemporary times, the acceptance of the huge financial implications of crewed space travel may depend on the emergence of a new era of the "militarization" of space and/or the drawcard of the possibility of commercial exploitation of off-Earth sources of rare or depleted materials on Earth. It might also emerge from an increase in "space tourism," currently very expensive (on the order of $20 million in the case of a visit to the ISS), but with the cost likely to reduce significantly with the increase in private-sector involvement (e.g., SpaceX, Virgin Galactic, and Blue Origin).

In spite of all the above known and yet to be revealed difficulties, the similarity of the climatic and geological conditions on Mars, and the prospect of paraterraforming small parts or all of the surface to enable long-term occupation, make it the most attractive option for settlement ex-Earth, with Titan as a potential second option. Habitation of one of the other moons or asteroids in the Solar System will be much more difficult due their remoteness from Earth and/or much harsher local conditions.

If and when long-term, multiple-generational settlement of Mars and/or Titan does take place, the long distances from Earth (especially in the case of Titan) may mean that members of those settlements have very limited or no physical contact with Earth. Thus, after many generations, they are likely to evolve in genetically distinct ways in response to their local adaption to the lower gravity, temperature and pressure differentials, lifestyle, and specifically different chemistry of the atmosphere and food sources (Wohlforth and Hendrix 2016). This differentiation may eventually develop to such an extent that the inhabitants may no longer be able (or want) to return to Earth. By such a process, like it or not, we may have contributed to the evolution of a new species of *Homo*.

This subject is explored in more detail in chapter 13.

CHAPTER SEVEN

Options for Settlement Outside Our Solar System

There is no way back into the past; the choice, as Wells once said, is the universe—or nothing. Though men and civilizations may yearn for rest, for the dream of the lotus-eaters, that is a desire that merges imperceptibly into death. The challenge of the great spaces between the worlds is a stupendous one; but if we fail to meet it, the story of our race will be drawing to its close.
   —ARTHUR C. CLARKE, IN *INTERPLANETARY FLIGHT*, 1950

If it is complicated and daunting to establish settlement platforms within the Solar System, including the Moon and Lagrangian points in the near-Earth environment, things certainly get much more challenging when one ventures outside the Sun's domain into neighboring parts of the Milky Way galaxy.

This chapter focuses on identifying and assessing practical options for the location of a settlement target outside the Solar System and reserves for part III a discussion of the manyfold technical, physical, social, political, psychological, and spiritual challenges embodied within a journey to such a destination.

THE EXOPLANET "EXPLOSION"
It wasn't so long ago that the existence of planets orbiting other stars than the Sun was hypothetical and solely the stuff of theoreticians and science

fiction writers. However, as described in chapter 1, following the discovery of the first exoplanet in 1992, the development of much more sensitive telescopes (on Earth, in orbit, and farther out) and massive enhancements in our ability to extract new and more-detailed information from the analysis of the spectra of stars and their environments has resulted in the discovery of many thousands of planets, protoplanets, and accretion discs around stars across the galaxy. Through these advances, as of April 18, 2024, there are 5,612 confirmed exoplanets in 4,170 planetary systems, with 948 systems having more than one planet (figure 1.12). It is now very clear that planets are ubiquitous in the galaxy as a natural outcome of the accretion of gas and dust during the formation of the stars themselves.

The majority of the exoplanetary stellar systems discovered and confirmed so far are profoundly different from our Solar System (figures 1.13 and 1.14). The characteristics of less than 100 of the planets in these systems are closely similar to Earth; that is, ones with a size, mass, and composition like our home planet, orbiting within its host star's habitable zone (Schulze-Makuch et al. 2020), apparently in support of the Rare Earth hypothesis.[1]

On the other hand, statistical estimates of the number of Earth-like planets in the *entire* Milky Way galaxy (rather than just those within observational reach of our best telescopes) may be as high as 40 billion (Petrigura et al. 2013), and there may be an even larger number of free-floating planetary-mass bodies that orbit the galaxy directly, unconnected to a specific star. These very large statistical estimates are consistent with the mediocrity principle (chapter 1), which predicts that planets like Earth should be common in the universe.

The conflict between these two assessments of the occurrence of Earth-like planets as rare or commonplace (in shades of Fermi's paradox) has not yet been resolved, for some of the reasons outlined in the next section.

Exoplanet Statistics and Properties

The statistical significance of the observed characteristics of exoplanets is constrained by the fact that the detection methods for these planetary systems are differentially sensitive to their spectral and other properties.

For example, planet-search programs have tended to concentrate on stars roughly like the Sun—that is, main-sequence stars of spectral categories F, G, or K^2—so naturally most known exoplanets orbit such stars. Also, statistical analyses indicate that lower-mass stars (red dwarfs, of spectral category M) are less likely to have planets massive enough to be detected by the radial-velocity method (Cumming et al. 2008).

In contrast, many exoplanets have been discovered by the transit method, which is able to detect smaller planets. Stars of spectral category A typically rotate very quickly, making it difficult to measure the small Doppler shifts from orbiting planets because the spectral lines are very broad. Observations of extremely massive stars of spectral category O are much hotter than the Sun and produce a photo-evaporation effect that inhibits planetary formation by dispersing gas away from the star, while stars that "go" supernova may push any planets it had out into the void.

Stars with a higher metallicity (that is, with a small but significant proportion of heavier elements, along with the usual light elements hydrogen and helium) are more likely to have planets, especially giant planets, than stars with lower metallicity (Buchhave 2012). However, metal-rich stars tend to be larger, making it more difficult to detect smaller planets, thereby producing another bias in detection rates for these stars and their planets.

Of course, overlaying all these limitations is that exoplanets naturally become more difficult to detect the farther away they are, so the population of stars from which the exoplanet statistics are drawn is restricted to stars located relatively close to the Sun. At the time of this writing, the most distant exoplanet discovered is SWEEP-11/04, around 30,000 light years away (Wikipedia 2024e).

It has long been suggested that multiple-star systems (that is, ones in which the stars are gravitationally connected and are revolving around a common center of mass) are unlikely to harbor planets because of an inherent instability in the orbits of these planets. Despite this, more than 100 exoplanets have been discovered orbiting at least one member of a binary star system and several planets have been discovered that orbit around triple star systems. In many cases it is uncertain around which star in the system the exoplanet orbits, and so the parameters of these transiting planets could be significantly in error. Finally, only a few planets have been discovered orbiting stars in open clusters, probably for the same reason of orbital instability.

SELECTION OF EXOPLANET TARGETS

Notwithstanding the large number of exoplanets that have been discovered in recent years (figure 1.12), when the characteristics of these planets as a population are compared to the corresponding parameters of Earth (figures 1.13 and 1.14), our own planet is seen to be very much an outlier in terms of radius, orbital period, mass, density, and irradiation from the host star. Although we have only one example of intelligent life to act as a benchmark, this "outlier" status of Earth may be an indication of the rarity of technologically advanced life and the complex, difficult and long, stable planetary history that is needed to give rise to, and to sustain, life of this sort.

The criteria for the emergence and flourishing of intelligent life (chapter 1) are very relevant to the selection of a target exoplanetary system for the establishment of a potential human settlement in the (presumably far distant) future. If the exoplanet has characteristics that are incompatible with even simple, let alone complex, life, then this environment will not be a sensible place to land humans. Indeed, these conditions may be so detrimental to the existence of life that they will significantly reduce the prospects for successful terraformation or paraterraformation of the planet.

Thus, the characteristics of an exoplanet that might be favorable for its selection as a viable landing platform for a settlement are likely to be similar to those required for the evolution of complex life on Earth (chapter 1), namely:

1. A host star with an energy output that is stable, absent the emission of dangerous radiation, and one that is located outside densely populated regions of the galaxy (to distance itself from the radiation of novae nearby).

2. Part of a single-star system to increase the likelihood of a stable planetary orbit.

3. Not so close to the host star such that the rotation period of the exoplanet is "locked" to its orbital period with one side always facing the star, thereby avoiding extreme variations in surface temperature over entire hemispheres of the planet.
4. Location in the Goldilocks zone of that star to enable water to exist in the liquid state.
5. Solid/rocky surface with a chemical composition necessary for life to be sustained through the growing of food (i.e., carbon and nitrogen, at least) and the extraction and processing of resources for the construction of dwellings, manufacturing facilities, energy production, etc.
6. A nontoxic atmosphere thick enough to preserve water in the liquid state (through the prevention of evaporation due to inadequate surface pressure) and to absorb harmful radiation and reduce meteorite influx.
7. The presence of a strong magnetic field for protection from harmful solar and cosmic rays, largely determined by the planet's size and its differentiation into a liquid metallic core and rocky crust.
8. The absence of a large remnant accretion disc around the star, together with the presence of other larger planets in the system to clear the planet's neighborhood of objects as insurance against routine catastrophic, extinction-level bombardment by other bodies.

The characteristics of targets within the Solar System such as the Moon, Mars, Titan, Ceres, Enceladus, Callisto, and several asteroids have already been the subject of intense scrutiny, to varying degrees, by telescopes of all types, spacecraft flybys, orbiters, landers, and rovers (chapter 6). Much is known about several of these targets, especially the Moon and Mars, and this information will be invaluable for the planning and execution of future settlement expeditions to these targets.

The same detailed characterization missions will need to be undertaken on potential exoplanets before decisions about their suitability for habitation can be made. It might be expected that with the continuation of recent spectacular improvements to telescopic imaging and spectrum analysis, many of the critical parameters will be able to be determined remotely for exoplanets, at least to some degree. However, this will be no easy task because of the massive distances involved, the overwhelming dominance of the spectrum from the host star, and the paucity and uncertainty of even the basic properties of the closest of these targets (the Proxima Centauri system at just 4.3 light years away). Obviously, the risks of getting these characteristics wrong are much higher than for the much closer settlement options within the Solar System. This underlines the critical role of preliminary "scouting" expeditions to potential platforms by automated probes.

In chapter 1 mention was made of several hypothetical exploration objects/craft, including von Neumann and Bracewell probes, that might have been dispatched by other civilizations to explore for and/or communicate with other civilizations (i.e., us) in the galaxy. Humans will need to dispatch similar craft, potentially in large numbers, to assess the prospects of specific exoplanet options for which the Earth-based remote sensing has provided positive indicators before committing crewed spacecraft to the journey.

An early proposal to develop such a fleet of robotic spacecraft to scout nearby stellar systems was Breakthrough Starshot,[3] announced in April 2016. This US$100 million program intends to develop a fleet of 1,000 tiny spacecraft capable of making the journey to Alpha Centauri (and perhaps including a flyby of Proxima Centauri b) at 20 percent of the speed of light (60,000 km/s) within the lifetime of a human being. At this speed, it would take the spacecraft about 20 years to get there, but "only" 4.3 years for a return signal to notify Earth of a successful arrival. Using small onboard cameras and perhaps other sensing equipment, the project hopes to provide images of exoplanet surface features, in pursuit of the objective of providing information about the potential existence of extraterrestrial life, and whether or not we need to suss out further properties of the platform to determine its suitability for habitation.

Each Breakthrough Starshot vehicle is proposed to be propelled during the initial stages of its journey by ground-based lasers with a power of up to 100 gigawatts, by no means an insignificant technical achievement in itself. The fleet would comprise very small centimeter-size craft weighing only a few grams, and each would transmit data back to Earth using an onboard laser communications system.

A proof-of-concept initiative in this area of robotic exoplanet scouting is LightSail, a crowdfunded project developed by the Planetary Society[4] to demonstrate controlled solar "sailing" using the pressure of light photons acting on extended "sails" attached to the craft. The half-shoebox-size *LightSail 2* cubesat[5] spacecraft was launched in June 2019 and remained in low Earth orbit through to November 2022, successfully using sunlight alone to change its trajectory around Earth (see chapter 11 for further details). Given that light sails are one of the prime candidates for interstellar travel propulsion, this was a significant achievement in the development of technology to achieve the Breakthrough Starship goal.

While these miniature spacecraft have the potential to be accelerated to a non-trivial fraction of the speed of light and may provide critical habitation-related information on the target systems within a few decades of launch, even this relatively short flight time will add significantly to the time delays in embarking on an interstellar expedition with humans on board. Moreover, the much higher mass of follow-up spacecraft required for human transportation is very unlikely to permit the achievement of speeds that are a significant fraction of the speed of light for these settlement missions. Part III specifically addresses these transportation and other challenges.

Exoplanets Closest to Earth

While transportation and scouting plans for habitation outside the Solar System are being developed, the exoplanets that present the greatest potential opportunities for first-mission targets are those that are closest to Earth. But this depends on whether preliminary investigations show that one or more of these "close" neighbors has characteristics that are consistent with (or at least not completely adverse to) human occupation.

The observed or inferred physical characteristics of the 18 confirmed exoplanets located within 12 light years of Earth (as of January 2023) are provided in table 7.1. Very few of these are considered to be habitable on currently available observations, with Proxima Centauri b considered the most likely, according to the Habitable Worlds Catalog (Planet Habitability Laboratory, n.d.). Let's analyze the properties of this nearest exoplanet first.

Table 7.1. Exoplanets lying within 12 light years distance from Earth (as of 2022).

Name of Host Star	Distance (light years)	Host Star Mass (rel. to Sun)	Exoplanet Label	Exoplanet Mass (rel. to Earth)	Semi-major Axis (AU)	Orbital Period (Earth days)	Discovery Year
Proxima Centauri	4.25	0.123	d	≥0.26	0.029	5.12	2022
			b	≥1.07	0.049	11.19	2016
Lalande 21185	8.30	0.46	b	≥2.69	0.079	12.9	2019
			c	≥13.6	2.94	2,946	2021
Epsilon Eridani	10.49	0.78	Ægir	242	3.53	2,689	2000
Lacaille 9352	10.724	0.49	b	≥4.2	0.068	9.26	2019
			c	≥7.6	0.12	21.8	2019
Ross 128	11.01	0.17	b	≥1.40	0.050	9.87	2017
Groombridge 34A	11.62	0.38	b	≥3.03	0.072	11.4	2014
			c	≥36	5.4	6,600	2018
Epsilon Indi A	11.87	0.76	b	941	11.1	15,700	2018
Tau Ceti	11.9	0.78	g	≥1.75	0.133	20.0	2017
			h	≥1.8	0.243	49.4	2017
			e	≥3.9	0.538	163	2017
			f	≥3.9	1.33	640	2017
GJ 1061	11.9	0.11	b	≥1.37	0.021	3.2	2019
			c	≥1.74	0.035	6.7	2019
			d	≥1.64	0.054	13.0	2019

Source: Data extracted from https://en.wikipedia.org/wiki/List_of_nearest_exoplanets

Proxima Centauri b

The nearest star to the Solar System, Proxima Centauri, is a red dwarf[6] star just 4.3 light years away with a family of three (one disputed) orbiting exoplanets (table 7.1). One of this family, Proxima Centauri b, orbits at a distance of about 0.05 AU (7.5M km), much closer than Mercury is to the Sun (0.4 AU, or 58M km), with an orbital period of approximately 11.2 Earth days (Faria et al. 2022). It is understood to be a potentially Earth-like planet with a minimum mass of at least 1.07 times that of Earth and with only a slightly larger radius.

The planet orbits within the habitable zone of its star, receiving about 65 percent of Earth's insolation, but it is not known whether it has an atmosphere. Unfortunately, Proxima Centauri is a "flare" star[7] with intense emission of electromagnetic radiation that probably has already stripped or could in the future strip an atmosphere off that planet.

The ubiquity of red dwarfs in (at least) our part of the galaxy means that the suitability of this type of star as a host for habitable exoplanets is important, but questionable, and quite controversial (Waltham 2017; Gilster 2012). It is, therefore, instructive to consider the general characteristics of red dwarfs that might be used to determine whether a visit to our nearest exoplanetary neighbor, or any other exoplanet in our vicinity that is orbiting a red dwarf star, is justified (Ritchie et al. 2018).

- Red dwarfs are much cooler than other star types,[8] so the habitable zone is much closer to the star (figure 1.2) and thus there is a high chance that any planet in this zone will be tidally locked, that is, always presenting one face toward the star. This would create permanent extremes of surface temperature over the majority of each hemisphere that would not be conducive to the emergence and support of life as we know it (akin to one of the difficulties extant on Mercury).
- The rapid and extreme changes in emissions (especially ultraviolet radiation) from the star due to its flare activity would be likely to hinder the support of life through radiation damage to the large molecules that are critical to the emergence and maintenance of life as we know it on Earth.

- Even if the planet has an atmosphere, its close proximity to the host star means that the stellar wind is much stronger than Earth's and this may have accelerated the loss of atmospheric gases, including the dissociation and loss of water molecules (perhaps as took place during the early history of Venus).

- Depending on the eccentricity of the exoplanet's orbit (around a star of any type), it may lie outside of the habitable zone for significant parts of its orbit, even though the average distance lies within the zone.

- On the other hand, the extremely long lives of red dwarfs like Proxima Centauri provide plenty of time for life to establish itself and to evolve (but also, of course, ample time for whatever life there is to be extinguished).

Other Potentially Habitable Exoplanets

The NASA Exoplanet Archive operated by the California Institute of Technology, under contract with NASA under the Exoplanet Exploration Program, reports that as of March 23, 2024, the current number of confirmed, potentially habitable worlds (regardless of distance from Earth) is 63 (table 7.2). The estimate includes only the likely rocky planets with

Table 7.2. The 63 confirmed potentially habitable exoplanets that have been catalogued up to March 2023, divided into three classes of mass and radius and named in relation to equivalent planets in the Solar System.

Name	Number	Mass rel. to Earth	Radius rel. to Earth
Subterran (Mars size)	1	0.1–0.5	0.4–0.8
Terran (Earth size)	23	0.5–3.0	0.8–1.6
Superterran (super Earths/ mini Neptunes)	39	3–10	1.6–2.5

Source: NASA Exoplanet Archive, https://exoplanetarchive.ipac.caltech.edu

dimensions up to 2.5 Earth radii or 10 Earth masses orbiting within an optimistic stellar habitable zone. These exoplanets fall into three subsets relatable to corresponding planets of our own Solar System.

The higher proportion of large exoplanets in table 7.2 and overall in figure 1.13 (whether estimated to be habitable or not) is likely to be a result of detection efficiency rather than reflecting the fraction of large exoplanets in the general population; the smaller a planet is, the smaller the effect it has on several of the detection procedures, especially the radial velocity and transit photometry methods (chapter 1), and, therefore, the less likely it is to have been observed.

The most distant potentially habitable planet *confirmed to date* is Kepler-1606b, at 2,870 light years away, discovered in 2016 (NASA, n.d.). This planet is a "super Earth" with a mass of 4.94 Earths, orbiting a G-type star every 196.4 days at a distance of 0.6421 AU. There are other suggested exoplanets in the Milky Way galaxy more distant than Kepler-1606b, but they are unconfirmed as yet. There have also been several indirect gravitational microlensing observations of planets at much greater distances, in other galaxies, but they have not been confirmed. Unless there is a truly transformational breakthrough in spacecraft transport methods, these more distant exoplanets will remain out of practical reach for both uncrewed and, especially, crewed spacecraft.

SUMMARY AND CONCLUSIONS FROM CHAPTER 7

Platforms for human settlement outside the Solar System are limited to Earth-like planets in orbit around stars other than the Sun—the so-called exoplanets. Given the number of stars in the galaxy, the possibilities of finding a suitable such location on which we might settle might seem to be limitless at first glance, but this is only the case if we are able to solve all of the multifarious challenges associated with long-distance space travel by fragile human beings. Furthermore, while the number of known exoplanets orbiting stars in the galaxy continues to increase unabated, the farther away from Earth they are located, the more difficult it is for our current technology to determine their characteristics and suitability for settlement.

What we know about the ones discovered so far indicates that only a small proportion possess the two primary conditions for human occupation, namely, a "rocky" surface to provide a solid platform for habitation, and a location in the Goldilocks zone around the planet's host star where water can exist in the liquid state. Even if these two conditions are met, there are a host of other characteristics that are at the very least desirable, and potentially necessary, for occupation: a stable orbit and host star not prone to outbursts of deadly radiation, a nontoxic thick atmosphere, the presence of a protective magnetic field, and an orbital path clear of the threat of collisions with other large bodies.

Before sending humans to any putative destination, there will need to be a series of exploration visits by small robotic spacecraft to determine their habitability from closer quarters and with greater certainty. These visits, even at significantly enhanced speeds relative to current spacecraft (say, up to a very optimistic 10 percent of the speed of light), will require many decades if not centuries to complete, even for the closest of the known prospective exoplanets. This requirement for robotic reconnaissance missions will push out any human mission by at least a similar amount of time, but this will have the advantage of perhaps allowing technology development to provide solutions to the negative impact of spaceflight on the human condition.

Conclusions to Part II

A small number of options are emerging for the location of potential settlement platforms within the Solar System. With the exception of those in low Earth orbit (with ready access to support from Earth), all of these options present several very serious challenges for any putative community of humans, including greatly reduced or zero gravity; absence of breathable air; extreme cold or heat; the need to generate water, food, energy, and shelter in situ from local resources; and/or little if any protection from cosmic radiation, meteorites, and the solar wind. Short-term solutions to these manifest problems may, perhaps, be affected by paraterraformation of the local environment, namely, the construction of small contained environments (like those on space stations) that create and sustain conditions amenable to life on Earth.

In due course, these fabricated, local, Earth-like environments either underground or on the surface of the planet or moon might be expanded to encompass the entire celestial body through a process of its full terraformation. However, there are huge financial, energy, and materials costs involved in the creation of a remote platform to house a self-sustaining population of humans that is much larger than just a small group of individuals. This expenditure may not be allocated a social license from the citizens of Earth when so many issue remain unresolved "at home." Furthermore, it is very unclear whether all the technology necessary for (para) terraformation exists currently or is likely to be created anytime soon.

In terms of settlement platforms farther afield, we know very little about even the closest planetary systems outside the Solar System, and the technical means to get there are still not possible in practice and seem very unlikely to be developed for decades, if ever. Even if we could determine that one of the potential platforms looked suitable for occupation, getting to it with anything other than a very small uncrewed spacecraft is likely to require very (impractically?) long travel times and would be plagued by a host of other potentially insoluble physical, emotional, and political challenges.

Part III provides a discussion on all these issues, and more.

PART III

THE JOURNEY OFF-EARTH

The Earth is the cradle of humanity, but mankind cannot stay in the cradle forever.
—Konstantin Tsiolkovsky, in *The Exploration of Cosmic Space by Means of Reaction Devices*, 1895

Space, the final frontier . . . to boldly go where no man has gone before.
—Gene Roddenberry, *Star Trek*, 1966

As we have seen in part II, human travel to a destination within the Solar System is difficult enough (chapters 5 and 6), but the establishment of a settlement outside the Solar System (chapter 7) carries with it a plethora of even more serious, if not ultimately insurmountable, challenges, as follows:

- We do not yet have the technology to travel at anywhere near the speed of light, so interstellar travel will take a lot longer than a single human's lifetime (at least as measured by the people left behind). If we could develop technology to get close to light speed (see chapter 10), time, and with it human metabolism and aging processes, would slow down significantly for the individuals on board, such that they may then be able to make the journey within their lifetime. However, it takes impossibly large amounts of energy to accelerate a spacecraft large enough to accommodate humans and their cumbersome life-support systems, even to a small fraction of the speed of light.

- We do not yet understand the full effects of long-term zero or near-zero gravity on the human body. Travel outside the Earth-Moon system to the outer regions of the Solar System, let alone to other star systems, will involve much longer exposure to zero gravity, with largely unknown consequences (see chapter 9). In addition, human DNA is damaged by exposure to cosmic rays and the doses received on long-duration spaceflights will be orders of magnitude larger than those accumulated during the 12 days or less of exposure on the Apollo missions to the Moon.

- Human travel to destinations within the Solar System and farther afield will require enough oxygen, water, and food to sustain the travelers over the entire journey, and also for at least the initial (potentially lengthy) period of residence at the mission destination while sustainable support systems are being established. These resources will have to be generated on board the spacecraft using very efficient recycling and minimization of waste products. The alternative is to maintain all of the passengers in a state of "hibernation" in which their life-support requirements are significantly reduced or eliminated altogether, as proposed for the so-called sleeper ships or freezer ships of science fiction (see chapter 10 for details on these craft).

- In the outer reaches of the Solar System far from the Sun, not to mention interstellar space, solar insolation will be insufficient for the generation of electricity to power all the life-support and other equipment required on board. This will include temperature and humidity control, recycling processes, lighting, communications, powering of electrical equipment, refrigeration and heating of food, course corrections, possible need for rotation of the craft to generate artificial gravity, and many other things. For these long journeys (unlike the Apollo missions) fuel cells and batteries will not be sufficient, and power from the propulsion system will only be available during the acceleration and deceleration stages of the mission, not during the long cruising stage. Thus, as for the robotic craft that have been and are operating in and outside the Solar System, a large and (very) reliable onboard power supply will be required, most likely more than one nuclear reactor (to provide backup capability in case one fails). But even a nuclear reactor runs out of fuel eventually, as more and more half-lives of its radioactive constituents are consumed.

- A lot of additional resources and paraphernalia must be transported in a settlement ship, similar in many respects to the supplies and equipment needed by the first European colonists bound for Australia in the 18th century when they embarked on a three-to-four-month sea journey with very limited opportunities to replenish stocks on the way, and almost no chance of returning to their homeland. These include:
    - medical expertise and supplies;
    - technical equipment for repairs and life support;
    - the materials and equipment needed to build shelters and greenhouses and to extract and process the necessary resources from the local environment for food, clothing, and energy; and
    - materials and activities for intellectual stimulation and physical well-being.
- Psychological, social, and political issues will be very significant on the longer journeys. Without major advances in current propulsion technology, and in the absence of "sleeper" or "frozen" ships, thousands, if not tens of thousands, of human generations will have expired on board before reaching the destination. The ability to manage morale, emotional well-being, social interactions, and "governance" in a small population confined to a small spacecraft for thousands of years, with no chance of escape, will be sorely tested.
- Within a handful of generations after departure, none of the humans traveling in the spacecraft will have a real connection to Earth. For all generations except those on board as the spacecraft nears its destination, the travelers will also be disconnected from the target because they have no chance of reaching it.
- As the distance from Earth increases over the first few years of the journey, communication delays with home base will slowly

increase, ultimately rendering practical contact impossible, even for a journey to the closest planetary system in Centauri, from which light and radio signals will still take 4.3 years to be received from their source.

- The ability to assess the destination for human livability well before arriving there and perhaps to change destinations, if necessary, will be essential. Most of this would (hopefully) have been anticipated and actioned prior to departure through feedback from robotic scouting missions, but provision will nevertheless need to have been made for unanticipated changes of plan and/or emergencies.
- Also required is the means by which course corrections and changes to the orientation of the spacecraft in flight can be made along with its deceleration from cruising speed in preparation for a controlled landing. This would involve the carriage of at least as much additional fuel as was required to accelerate the craft to its cruising speed in the first place.

These challenges are discussed in chapters 8 through 11 of part III.

Chapter Eight

The Tyranny of Distance and Time

A manuscript I wrote on January 14, 1918 . . . and deposited in a friend's safe . . . speculated as to the last migration of the human race, as consisting of a number of expeditions sent out into the regions of thickly distributed stars, taking in a condensed form all the knowledge of the race, using either atomic energy or hydrogen, oxygen and solar energy. . . . [It] was contained in an inner envelope which suggested that the writing inside should be read only by an optimist.

—Robert Goddard, in
Material for an Autobiography, 1927

As outlined in the introduction to part III, space is a very dangerous place for humans, and any living creature, including but not limited to the obvious hazards arising from vacuum, weightlessness, and radiation. This applies to all spacecraft, whether in low Earth orbit, the local neighborhood of the Earth-Moon system, elsewhere within the Solar System, planetary systems on other stars, and the region between the stars (and galaxies). Another very substantial challenge that must be confronted in space travel is the immense expanse of "nothingness" between the start and end points of any such journey which can require many years to cross, even within the Solar System, let alone missions to other "nearby" stars and beyond.

Table 8.1 provides estimates of the time it would take, with a range of past and current propulsion systems, and at the speed of light, for a spacecraft to get to potential human settlement platforms within and outside

Table 8.1. Average speed and time to destinations of the Moon, Solar System planets and moons, and Proxima Centauri using a range of current spacecraft propulsion systems, compared to a motorcar and light/photons as extreme examples of travel speeds.

Craft (launch date)	Average Speed	Destination from Earth	Time to Destination
Motorcar	100 km/h	Moon	3,860 hours (5 months)
Apollo 11 (1969)	~5,400 km/h (1.5 km/s)	Moon	76 hours (3.2 days)
		Proxima Centauri (nearest star)	840,000 years
New Horizons (2006)	58,536 km/h (16.3 km/s)	Moon (on the way to Pluto)	9 hours
		Pluto (with Jupiter gravity assist)	9.5 years
Mariner 4 (1964)	7 km/s rel. to Mars and 1.7 km/s rel. to Earth	Mars flyby (July 1965)	229 days
Voyager 1 (1977) Fastest interstellar spacecraft to date	17 km/s (0.006% of light speed)	Jupiter flyby (March 1979)	580 days
		Saturn flyby (November 1980)	1,162 days (3.2 years)
		Proxima Centauri (nearest star)	78,000 years
Spacecraft traveling at 5% of light speed	15,000 km/s	Proxima Centauri (nearest star)	87 years
Light speed (communication time)	300,000 km/s	Mars	5 to 23 minutes, depending on orbital position
		Proxima Centauri (nearest star)	4.37 years
		Center of the Milky Way galaxy	30,000 years

the Solar System. In this table, a motorcar and light/photons are used as examples of the slowest and fastest ends of the spectrum of travel speeds.

To put these transit times in context, the record for the fastest speed at which any human has ever traveled (relative to the Earth) was set by *Apollo 10*, which reached 39,900 km/h (10.8 km/s) or about 0.0037 percent of the speed of light. The fastest unmanned craft ever launched to date is the *Parker Solar Probe*, which reached a speed of around 690,000 km/h (191 km/s) or 0.064 percent of light speed during a flyby of the Sun in December 2024, although the vast majority of this extraordinary speed is due to gravitational enhancement by the Sun.

For the lengths of the journeys summarized in table 8.1 to be shortened by any material amount, the kinetic energy of the spacecraft must be increased in proportion to the square of the required increase in average speed as demonstrated by equation (2):

$$K = \tfrac{1}{2} m v^2 \qquad \text{Eqn (2)}$$

where K is the kinetic energy, m is the mass of the craft, and v is its velocity.

To decrease the travel time of 78,000 years required by *Voyager 1* to get to Proxima Centauri to, say, 78 years, the average speed of the spacecraft would need to be 1,000 times faster, and the kinetic energy provided must therefore increase by a factor of a million. This is far beyond current rocket (and all other moderately practical future) propulsion technologies.

In cases where the mission seeks to land on another body rather than to just execute a flyby, the energy requirement of the mission doubles due to the need to expend an amount of energy in decelerating the spacecraft before arrival equal to that required to accelerate to its cruise speed. In cases where the rate of acceleration and deceleration is very slow (i.e., for a light sail propulsion system), significant time will be spent in getting to, and reducing from, the cruise speed, thereby requiring that the coast speed is much greater than the average. The associated massive energy requirements for a crewed spacecraft to make the trip to the nearest star have been estimated to be equivalent to the total annual energy output of the entire world, thereby causing some to consider it improbable that humans will ever be able to explore beyond the Solar System (O'Neill 2008).

The following sections consider in detail the current and future options for human travel to a range of specific destinations both within and beyond the Solar System.

WITHIN THE SOLAR SYSTEM

Human missions to Mars, an asteroid (perhaps Ceres), and the promising moons of Jupiter and Saturn discussed in chapter 6 will take between at least 7 months (Mars) and 3.2 years (Saturn), depending on the mass of the spacecraft and the chosen trajectory. These are not insignificant journeys, and they present many hardships and risks. As documented in chapter 4, the longest continuous period spent in low Earth orbit is by Valeri Polyakov, who spent 437 days on the Mir space station between 1994 and 1995. The longest total length of time spent outside the protection of the Earth's magnetosphere is only around two weeks, by John Young on his *Apollo 10* and *16* missions in 1969 and 1972, respectively. The lengths of these missions are well short of the times required to reach any destination outside the Earth-Moon system. This leaves a great deal of uncertainty about the risks that will be encountered in almost all mission-critical functions in a multiple-years crewed flight.

Furthermore, and importantly, uncrewed probes did all the preliminary exploration and evaluation for low Earth orbit and missions to the Moon prior to crewed missions being sent to the same environments. Likewise, robotic probes have been doing, and are continuing to do, all of the initial scouting in advance of human missions to Mars and farther afield in the Solar System.

OUTSIDE THE SOLAR SYSTEM

Robotic missions to deep space for scouting work in advance of direct human exploration will entail very long travel times with feedback loops that may be frustratingly and sometimes unacceptably long. For example, even the proposed Breakthrough Starship mission(s) to the closest exoplanet system to Earth, Proxima Centauri, will require around 22 years (traveling at 20 percent of the speed of light; table 8.1) to get to this star system and another 4 years for the return signal to get back to Earth.

It is more likely that preliminary assessments of targets outside the Solar System will be done remotely via the current and next generation of Earth-based and space-based telescopes. These telescopes will be able to identify, examine, and directly image planets orbiting in the habitable zone of stars to determine if they possess the appropriate conditions for life to be supported (as discussed in chapters 1 and 7). They will have been closely assessed for their size, orbital position, insolation rate, atmospheric composition, mean temperature, presence of liquid water, a rocky surface, and a magnetic field to protect life from deadly cosmic radiation. Only when the conditions have been established as sympathetic to life should an expedition be mounted with humans on board, possibly including rendezvous with "stepping-stone" celestial bodies or with robotic fuel and food supply ships that were sent off earlier at slower speeds.

Even if the appropriate conditions have been confirmed in advance on the target platform, the challenges for a spaceship containing humans traveling to, and landing on, another star system are nothing less than extremely daunting:

1. At a very optimistic average speed of 5 percent of the speed of light, the journey to the closest star system, Proxima Centauri, would require 87 years (table 8.1) and would likely require three generations of humans to occupy the spacecraft over the course of the journey. The energy required to accelerate and decelerate the spacecraft from this speed would be truly enormous. Moreover, the damage caused at this speed by impacts with even tiny interstellar dust particles could be terminal for the mission.

2. At more reasonable speeds similar to the fastest craft that have been launched from Earth to date (*Voyager 1*; see table 8.1), the journey to Proxima Centauri would require 78,000 years, or around 2,600 human generations. The many physical, social, and mental health issues associated with a journey of this length through the endless path of unimaginable, inescapable emptiness are outlined in chapter 9. After a handful of generations, the humans on board will have lost essentially all physical and emotional connection to the Earth, except as provided by audio and visual recordings. News from Earth throughout the journey via radio communications will require up to a 4.3-year one-way delay.

3. The very long travel times suggest that a strategy of using "stepping-stone" objects in the outer Solar System, the Kuiper Belt, and the Ort Cloud might be appropriate, rather than proceeding directly to the targeted final destination (Lucas 2004). This will increase the time taken to get to the end point but could be organized to allow replenishment of resources (through earlier storage of fuel, food, and other supplies) and therefore enable the overall size and weight of the spaceship to be minimized. This is one of the strategies suggested for getting to Mars, namely, to establish a permanent base on the Moon (or in Earth orbit) as a temporary repository for consumable and other supplies, from which the final trip to Mars can launch with a much lower energy penalty.

4. The so-called travel time or wait calculation problem in which the earliest "pioneer" spacecraft are overtaken by later, faster ones is a difficult issue. Specifically, why send out missions when it can be safely assumed that, with the development of better propulsion technology, later missions will arrive at the destination before them? However, there comes a time when, despite continued technological advances producing higher spacecraft speeds, the waiting time for that growth in technology is too long to make up the velocity difference between the earlier and later craft (Kennedy 2006).

5. While much research has been undertaken on the physical, social, and mental health of human spaceflight to date (see the discussion in chapter 9), the missions upon which this real-world evidence is based are very limited in length and number. The hugely extended travel times to other star systems leave many of the long-term physical and mental health issues unresolved, so it is likely that the true impact will only be discovered by executing the trip itself.

EXTENDING THE LIFESPAN OF HUMANS

At the time of this writing, the longest documented human lifespan was 122 years and 164 days, by Jeanne Calment who was born in 1875 and died in 1997. Even traveling at a speed of $0.1c$, this extreme longevity is insufficient to survive a return journey to an exoplanet more than around 6 light years away from Earth. In practical terms, the required spacecraft speed for a human passenger undertaking a journey to the nearest star system in the prime of their adult life (ages 20 to 50 years) is around 15 percent of the speed of light. This is unattainable with currently available propulsion technology and could only be achieved (perhaps) by one or other of the unproven theoretical systems discussed in chapter 11.

If travel speed cannot be increased significantly (at least for the time being), the maximum distance for survival of a human can be lengthened (a bit) by increasing their life expectancy.[1] At present, eight major processes are believed to contribute to aging and death (Bohan 2022): DNA damage and cancer-causing mutations; epigenetic changes (i.e., cell function disconnected from DNA); mitochondrial (cell energy production) dysfunction; the shortening of telomeres (repetitive sequences of specialized proteins that protect DNA from degradation); senescent cell (those that have ceased to divide) accumulation; dysregulation of protein function and nutrient sensing; depletion of stem cell reserves; and impaired cellular communication.

Delaying the onset of debilitation and death requires a suite of future medical breakthroughs in tissue rejuvenation using stem cells, regenerative medicine, molecular repair, gene therapy, pharmaceuticals, and organ replacement, collectively known as "strategies for engineered negligible senescence." While some experts are optimistic about our ability to eliminate age-related damage (Zealley and de Grey 2013), others are more cautious (Warner et al. 2005). In any event, progress in eliminating senescence must be achieved across all of these strategies before a material increase in human lifetimes can be achieved, and this is not going to happen anytime soon.

Playing with Time Itself
Special Relativity and Time Dilation

Einstein's theory of special relativity indicates that for passengers in a spacecraft moving relative to Earth, their clock will move slower than those left behind on Earth and they will experience "relativistic time dilation." Equation (3) shows the relationship between the rate at which time passes for someone on the spacecraft and the rate that it passes for the person left behind:

$$t' = t / [\sqrt{(1-v^2/c^2)}] \qquad \text{Eqn (3)}$$

where t' is the time experienced by a passenger on the spacecraft, t is the time experienced by the person left behind, v is the velocity of the spacecraft, and c is the velocity of light. Worked examples of equation (3) show that the time dilation (i.e., the speed with which time slows down) at a velocity of $0.1c$ is 0.5 percent, at $0.5c$ it is 15.5 percent, at $0.9c$ it is 229 percent, at $0.99c$ it is 709 percent, and at the speed of light time stands still and the dilation is infinite.

The practical effect of the slowing down of time for passengers on board the craft can be described by considering that *either* (1) the effective speed of the craft is increased in direct proportion to this time dilation *or* (2) the effective distance to the destination is reduced by the same proportion. Under both descriptions, the destination is reached in a shorter time for those on board, but not for anyone outside that frame of reference. The amazing extension of this algebra is that in the limiting case of travel between two objects at the speed of light, for example by a photon of light, no matter how far apart these objects are, the photon will arrive at its target at the same time as it left the source.

While everything appears quite normal on board the spacecraft, back on Earth time has passed at its "normal" rate, and so when/if the passengers ever return to Earth, they will be younger than their stay-at-home contemporaries. For example, in the case of a person on a 10-year return journey from Earth at $0.5c$, time seems to pass as normal, but for the person back home, they will have been away for 11.5 years.

Time dilation makes the concept of interstellar travel more attractive, at least from the perspective of the traveler. However, while this is an exciting prospect when the spacecraft speeds are near c, with current propulsion technology the achievable speeds are such a small proportion of light speed that the time dilations experienced in practice are minuscule. For example, after six months on the International Space Station (the normal length of an ISS mission), orbiting Earth at a speed of about 7.7 km/s (i.e., 0.0026 percent of c), an astronaut ages about 0.005 second less than everyone else on Earth.

On the other hand, unlike the benefits that accrue from time dilation, special relativity dictates that the mass of the spacecraft increases exponentially as it approaches light speed, with consequent dramatic effects on the fuel requirements. Like time dilation, the (relativistic) increase in the mass of the spacecraft is insignificant at low speeds, but at the speed of light the spacecraft has infinite mass. The additional fuel requirements on an already very massive spacecraft accelerated to, say, $0.5c$, will increase the mass of the craft by the same 15 percent as passenger time was dilated, creating an added burden on the achievement of high speeds, even if these speeds were achievable without mass expansion in the first place.

Another problem with time dilation is that it complicates communication between the spacecraft and Earth. Depending on how fast they were moving, the passengers will not experience the years and decades of travel but, perhaps, only weeks or months. This difference in elapsed time for each of the parties (on board and at home base) would introduce serious difficulties in coordinating messages, especially if the spacecraft was accelerating (or decelerating) rather than cruising with constant speed.

Furthermore, depending on the time that the message was sent from Earth, it might not have enough time to catch up to the spacecraft before it arrived, thereby introducing a kind of "event horizon"[274] for those on board, after which messages would only reach the spacecraft subsequent to its arrival at the destination. On the other hand, if the spacecraft was traveling at a significant proportion of the speed of light, signals sent *from* the spacecraft to the destination would not arrive until shortly before the craft itself got there.

Wormholes

Another method of dramatically reducing the time required to reach an interstellar target is to use "wormholes," the (currently hypothetical) shortcut through space proposed by Einstein and Rosen (1935). Wormholes are consistent with the general theory of relativity and are an outcome of a special solution of the Einstein field equations. They may be simply visualized as providing a reduced distance between two points on a 2D representation of space by folding that plane over onto itself. The path between the two points is then represented by a tunnel, or "wormhole," between the folded plane surfaces that can dramatically cut the travel time from one point to the other. It could potentially reduce the travel time between galaxies from millions of years to hours or minutes.

Wormholes might give the appearance of faster-than-light travel to those located outside it, but the speed of light is not exceeded locally at any time while traveling through the wormhole. Slower-than-light speeds are always used, but the wormhole shortens the distance between the start and end points of the journey so the time taken to traverse it could be less than the time it would take a light beam to make the journey outside the wormhole.

While wormholes do not require a violation of the laws of physics, they may require the existence of yet-undocumented matter and have not been observed in nature. They are a common theme in science fiction because they permit interstellar, intergalactic, and even inter-universe travel within the span of a human lifetime and have also served as an effective method for time travel, that is, transport to another era.

Relocation into the future would occur naturally whenever the speed of a person is greater than that of the environment (or frame of reference) to which they are returning. This is illustrated in the example provided above of a member of the crew of the ISS who, after six months in orbit, would return to Earth 0.005 second in the future relative to their colleagues on the ground. However, many in the scientific community believe that backward time travel is highly unlikely since any such circumstance would introduce potential problems of causality[3] (Bolonkin 2011).

SUMMARY AND CONCLUSIONS FROM CHAPTER 8

The hurdles confronting humans traveling to the outer Solar System (that is, the asteroid belt and beyond), let alone to other star systems, are very challenging, and some may be insurmountable for the longer of these journeys. Foremost among these difficulties are travel times of years for the outer Solar System and centuries or millennia for interstellar destinations; the impact of long-term exposure to weightlessness and cosmic radiation;

the need to carry a multitude of resources to sustain life on board (including a reliable source of energy) and materials for construction, repairs, and other activities at the destination; and the as yet largely unknown psychological, social, and political repercussions of living for very long periods in an isolated, confined, and inescapable "bubble."

The long travel times are perhaps the most serious, since they are responsible for most of the other physical and emotional challenges, and they are the ones over which we (currently) have the least control. Chemical rocket propulsion systems in present use are reaching the limit of their capability, and the next generation of technologies are as yet unproven in practice and/or at scale (see chapter 11). Options for time dilation (i.e., a relative slowing down of the passage of time in the frame of reference) for those humans on board the spacecraft by increasing travel speed to a significant proportion of the speed of light are therefore limited to negligible levels. Attempts to extend human life beyond its current level of around 120 years by medical (or any other) means are in their infancy and, in any event, are unlikely to be significantly helpful for all but the shortest of journeys (i.e., those limited to decades). Finally, other methods of utilizing distance "shortcuts" (wormholes, etc.) remain in the realms of theory and are very likely to remain so for the foreseeable future.

Whether, like many other scenarios depicted in science fiction, any of these methods of "playing" with the passage of time to enable interstellar travel ever come to practical reality remains to be seen. Further discussion of these perhaps insurmountable challenges of human travel outside the Solar System is provided in chapters 9, 10, and 11.

CHAPTER NINE

Physical, Social, and Psychological Challenges of Space Travel

> *The greatest unknowns, and perhaps the most dangerous are those we have not considered or are unaware of, colloquially termed the "unknown unknowns."*
> —J. D. POLK, CHIEF MEDICAL OFFICER, NASA
> (LinkedIn message to author, April 29, 2024)

In his book *Endurance*, Scott Kelly describes in agonizing and graphic detail his experience after returning to Earth after nearly a year on the International Space Station in 2015 (Kelly 2017). His accounts of the nausea, fever, disorientation, muscle atrophication, and pain arising from fluid movements (e.g., the blood in his body rushing to his legs) several days after reentering the Earth's gravity well were described as much, much worse than after his previous six-month mission. Some of the cognitive changes he experienced took months to return to something near normal.

These and a host of other post-mission outcomes documented by most of the long-term astronauts and cosmonauts highlight the fundamental problem for humans in their quest to venture deep into space: that we are not built for space travel and are unlikely to be able to evolve quickly enough under natural pathways to be so in the foreseeable future (Criscuolo et al. 2020; Des Marais et al. 2008). To build the required resilience to the rigors of space travel in the short term (that is, in a handful of reproduction generations) we would need to (1) generate adaptions to the conditions through the widespread use of human genetic engineering and/or synthetic biology to ameliorate or remove our inbuilt encumbrances and fragilities, (2) accelerate reproductive selection in favor of individuals who display greater resistance to one or more of the negative effects of space travel, or (3) progress away from a solely carbon-based structure, potentially to a cyborg[1] or ultimately completely inorganic, "machine" version of humanity.

Natural evolution has led to the existence of a host of species (admittedly only those of simple form, like bacteria) that can survive in extreme environments such as the high-pressure, boiling, metal-rich waters of the so-called black smokers (geothermal vents) along deep mid-oceanic plate boundaries, and the high radiation, near vacuum, and extremely low temperatures of low Earth orbit (see chapter 1). Criscuolo et al. (2020) have suggested that, like these simple creatures, humans may find ways, perhaps during very long-duration spaceflight (i.e., millennia), to adapt to these generally debilitating (if not fatal) conditions through the development of physiological modifications that reduce or eliminate their impact(s). The facilitation of the reproductive success of these "protected" individuals would be critical to the long-term adaption of the species to these alien conditions.

In the shorter term, the relevant protective changes might be expedited by genetic engineering and/or the use of synthetic biology that prevent the short-term physiological alterations, namely, nausea and bone density loss from low gravity, fluid movement in the body, cognitive degradation, retinal issues, radiation damage, etc. (see sections below).

The creation/evolution of an even more resilient, partly inorganic "architecture" for those individuals who undertake long-term missions might also be contemplated. This could be deemed to be merely an extrapolation of the increasing level of inanimate augmentation or replacement of functionality that is being incorporated into the human body right now. It runs in parallel with the rapidly increasing proficiency of the many robotic spacecraft, with their onboard decision-making (artificial intelligence), that are in current use within the Solar System, including all manner of orbiters, landers, and rovers.

Despite their advancing capability, none of these existing robotic systems are entirely autonomous since they all require periodic or continuous oversight and intervention by humans to make most of the critical decisions. Nevertheless, jumping forward for a moment in advance of the discussion of this unfolding human and machine transformation in chapter 13, robotic systems have huge potential as a means of avoiding exposing the fragility of the human biological form to the rigors of space. However, they will not provide a mechanism to guarantee the continuation of human civilization off-Earth until they have jettisoned their dependence on humans and have advanced to the point of acquiring and assimilating all the innate, biological characteristics of the beings that we use to define "humanity."

Current technology embedded in autonomous inorganic systems is improving rapidly, and recent developments in advanced generalized intelligence (AGI), such as ChatGPT, do contain some of the elements of human "reasoning," albeit derived from the raw statistics of language gleaned from the massive database that is the "internet." Having made that last statement in a somewhat dismissive tone, is not human learning from birth founded on just such a data accumulation (aka "machine learning") protocol?

Regardless of these spectacular achievements in AGI, the current state of this software in no way (yet) replicates the full range of human intelligence or functionality. In consequence, the diehard proponents of in-situ human engagement in these exploratory forays into off-Earth environments can maintain a case for continued funding of proposals for crewed spacecraft since only humans can (currently) provide reliable "on the ground" scientific and technical interpretations and timely response to emergencies.

While no humans have ventured outside low Earth orbit since *Apollo 17* in 1972, NASA, ESA, and other space agencies have continued to explore the Solar System with robotic missions, and the sophistication and capability of these robots have increased dramatically over those 50-plus years. Nonetheless, NASA, SpaceX, and other organizations are now pushing for a resumption of human-crewed missions to the Moon and Mars, rather than persisting solely with the development of the next generation of truly autonomous machines to "do the dirty work" in these hostile environments. It seems that both models (animate and inanimate) for exploration of other parts of the Solar System and outside it will be pursued until humans can evolve into a more resilient species (Criscuolo et al. 2020) or "the machines" can fully take over our roles.

Another option for the far future (perhaps) is one in which human consciousness alone is transported across space, with our physical presence remaining (perhaps not so) safely on Earth. Once again, chapter 13 provides much more discussion on this all-important subject. In the meantime, let's refocus back on resolving the many inherent issues for human spaceflight.

The Difficulties for Human Spaceflight

In the case of the transportation of humans off-Earth in their current, very fragile, wholly biological form, the challenges are manifold (Patel et al. 2020; Buckley 2006). These range from the physical impacts of low or zero gravity, including fluid shifts and muscle atrophy and a multitude of other problems described by Kelly (2017), to radiation damage to DNA, brain function/cognition and the nervous system, carcinogenesis, psychological effects such as isolation and confinement stress, sleep disorders, and social/behavioral disorders.

Ensuring that humans can survive and work safely in space for relatively short periods of weeks or months, let alone potentially indefinite periods, has become the greatest challenge, aside from mission funding, for human space exploration to resolve. The very sophisticated and complex technology of spaceships and spacesuits has enabled these travelers to survive the harsh conditions of vacuum and cold (near absolute zero, or −270°C) in outer space; has provided breathable air, water, and food; and has dealt with the disposal and recycling of waste products.

Some, but by no means complete, protection from radiation and micrometeorites has been provided by relevant shielding, but massive and unexpected solar flare events have not had to be dealt with yet, and they can deliver fatal doses in minutes. The ISS has special designated locations on the spacecraft, protected by water reservoirs, that provide extra shielding[2] and to which occupants can retreat when advance warning of solar flares and coronal mass ejections is provided by the SOHO spacecraft and other monitoring stations on Earth. Despite this progress, the risks of long-term exposure to zero or near-zero weightlessness and the social and psychological impacts of confinement and isolation have yet to be tested meaningfully.

At the time of this writing, more than 60 years of human experience in space has been accumulated since the first flight of the Soviet Union's Yuri Gagarin in April 1961. Much has been learned through studies of the medium-term (multiple-month) effects of the zero-gravity environment of missions in the Skylab, Salyut, Mir, and ISS low Earth orbit bases. However, so far, only 11 humans have spent more than 300 consecutive days in space (see chapter 4 for details), and only 28 humans have each spent more than 500 days in space spread over multiple missions, the longest total time of which was 878 days (five missions) achieved by Russian Gennady Padalka. His accumulated mission time off the Earth's surface was longer than it takes to complete a round-trip to Mars (400 to 500 days of zero gravity), even allowing for a significant period (at 38 percent of Earth's gravity) spent resident on the surface at the destination, while waiting for the celestial mechanics of Mars and Earth to permit a Hohmann transfer trajectory for the return to Earth.

Thus, nothing is known of the impact of continuous space travel for periods longer than around 15 months. Furthermore, and important for long-term settlement missions, nothing is known about the effects of this travel on pre-adult humans or on adults who have not been heavily screened and preselected for resilience prior to the mission. The uncertainties will multiply significantly during the extended human missions proposed for an orbiting lunar space station, a "permanent" outpost on the lunar surface, and the journey to Mars. Destinations farther afield in the Solar System and extended residency on the Moon and Mars will be even more challenging.

It is worth remarking at this juncture that Captain James Cook's journey to Tahiti from Britain to observe the 1769 transit of Venus and subsequently to search the Southern Ocean for the presence of *Terra Australis Incognita* (ultimately leading to the settlement of Australia by the British in 1788) took eight months. This is about the same time as it takes spacecraft to get to Mars from Earth. Indeed, these early sea voyages by single or small groups of wooden sailing ships to the far-flung and uncharted reaches of the world in centuries past were, at some levels, as dangerous and daunting as a future human journey to Mars will be.

Perhaps they were even more difficult in some respects, since in Cook's case there was no possibility of receiving reassuring conversations and technological and operational advice from "mission control" back in England, while in the case of Earth-to-Mars missions, this comfort and advice can be received with just a one-way time delay of between 3 and 22 minutes, depending on the relative positions of the two planets. Also, navigation on the long sea voyages across vast uncharted and dangerous waters was very uncertain because, without accurate clocks, there was no means of reliably and accurately determining longitude. Moreover, wild storms arose frequently without warning, and Cook and his crew did not have the advantages of modern nutrition and medical treatments. They had to carry supplies of food and water sufficient for dozens of men for long periods of open-water sailing between the uncertain availability of top-ups when they eventually did reach landfall. They were also subject to potential attack from unfriendly local populations with which they might make unforeseen contact.[3]

Cook's ship the *Endeavour* was a mere 29.8 meters long and 8.9 meters wide and carried a complement of 95 men. For comparison, the ISS is 94 meters long with a habitable zone of 388 cubic meters, which can accommodate up to just 7 people (not counting temporarily docked spacecraft from Earth).

Despite this, Cook arrived at Tahiti with the loss of only 5 of the initial 94 crewmen while rounding the treacherous Cape Horn, and the suicide of one other while crossing the Pacific Ocean. By the seafaring standards of the time and the usual loss of sometimes up to 50 percent of the crew from the effects of scurvy, this was a quite remarkable achievement.

The expedition's success was due largely to Cook's masterful navigation (for the day) and leadership abilities and the fact that he carried a variety of experimental foods on board, including sauerkraut and malt wort, which turned out to provide the vitamin C needed to prevent the debilitating and deadly scurvy. Cook's steadfast faith in these supplements were put to the test and may have been an abject failure had he not administered harsh punishments for those of his crew who did not fancy the taste of these vegetables. Modern management practices may not permit the use of such coercive tactics when the crew objects to orders "from above" during spaceflight.

The individual, interactive, and combined hazards of long-term spaceflight, especially those missions that venture outside low Earth orbit and the protective envelope of Earth's atmosphere and magnetic field, are provided in Table 9.1. They include NASA's "red risks," namely, space radiation, isolation and confinement, distance from Earth, gravity fields, and hostile/closed environments, identified as having the greatest likelihood of occurrence, the highest impact on crew health and performance, and the possible contribution to failure of mission objectives (Lewis 2023).

Table 9.1. Summary of the physical and psychological/social impacts of space travel on humans.

Physical	Psychological/Social

Low or zero gravity:
  Muscle atrophy
  Loss of bone mass
  Balance disorders
  Fluid shifts
  Ocular degeneration
  Cardiovascular deconditioning
Radiation:
  DNA damage (genome instability)
  Carcinogenesis
  Nervous system damage
  Cardiovascular disease
  Compromised immune system
Fatigue:
  Reduced task performance
	Isolation:
	  Quality of life perceptions
	  Behavioral impacts
	  (lethargy, hostility)
	  Personality clashes
	  Anxiety (medical, safety, remoteness)
	  Depression (no chance of escape)
	Noise:
	  Sleep disorders
	Cognitive and performance decline:
	  Negative team dynamics
	  Boredom (fixed routine)
	Resource limitations:
	  Poor food variety
	  Lack of space for recreation and "novelty"

Some of the lower-level risks are ranked as such because they are of low probability and/or can be mitigated by in-flight or post-mission actions. However, interactions between the individual physical and psychological/social issues can exacerbate their impact when exposed in the confines and isolation of a spacecraft.

In a benchmark experimental study by NASA of some of these effects, Scott Kelly's mission of close to a year, double the length of a typical ISS mission, involved comparisons of the impact of the flight on a wide range of parameters including, inter alia, his cognitive abilities and changes to his genes. This was part of a unique "twins study" that tracked his condition relative to his identical twin Mark Kelly, was was/is also an astronaut (Garrett-Bakelman et al. 2019).

Scott's degraded physical condition soon after his return to Earth was documented in the early part of this chapter, although this account appears to contrast with the condition of Valeri Polyakov, who was reported to have walked unassisted from his capsule upon landing at the conclusion of his record-breaking 437-day mission in 1994–1995. Of concern was that Scott's cognitive scores took more than 18 months to recover fully to their pre-flight state and there were changes in his genome and epigenome that occurred in the last half of the flight. However, solid conclusions were hard to draw from the study because several factors could have contributed to this outcome and the sample size is just one person.

The impacts of zero or minimal exercise on heart function were measured in experiments on Earth in the mid-1960s as part of the Dallas Bed Rest Study (Mitchell et al. 2019), where subjects were not allowed to exercise for 21 days, after which they were subjected to intensive endurance regimes. The results showed that their maximal oxygen uptake declined 27 percent after bed rest and increased 45 percent after training, while average maximal cardiac output declined by 26 percent after bed rest but increased by 40 percent with training. Much of current knowledge about the adaptive capacity of the cardiovascular system can be traced back to this study and its follow-ups over the ensuing 50 years. It had important clinical implications for how heart attack patients were treated, which almost immediately changed from strict bed rest to early ambulation and use of cardiac rehabilitation. However, zero or minimal activity is not entirely analogous to the zero or low gravity situation experienced by space travelers.

The following sections discuss details of the various specific physical, social, and behavioral challenges and their associated case studies (Patel 2020, and references therein).

Physical Challenges
Low and/or Zero Gravity

In response to weightlessness, some physiological systems start to change and, although most of these changes are temporary, some have a long-term impact. One that can occur early in any mission, and which affects close to 50 percent of all space travelers, is the adjustment of the vestibular (inner ear) system, causing nausea, vomiting, and balance issues. Longer-term exposure leads to loss of muscle mass, including debilitating effects on the performance of the cardiovascular system. This is because in zero or microgravity environments skeletal muscle is no longer required to maintain posture, and the muscle groups used in moving around in this environment differ from those required for terrestrial movement. The types of muscle fiber in muscles also changes from slow-twitch to fast-twitch (Tischler and Slentz 1995).

Bone metabolism also changes due to the decrease in mechanical stress, resulting in a progressive loss of bone mass and increased brittleness resembling that which accompanies osteoporosis (for a review, see Baran et al. 2022). The resultant elevated blood calcium levels can result in calcification of soft tissues and the formation of kidney stones. While bone density is eventually recovered upon return to Earth, this can take several years. To combat these bone and muscle issues, the space travelers are required to exercise on board for at least two hours every day on specially designed resistive treadmills and other apparatus.

In spaceflight, the fluid volume of the body reduces, with a lowering of blood volume by up to 22 percent (Alfrey et al. 1996). With less blood to pump, the heart atrophies, causing low blood pressure and associated fainting and dizziness. With the removal of gravity, body fluids no longer have to compensate, resulting in a general accumulation into the upper part of the body. This causes balance disorders, distorted vision, and sometimes a loss of taste and smell.

Vision impairment in spaceflight has been observed for more than 15 years, with up to 50 percent of long-flight space travelers reporting diminished vision because of increased intracranial pressure. It is primarily observed as swelling of the optic disc, folds in the retina, shifts in refractive error, and flattening of the eyeball, only some of which can be corrected with appropriate lenses (Paez et al. 2020). So far, none of the observed eyesight effects have persisted after return to Earth, and methods to minimize their appearance in flight include the aforementioned resistive exercise, high-sodium dietary intake, and high carbon dioxide levels. However, such problems would be a major concern for deep space flight missions since there would be no immediate opportunity for a return to Earth to recover.

Artificial gravity, achieved by slow rotation of large "pinwheel"-shaped spacecraft with personnel occupying the outer regions, farthest from the center of rotation, or a tethered pair of craft rotating around their center of mass (Zubrin 2011) have been proposed as solutions to the weightlessness problems, especially during long periods in transit from Earth. However, while some spacecraft have been rotated slowly in orbit and on missions to the Moon to "spread" the heat load of exposure to sunlight, the engineered simulation of gravity has yet to be demonstrated at scale and its long-term effects have not been investigated.

These include the need to generate the relevant (outward) centripetal force perpendicular to the rotation axis to provide the desired pseudo-gravitational force g by a delicate balance between the radius and speed of rotation. Keeping the radius of rotation as small as possible has the advantage of decreasing the size and mass of the spacecraft but produces a smaller g for the same rotation speed. Increasing the speed of rotation produces a higher g for the same radius, but if increased above around 2 rpm may then cause inner ear problems due to the Coriolis effect.[4] In any event, achievement of the full gravitational force experienced on the Earth's surface, or even on Mars (38 percent of that), is unlikely to be fully realized by this method due to limits on spacecraft size, mass, and fuel usage.

The limitations arising from the need to keep the spacecraft mass as low as possible also applies to the automatic production of artificial gravity during the (hopefully uniform) acceleration of the craft up to, and down from, its cruising speed. The level of acceleration would need to produce the equivalent of around 50 percent of Earth's gravity and would have to be applied for the majority of the total journey in order to achieve the reduction or elimination of the negative effects of low gravity. The implications for the mass of fuel required to maintain this level of acceleration over a large portion of the journey are obvious and probably render this option impractical except, in theory, for the cases of light sail, nuclear, and ramjet propulsion systems (see chapter 11).

The same physical and emotional health and performance risks associated with weightlessness during the journey to a distant target platform apply to the potentially extended periods of lower gravity experienced by humans located in long-term bases/settlements on the surfaces of Mars or the Moon, where gravity is only 38 percent and 17 percent, respectively, of Earth's.

Radiation

Exposure to space radiation outside low Earth orbit is one of the most significant risks to organic life-forms. Damage caused by galactic cosmic rays (high-charge and high-energy ions, high-energy protons, neutrons, and secondary particles) and solar particle streams (low- to medium-energy protons) is much more destructive to biological tissue than that produced by X-rays and gamma rays because the cosmic rays and other particles have mass. The highly energetic high-charge and high-energy ions, for example, produce double strand breaks and oxidative damage to DNA that are difficult to repair and are associated with altered cellular behavior and signaling patterns that may lead to disease outcomes (Patel et al. 2020).

Evidence for radiation carcinogenesis comes from epidemiological studies of humans exposed to atomic bomb radiation (e.g., Ozasa et al. 2019) and, in the case of simulated space radiation, from ongoing studies of rodent and advanced human cell-based model systems (Pariset et al. 2021). So far, the only humans to have spent time outside the protective bubble of Earth's magnetic field are the 27 astronauts who flew there (three of them twice) in the Apollo program between 1968 and 1972. There were no genomic studies of these astronauts at the time and, in any event, the longest exposure to cosmic radiation was around two weeks, by John Young on his *Apollo 10* and *16* missions in 1969 and 1972, respectively. Much longer exposure experience will be needed to properly assess the effect of radiation on human genes.

Terrestrial radiation exposure has been associated with a range of degenerative tissue effects, including cardiovascular and cerebrovascular diseases, cataracts, immune system degradation, and respiratory dysfunction. These effects may be more serious during spaceflight radiation exposure, with research focusing on finding the mechanisms for damage, the elucidation of diagnostic and therapeutic approaches, and the determination of practical exposure limits and disease-free survival years (Patel et al. 2020). Meaningful epidemiological studies of these effects are thwarted by overlap with lifestyle and genetic factors and the notoriously small sample sizes for space traveler cohorts.

Risks to the central nervous system are also major concerns for the exploration of space (Chancellor et al. 2014). These include impaired neurocognitive and motor function, along with behavioral changes, observed in both humans (exposed to high doses of gamma rays and protons) and a host of animal models (exposed to high-charge and high-energy ions). The effects are expressed in dementia, cognition impairment, memory loss, depression, "fear extinction" (i.e., reduction of fear after repeated experiences without adverse effects), and anxiety. Nervous system effects have yet to be observed as significant outcomes of long-duration ISS missions, but changes in brain shape and volume and gastrointestinal issues have been noted. Much further experimentation is needed, not least because it is unclear how individual components in a profusion of mission hazards interact with each other and the overall radiation risks.

To add to the challenges, the radiation exposure for humans in orbit or in transit is not reduced when they land and occupy bases/settlements for long periods on the surfaces of the Moon, Mars, or any other body with little or no atmosphere or magnetic field to protect them. Data from radiation measurements by the *Curiosity* rover on Mars suggest that doses on this planet might be problematic for humans (Gelling 2013). Neither the Moon nor Mars has a magnetosphere to shield visitors from solar and cosmic radiation; the Moon has no atmosphere at all, and Mars's atmosphere was long stripped away by the solar wind and is now only about 1 percent of Earth's. Thus, settlements on these bodies will need to build robust accommodation structures on the surface or be in sites well below the surface (e.g., lava tubes) to minimize the radiation risks (i.e., use paraterraforming protocols).

In anticipation of parts of the discussion in chapter 13, it is noteworthy that almost all kinds of physical damage experienced by humans through reduced gravity and radiation will not be experienced by robots. Indeed, doing away with flesh and replacing it with silicon, titanium, and other metals and ceramics, etc., along with computer-based artificial intelligence (potentially massively superior to humans), imbued with all or much of the information, history, and culture of humanity, might allow "us" to live indefinitely off-Earth.

However, robots have their own downsides. Unless adequate radiation protection is provided, the control software in these robots, not unlike the human brain, can be damaged by cosmic rays, as evidenced by the breakdown of communication satellites in low Earth orbit during periods of strong solar flare activity. Furthermore, like human flesh, the inorganic materials (i.e., hardware) from which these robots will be built may also be damaged (although in different ways from biological beings, and at much reduced levels) by very prolonged, intense radiation exposure.

Social and Behavioral Health

It is natural to expect that there are strong interactions between the physical/mechanical and social/political domains of spaceflight due to the impact of biological, social, and psychological factors on each individual and the polity, depending on the specific environmental conditions in play. How the human mind copes with the isolation, loneliness, and "no-escape" aspects of long-distance space travel may be the biggest and most potentially unsettling unknown.

An exploratory trip to Mars and back will most likely occur aboard a ship smaller and with fewer crew than the ISS, and there will be increasingly delayed two-way communications with Earth as the occupants get farther and farther away. It will be a long, lonely, cramped journey with monotonous, processed, and packaged food; the continuous noise from instrumentation and the fans needed for cooling of electrical equipment and the forced circulation of breathable air necessary in a weightless situation; unnatural night/day sequences disrupting circadian sleep cycles; and no exit option. It remains unknown what happens to people's minds when those conditions last for years.

There is considerable evidence that psychosocial stressors are among the most important impediments to optimal crew morale and performance (Kanas 2023). A wide range of studies have been conducted in (imperfect) space-analogue environments, such as in Antarctica, Biosphere 2, submarines, and land-based and submersible simulators, and from studies of hypodynamia (confined bed rest), many lasting months to a year or more (chapter 6 and earlier in this chapter), but the relevance/importance of such analogue studies for manned space missions is not clear.

There is evidence that crew anxiety is expressed differently because of the much lower degree of danger in isolated environment experiments on Earth (or even in low Earth orbit) compared to those that will be experienced in space. Crew tension in the on-Earth examples seemed to be related to crew heterogeneity, as reflected in gender, cultural background, native language, level of career motivation, lower levels of crew autonomy, and the leadership skills of the crew members in charge.

In studies of small, closed communities in Antarctica, Wood et al. (2005) found that there was not just a simple decline in psychological well-being over time and that many of the changes in mental state occur in response to specific events, rather than to the isolation alone. The psychological changes experienced by members of the community also depended on the emotional history that they brought with them, how they interacted with the other people with whom they are isolated, and what kind of events they experienced while isolated. They were found to have significantly different experiences on different expeditions, depending on the events and people they encountered.

A stark example of the instability of the mental state that can arise in members of isolated groups was encountered by Sir Douglas Mawson during his leadership of the Australian Antarctic Expedition (AAE) from

December 2, 1911, to February 26, 1914. As one of the last scientific and mapping excursions before the planned return to Australia in January 1913, Mawson led an ill-fated Far Eastern Expedition from his base in Commonwealth Bay approximately 400 kilometers east across King George V Land and then back (Mawson 1915; Ayres 1999). Mawson was the only survivor of this three-month expedition from November 10, 1912, to February 8, 1913, the other two members, Belgrave Ninnis and Xavier Mertz, perishing en route through accident and illness, respectively.

Mawson arrived back at Commonwealth Bay alone on February 8, nearly a month after his scheduled return and just hours after the departure of the supply ship *Aurora*, which was scheduled to extract the entire complement of the AAE back to Australia. The *Aurora* had delayed its departure as long as it could before winter set in when it, too, would then be stranded at Commonwealth Bay for the winter. A relief party of six men under the leadership of Cecil Madigan, five of whom were members of the AAE and one of which (a radio operator, Sidney Jeffryes) had volunteered to be transferred from the ship, remained behind for another year to search for and hopefully ascertain the fate of Mawson, Mertz, and Ninnis.

All was going well with the small group until July, when Jeffryes began to display intermittent symptoms of hypochondria, melancholia, suicidal desires, "foolish cant," and paranoia, and threatened to expose his comrades upon return to Australia for "contemplating murdering him." After receiving an "intimidation" to the effect that Jeffryes, "believing his life to be in peril," "therewith resigned from his position as a member of the expedition" (Ayres 1999), Mawson decided that he needed to expose this behavior in a speech to all members of the expedition in the hope that it would (1) encourage Jeffryes to acknowledge that he was unwell and (2) prevent the "melancholy and madness" from spreading to the other members. Part of this speech is reproduced as follows from Mawson's papers:

> *[I]t is all a vapour of the mind because as every thinking person knows, there is no such thing as "resignation" on an enterprise of this kind. . . . every member of the expedition signed articles just as binding and onerous as in the case of military service. . . . What I desire is that he shall recognize that he was ill for a time and to continue as a full*

member of the expedition, then the matter will be forgotten. . . . In the event of Jeff convincing people by some subtlety that he was not ill then the excommunication [in an ice cave separate from the main accommodation] *comes into force. (Ayres 1999)*

The situation improved somewhat after this intervention by Mawson, but Jeffryes's ongoing fluctuating moods and mania had to be tolerated, as he was the only one who could operate the wireless. They moderated as spring and summer arrived, and by the time the *Aurora* returned to Commonwealth Bay on December 13, 1913, he was "almost normal."

The AAE example of a breakdown in the mental condition of one of the members of the expedition as the darkness and confinement of winter descended, the absence of work to fully occupy the mind, and the thought of long months ahead took hold, has some close similarities to the physical and social situation of a spacecraft. In particular, it highlights the importance of strong leadership, an enduring commitment to the overall goals of the mission, the need to keep everyone usefully occupied, and the availability of some form of "refuge" to which relocation of an individual might be consigned for the protection of the others.

The effects of confinement and isolation in Earth-based analogue environments have been measured in the 520-day simulated mission to Mars in an isolation chamber at the Moscow Institute of Biomedical Problems (Basner et al. 2013). Once again, these experiments are not exact parallels with space travel since the participants know that they are on Earth and that they can be rescued in an emergency or for other reasons at almost any time, but they confirm the impact of multiple stressors, as revealed by the completion of regular behavioral questionnaires.

One of the key findings from these experiments and in-orbit monitoring studies (Kanas et al. 2000; Kanas et al. 2007) is that in autonomous (i.e., self-managed) conditions, the crews become less dependent on mission control, undergoing psychological "autonomization." Also, the crews tend to increase their cohesion when crew members become closer and more "similar" to each other, despite their initial personal, cultural, and other differences. These phenomena look promising for future solar system exploration. Indeed, surveys of cohorts of astronauts and cosmonauts after they returned to Earth have revealed that they had developed an enhanced positive appreciation of Earth's beauty, other people, nature, spirituality, and power.

The generalized themes that can be extracted from these studies are the characteristics of personality types that make up a good crew, the kinds of problems (such as stress) that can be prevented, and the kinds of countermeasures that would make life easier for people in isolated settings. This might include preselecting crew members for high emotional intelligence (EQ), absence of mental health issues, good interpersonal skills, and high-level leadership qualities in those tasked with these responsibilities; an allowance for crew autonomy; and ensuring that there is regular support and contact with family and friends at home (where possible and practical). Pagnini et al. (2023) provide an instructive summary of these and other issues in deep space exploration and propose potential countermeasures that might be pursued to resolve them.

With a round-trip to Mars requiring up to 18 months in transit time alone, further comprehensive and multiple studies of spaceflights of at least this duration are required before all the risks and their impacts on the social and behavioral health of humans can be determined and understood and countermeasures developed. This understanding might best be obtained in lunar orbit, where outside intervention can still occur (albeit not immediately) when/if things go wrong, but there may be a greater sense of isolation and danger than for a corresponding study in low Earth orbit. Even so, every process and anticipation of intervention is likely to contaminate the results by reducing some of the stressors that cause or accelerate the negative impacts. Only an actual trip to Mars (with all its inherent known and unknown dangers) is likely to provide real-world data on the impact of long-term isolation on humans.

Nutrition

The history of major exploration missions on the land and oceans of Earth has underlined the importance of food availability and nutrition for the success or failure of the mission. The case, described above, of Captain James Cook's use of vegetables containing vitamin C (the specific beneficial ingredient was, of course, not known to him at the time) in the prevention of scurvy on long sea voyages in the mid-1700s is but one example of the importance of nutrition, especially in the case of a closed food system like that in space flight.

Antarctic explorers in the 20th century were also occasionally confronted with serious food shortages, a case in point being Sir Douglas Mawson's Far Eastern Expedition in 1913 (described in a different context above). After the accidental loss, down a crevasse, of a dog sled containing most of the food provisions for the small party of three, Mawson, after the deaths of his companions, was forced to survive in part on the offal of the remaining sled dogs, nearly killing him in the process due to vitamin A poisoning.

Some afflictions on Earth are similar to those experienced in space travel (e.g., nausea, bone density loss, etc.) and can be ameliorated by adjustments to nutrition. Pre- and post-flight biochemical analysis of the blood and urine of ISS astronauts are being undertaken to understand the relationship between these indicators and bone loss, visual impairments, body mass, radiation protection, exercise regimes, fluid distribution in the body, genetics, and, ultimately, crew cognition, performance, and morale (Smith et al. 2005). With ongoing research in this area, many of these deleterious effects of space flight might be able to be reduced or eliminated altogether by administering to crew members the full complement of their nutritional requirements (Patel et al. 2020).

GOVERNANCE
International Space Treaties
Current treaties involving the legal governance of outer space were discussed in chapter 5. Since these agreements/protocols were contemplated, the number of private commercial space actors has increased substantially and many have developed their own launch capabilities which enable them to undertake activities in outer space, independent of national interests. The existing treaties have not kept pace with the recent increased activity in private space exploration, the likelihood of the establishment of semipermanent "bases" on celestial bodies, and the potential for off-Earth resources to be exploited by the entities that get there first.

Atkins et al. (2022) claim that these governance treaties are now largely obsolete and do not provide a clear global space regulatory framework to deal with property and ownership rights of equipment and resources, liability in the event of a collision, management (and salvage) of space debris, resolution of disputes, investment protocols, and uncertainty about who is responsible for taking action to ensure a safe and effective operating environment. To fill this void, individual nations have created their own distinct space legislation and policies which are likely to lead to an increased occurrence of disputes and conflicts with the potential to undermine scientific and commercial activities in space altogether.

Small Groups

Similar uncertainty arises in the case of spacecraft "social governance" in the broader sense of the assigned authority, hierarchy of decision-making, reporting lines, and roles and responsibilities of the individual crew members. To date this has not been a serious issue, largely, if not entirely, because the missions have been relatively short, with only small numbers of individuals in the "teams" and with the hierarchy of management and responsibility set well in advance and subject to real-time monitoring and intervention by mission control.

In the case of the Apollo and Soyuz missions, roles and tasks were determined and assigned to individuals during mission design and training, and even on the longer ISS expeditions, there was no suggestion that these authorities, roles, and duties would "evolve" during the flight. With very few exceptions (Neil Armstrong of *Apollo 11* being one), the crew were military personnel rather than civilians and so they were very used to management arrangements of this sort.

As described above, the concept of "autonomization," in which the crews become more independent of home base, has been observed in some on-Earth and in-orbit experiments (Kanas et al. 2000; Kanas et al. 2007). An extreme case in point occurred during the *Apollo 7* flight in 1968, an 11-day "shakedown" mission for an eventual trip to the Moon. It was the first crewed Apollo flight after the deadly *Apollo 1* fire in January 1967, the first time three men flew in space together (Wally Schirra, Donn Eisele, and Walter Cunningham), and the first time NASA broadcast a television feed from space. The near-fatal explosion on *Apollo 13* that resulted in the abandonment of its lunar orbit and landing mission demonstrated that very complex technical problems can be solved when crew and ground controllers cooperate, but *Apollo 7* showed that when disagreements emerge between the crew and mission control, anything can happen.

On *Apollo 7* there were arguments over whether to launch at all, conflicts over the television broadcast, complaints about the food, and unhappiness with spacesuits that required 30 minutes for astronauts to go to the toilet (Niiler 2018). All of this discontent played out in the context of the launch pad fire that killed three astronauts 20 months prior (one of whom, Gus Grissom, was a close friend of Wally Schirra) and the associated uncertainty about whether the redesign of the spacecraft that emerged from close analysis of that accident had eliminated all the safety concerns. The *Apollo 7* mission also carried the added weight that it had become the dress rehearsal for *Apollo 8*, a mission brought forward as a circumnavigation of the Moon when evidence emerged that the Russians were getting close to a crewed mission to the Moon.

It was also significant that Schirra, as mission commander, felt a heightened sense of responsibility for the crew, had already determined that this was to be his last flight, and had a bad head cold. Schirra is reported as saying that nobody on the ground is taking the risk that he and his crew was. Things came to a head near the end of the mission when Schirra and the crew refused to don their space suit helmet during landing, a clear violation of safety protocols. Ultimately, mission control agreed to allow them to leave their helmets off, and the crew landed safely. None of the three crew of *Apollo 7* flew again with NASA.

Boyd and Richerson (2009) argue that the *Apollo 7* situation is an example of the rapid evolution of an onboard "culture" as a local adaption to the local environment of (and background to) the flight. The degree of cooperation and sense of an onboard "team" that emerged on this mission might reflect the competitive advantage that enabled humans to produce more-cooperative and larger societies than their competitors over the past two million years on Earth. However, it can also cause discoordination of actions and give rise to factions and subgroups that have the potential to be massively disruptive to a mission.

Discontent within spaceflight teams has been associated with high levels of contact with "home," lack of contact with the outside world, fatigue, lack of clear goals and responsibilities, an unsafe environment for communication, unequal workload, low autonomy, and few creative opportunities. While diversity of background, outlook, and expectations are often a source of strength in the group, they can also be a source of conflict and disharmony (Hewer and Sleek 2018). The performance of teams can be improved by preselecting members for sociability, a low need for external stimuli, behavior moderation, collaboration and teamwork (rather than solely for individual qualifications), technical expertise, and designated roles. Improved outcomes can also be achieved with regular "debriefing" sessions involving input from multiple members and a strong focus on the specific goals of the mission (Hewer and Sleek 2018; Riley 2021).

Large Populations in Space
Things will get far more complicated in the case of medium- to long-term missions, especially those in which multiple generations of humans are involved and where the distances from "home" are much larger and the (physical) lines of accountability to the mission stakeholders and mission control become stretched.

For the much larger spacecraft populations occupying a "generation starship" or "worldship" (see chapter 10 for a description of these craft) required for a major settlement mission to Mars or an outer planet or moon of the Solar System, or ones even farther afield, the social, political, and management challenges increase significantly. These populations can be described as, and need to be, genuine "societies" in all meanings of the word. If they are fleeing Earth in the face of an impending extinction event, they will have been created to include representatives of a host of cultures, languages, ages, education levels, skills, and authority, and will be required to maintain all of the life-support systems for the society, along with progeny to maintain population numbers through hundreds if not thousands of human generations (see chapter 12 for further discussion on the selection of who makes the journey).

To suggest that this society will be stable, cooperative, and productive without the emergence of factions/"tribes" and serious conflict at some point, and probably very soon after launch, would be a classical triumph of hope over experience. Given the experience of social life on Earth over the past two million years, exacerbated by the isolation, loneliness, and "sterility of purpose" (i.e., to breed further "platformians" who themselves would be subjected to a similar sterile existence), it is almost guaranteed that any initial harmony will be short-lived.

On the positive side, if the composition of the society has been structured optimally at the beginning, each member or group of members will have an important role to play in the overall societal environment, looking after energy or food production, spaceship maintenance, entertainment/leisure organization, education of the children, breeders of the next generation, enablers of the democracy or other social management system, etc. Thus, short of the emergence of widespread insanity, it might be expected that most of the population would see the advantages of cooperation among these groups of roles. This would be consistent with the observation (Boyd and Richerson 2009) that one of the distinguishing features of humans is that they cooperate on a larger scale with unrelated individuals (in the genetic sense) than most other mammals.

In addition, just like the hoped-for adaptive physical responses to space flight of the humans themselves, discussed in the introduction to this chapter, there is likely to be ongoing adaptive social responses to the local conditions. The structure of the society itself would evolve, perhaps through trial and error, in ways that consider the fundamental fact that individuals in the population have no escape, except by death.

The optimal sociopolitical structure that provides the practical foundation for a large population/society of humans of this kind moving through space in a starship over the long time periods associated with interplanetary and interstellar travel has been discussed by Turner (2013). He presents an analysis of the social, political, cultural, genetic, and "cybernetics" (communication and control system) characteristics and the issues that might/will confront this society through the operation of a *gedanken* (thought experiment) that provides tools to design and optimize this ecosystem. In this model, simulations of ecology, economy, and polity are provided by an "exovivarium," a closed ecosystem in Earth orbit maintained from Earth remotely and democratically governed by a global community.

The characteristics of the worldship/exovivarium community are defined and compared to (1) isolated hunter-gatherer societies on Earth (e.g., Papuans), (2) "spaceship" Earth itself, (3) insular modern societies on Earth (e.g., North Korea), and (4) space stations such as the ISS and Mir. Drawing upon the work of Cohen and Rogers (1995), it is assumed that democracy is the "least-worst" system for implementation in a worldship, and that the biggest threat to its successful operation is "the mischiefs of factions": geographic, class, occupational, racial/ethnic/clan-based/tribal, industry-oriented, and ideological. The selection of reproductive mates and the prevention of inbreeding are also of major concern for the health and longevity of the society.

Cohen and Rogers (1995) and Turner (2014) propose that for the democracy to survive, it needs to retain its basic virtues of popular sovereignty/authority (goals, plans, and enforcement); political equality (no castes or second-class citizens); distributive fairness (compensation for individual advantages); civic consciousness (regard for the society as a whole yields net benefits for the individual); economic performance (maintain acceptable prosperity); and state competence (foster public confidence and stabilize the democracy).

Cohen and Rogers (1995) suggest that the ideal form of democracy can be approached by (1) limiting the power of the state, without strengthening the power of the already advantaged; (2) promoting a strong state but insulate it from factions; and (3) limiting factional dominance by giving all factions an equal voice. It is claimed that some of the more economically and politically successful societies have implemented something approaching item 2 above and that this might also be the best choice for worldship societies.

The virtual exovivarium experiment would allow comprehensive testing and optimization of these models for a worldship sociopolitical system to be undertaken at much lower cost, and with much less risk to health and well-being, than a real-life experiment using a "best guess" of the composition of the human population and its operating structure selected for the first mission. As a side benefit, what is learned from the experiment(s) may provide vital information to, at the very least, improve our lot on Earth, if not save the world from extinction.

Summary and Conclusions from Chapter 9

I am always jealous of my engineering colleagues—as they can take their system to failure to bound the risk. But the human system is the only system in engineering that you can't take to failure.
—J. D. Polk, Chief Medical Officer, NASA
(LinkedIn message to author, April 30, 2024)

The above quotation highlights the challenge that confronts space agencies in their quest to reduce enough of the inherent biological and behavioral weaknesses and fragilities of humans to enable them to survive the extended length of journeys that reach beyond the Earth-Moon system to another planet or asteroid in the Solar System, let alone the much, much longer voyages to other star systems. It highlights the view of Kelly (2017), who identified the human body and mind as the weakest link in the chain that makes spaceflight possible.

Opinions vary widely about both the seriousness and the ability to address the task: some claim that everything is soluble by technology or adjustment of habitat operating procedures, or by real-time adaption in-flight or at the remote location, or that the effects will ameliorate to an acceptable level after recovery periods back on Earth (e.g., Zubrin 2011; Kaku 2018). Others are less optimistic (e.g., Wohlforth and Hendrix 2016; Weinersmith and Weinersmith 2023, and references therein).

It might be that the health and well-being risks for humans on long journeys will remain too high for them to be accepted. Perhaps even more likely, the financial and political costs of human spaceflight will became too high to be acceptable to Earth's population at large, just as they were after the euphoria of the first Moon landing abated (at least in the United States) in the early 1970s when the focus of society turned to ending civil unrest and the war in Vietnam. It was another 50 years before plans for a human presence on the Moon (and Mars) became mainstream again, driven largely by (1) the increasing use of private-sector funding that removed the sole burden from governments, (2) the significant reduction in launch costs through the development of technology to reuse boosters and other equipment, and (3) advocacy of the perceived inherent need/desire for humans to settle other worlds (especially Mars) by influential individuals with the financial means to play a major part in achieving this goal.

However, an emerging snag to the achievement of human-based space travel, or indeed the continuation of major commitments to robotic space travel, is the acceleration of global attention on providing effective action to ameliorate and/or mitigate global warming and climate change. Thus, humans may turn their attention away from the preservation of life through the settlement of an off-Earth platform and focus their attention instead on keeping Earth itself livable. While this refocus of attention may lead to the elimination of one of the extinction scenarios described in chapter 2, namely, runaway global warming, it will leave all the other scenarios still in play. One very positive outcome from this path (aside from the obvious removal of one extinction scenario) is that it might accelerate the development of technology suitable for terraformation (of Earth, literally, in this case) that was to be part of the preparation for the off-Earth settlement goal, from which they will have turned away.

But, getting back to the main thrust of the extinction survival imperative outlined in chapters 1 and 2, it may be that the best chance we have of preserving human *civilization* is to entrust/embed/encapsulate it, lock, stock, and barrel, in much more physically robust nonbiological "beings." This opens the need to define what constitutes, specifically, the very "humanity" and "human civilization" that it is being attempted to preserve. This proposal is considered in depth in part IV.

Meanwhile, the following chapter examines the types of spacecraft that have been proposed to carry settlers to other platforms both within and outside the Solar System. Some are based on current technology, but others are quite theoretical and unlikely to be realized in the foreseeable future.

Chapter Ten

Spacecraft Options for Human Transportation

Apollo 8 has 5,600,000 parts and one and one half million systems, subsystems, and assemblies. Even if all functioned with 99.9 percent reliability, we could expect fifty-six hundred defects.
—Michael Collins, in *Carrying the Fire: An Astronaut's Journey*, 2019

All options for off-Earth settlement depend on the identification of a habitable or near-habitable (i.e., terrestrial) planet or moon as the desired platform. For potential targets identified within the Solar System (i.e., the Moon, Mars, Titan, Ceres, etc.—see chapter 6) the conditions there have already been determined, at least partially, by telescopic observations and/or data obtained at closer range by orbital or landing spacecraft. While some properties of these systems are known with a degree of certainty (especially for the Moon and Mars), others are less well characterized.

Settlement options for extremely remote platforms outside the Solar System (i.e., exoplanets) have only been able to be gauged (or, more accurately, inferred) from measurements of their influence on the motion and/or light curves of their parent star (chapter 1). This has provided information about their locations within the Goldilocks zone, and perhaps estimates of their size and density and the possible presence of a rocky surface. These characteristics are by no means detailed or robust and will need to be quantified much more tightly by uncrewed reconnaissance missions undertaken by small autonomous spacecraft (e.g., light sails). These craft will need to travel at a significant fraction of the speed of light to provide intelligence on their suitability for habitation on time scales that will allow decisions to be made about, and actions commenced on, the development and construction of the necessary human transport and accommodation technology.

While these exploratory robotic missions are being executed, it is likely that further details of the suitability of the putative platform will be able to be determined from detailed observations of the planetary systems using terrestrial and orbiting telescopes (such as the James Webb Space Telescope) equipped with even more powerful, next-generation sensors and spectroscopic analysis. Specific properties of the exoplanet, namely, the presence of a magnetosphere, the composition and extent of its atmosphere, and whether water and other biosignature gases are exhibited, will able to be determined with much more confidence. These observations might even be able to infer directly the presence of simple or complex life.

If some conditions on the exoplanet have been established as currently unfavorable to habitation, consideration might be given to the prospect of terraforming the planet to remove these impediments (see chapter 6). Only after these assessments by telescopic observations and/or reconnaissance spacecraft missions have taken place and positive prospects for settlement have been established beyond reasonable doubt could the green light be given for the human transport mission to get underway.

While all of this interrogation of the suitability of the target is taking place, another fundamental question needs to be answered, namely: "What is the size, mass, propulsion system, and other characteristics of the spacecraft that will enable the transport of potentially large numbers of humans (alive and in good health, of course) along with their life-support materials and equipment and all of the relevant supplies and other gear that will be needed to set up the settlement on the selected target platform?" The question of the propulsion system that will enable the journey to take place is reserved for discussion in chapter 11—but let's concern ourselves in the meantime with considering the options for the type of vehicle itself.

With current propulsion systems, trips to the outer Solar System, and especially to nearby star planetary systems, are necessarily very lengthy—from several years to tens of millennia (table 8.1). Time dilation is not relevant to the discussion at this juncture since existing propulsion technologies deliver speeds less than 0.02 percent of the speed of light, so we are stuck with very long journey times. The issue then boils down to what sort of vehicle can sustain humans alive over these long periods. The options are considered roughly in order of increasing technical difficulty and diminishing likelihood in the following sections.

Generation Ships

A "generation ship," or "settlement/world ship," is a spacecraft in which the crew is living on board for hundreds or thousands of years and therefore spans multiple generations of humans (Hein et al. 2012). This concept of interstellar craft was first proposed by rocket pioneer Robert H. Goddard in his book *The Ultimate Migration* in 1918 (Caroti 2011). A variant of this model in which the occupants are held in a state of suspended animation for much of the journey is discussed as a separate model for interstellar travel in the section titled "Sleeper Ships" below.

The most technically feasible option for a generation ship is to breed sequential generations of humans on board to overcome the limitation that the human lifespan places on the distances and times relevant for each individual. This might be done in combination with some form of artificially slowed aging or an extended "human" version of hibernation that slows metabolic processes to a level that requires much less material support (see chapter 8 and the discussion below).

Quite aside from the difficulties mentioned below, it is hard to imagine that anyone would volunteer to spend what remains of their lifetime encased in a spaceship, for the sole purpose of breeding further "platformians," who themselves would be subjected to a similarly sterile existence for their entire life, without any say in the matter and without any means of escape.[1]

The main challenges for generation ships are related to (1) the absence of any studies of human reproductive success in space, including fetus and mother health during gestation—a fairly fundamental absence of data that is critical the very nature of a generation ship; (2) the need to establish and maintain a closed-system (the living environment) on a spacecraft that is capable of producing sufficient food, nutrients, and oxygen for the not insignificant number of inhabitants of that system in a controlled and sustainable manner; (3) the need for the spacecraft to be of sufficient size and robustness to accommodate this massive and complex habitation system, with the related huge imposts on the amount of fuel required for acceleration from its original location (Earth or Moon) and corresponding deceleration when the target (exoplanet) is reached; (4) protecting the occupants from exposure to cosmic radiation and the effects of low/zero gravity for the entirety of their lives; and (5) effective and stable management of sociopolitical governance, professional and leisure activities, and group dynamics: psychological well-being, conflict, and cooperation.

Some of the psychological health and crew dynamics issues of confinement and isolation have been studied at length for scientists overwintering in Antarctic research stations and in long-duration Mir and International Space Station missions (see discussion on social and behavioral health in chapter 9). Other extended, on-Earth simulations of humans in isolated and fully confined environments specifically targeted for spaceflight have been undertaken, with mixed results. The largest and most comprehensive of these simulations were the Biosphere 2 (Bahr 2009) and Mars-500 (Basner et al. 2013).

Biosphere 2

Biosphere 2 was named as such because it was meant to be the second fully self-sufficient biosphere (after the Earth itself) to test the viability of closed ecological systems to support and maintain human life in outer space. It was a 1.27-hectare structure, proposed in 1984 by philanthropist Ed Bass and systems ecologist John P. Allen, with Bass providing US$150 million in funding until 1991 (Bahr 2009). It was built in Oracle, Arizona, between 1987 and 1991 and was specifically designed to explore the interactions between life systems and the challenges/viability of life on board an interstellar spacecraft. It contained several "biomes" (areas in which specific species live), living quarters for eight people, and an agricultural area and workspace. It was only used twice for its original intended purpose as a closed-system experiment: from 1991 to

1993, and then March to September 1994. A comprehensive assemblage of collected papers and findings from the Biosphere 2 experiments was provided in a special issue of the *Ecological Engineering* journal edited by Marino and Odum (1999).

Both of the Biosphere 2 closed-system experiments ran into problems of insufficient amounts of food and oxygen, die-offs of many animals and plants, group dynamic tensions among the resident crew, outside politics, and a power struggle over the management and direction of the project. Notwithstanding these difficulties, the experiments set records for closed ecological systems, agricultural production, and health improvements associated with the high nutrient and low caloric diet of the crew. They also provided insights into the self-organization of complex biological systems and the interaction of these systems with the closed atmosphere, including significant seasonal variations in its oxygen and carbon dioxide levels. Unlike the 1991–1993 experiment, the shorter 1994 trial achieved total food sufficiency and did not require injection of oxygen.

Overall, the Biosphere 2 experiments provided important scientific contributions in the fields of biogeochemical cycling and the ecology of closed ecological systems, and it has been described as "the first unequivocal experimental confirmation of the human impact on the planet" (Nelson 2018). Much of the learning from Biosphere 2 is likely to inform efforts to design a future interstellar generation ship.

Mars-500

The Mars-500 mission was a multistage psychosocial isolation experiment conducted between 2007 and 2011 by Russia, the European Space Agency (ESA), and China in preparation for a potential future full-length crewed spaceflight to the planet Mars (Basner et al. 2013). It was located at the Russian Academy of Sciences Institute of Biomedical Problems (IBMP) in Moscow.

Three different crews of volunteers lived and worked in a mockup spacecraft at IBMP, with the final stage of the experiment, designed to simulate a 520-day crewed mission, conducted by an all-male, six-member, mixed nationality crew with combined experience in engineering, medicine, biology, and human spaceflight. The facility simulated an "Earth-

Mars shuttle spacecraft, an ascent-descent craft, and the Martian surface" and imposed an average communication delay of 13 minutes with the "outside world" to simulate the transmission time to and from a real-world Mars-bound spacecraft.

The goal was to study the psychological, physiological, and technological challenges of long-duration space flight. This included studies of habitat design, the management of limited resources, the effects of isolation in a hermetically sealed environment, hypokinetic (reduced bodily movement) disorders, and the effect on the human body of the use of a fire-resistant oxygen-nitrogen-argon atmospheric mixture. As for Biosphere 2, conditions such as weightlessness and cosmic radiation could not be simulated.

After an initial 15-day technical and operational procedures test stage (in November 2007) with no occupants, a second, 105-day stage of the experiment began in March 2009, when six volunteers started living in the experiment's isolated living complex. The 520-day final stage of the experiment with the six-man international crew began in June 2010. This stage included a simulation of a crewed Mars landing and three simulated Mars walks.

In an assessment of the results, Basner et al. (2013) reported that four of the six crew members had considerable problems sleeping, leading to increased sleep and rest times (a behavior comparable to animal hibernation), disruption to their circadian rhythms, and some "psychological issues." Friendly and constructive communication existed throughout the experiment, with the crew using recreational activities as an opportunity to discuss films and interact socially. No language, social, authority, or cultural barriers were observed. The crew's activity levels dropped steeply in the first three months and continued to fall for the next year. On the simulated return journey, they spent 700 hours longer in bed than on the outward journey.

NASA Mars Habitat Simulation (Ongoing)

NASA established Mars Dune Alpha, an isolated 158-square-meter simulated Mars habitat in April 2023 at the agency's Johnson Space Center in Houston. It is occupied by two men and two women with expertise in microbiology, biology, engineering, and medicine. They will spend more than a year confined to the facility in preparation for life in a lunar base and human exploration of Mars. The mission is the first of three planned in NASA's Crew Health and Performance Exploration Analog (CHAPEA) habitat (NASA 2024), a series of missions that will simulate year-long stays on the surface of Mars.

Crew members will undertake simulated spacewalks, robotic mission operations, habitat maintenance, exercise, and crop growth. A mimic of Martian soil will be used as a resource for construction materials, and communications with the "outside world" will be delayed by 20 minutes, as would occur on a real Mars base. The crew will have to deal with deliberate environmental stressors such as resource limitations, isolation, and equipment failure, to evaluate the impact of these factors on mental health and stamina.

All these specific spaceflight and on-platform settlement simulations and on-Earth experiences of isolated and confined human populations are sure to provide (and have already provided) valuable information for the technical and operational design of generation ships. However, none of them address three of the most important factors that will apply on a real ship or settlement: (1) low or zero gravity, (2) cosmic radiation, and (3) the absence of the "no escape" constraint. These factors will only be able to be addressed properly during the execution of a "real" flight to Mars or farther afield.

Sleeper Ships

Without a dramatic increase in spacecraft speed, travel times for interstellar journeys will be measured in the tens of thousands of years, making some form of "life extension" necessary for the departure crew to live to see their destination. One of the proposed options is a "sleeper ship," or "suspended animation ship," a hypothetical craft in which most or all the crew spend the journey in some form of cryopreservation or other suspended animation, incorporated as a modification of the generation ship concept (Ventura 2019). The methodology is commonplace in fictional interstellar or intergalactic travel, most recently in the movies *Passengers* and *Alien*, and in shorter-length interplanetary travel within the Solar System, as in the films *2001: A Space Odyssey* and *Avatar*.

In contrast to generation ships in which the full passenger complement is alive and active throughout the entire journey, suspended animation ships have the major advantage of significantly reducing the need for oxygen, food, and other resources by the "suspended" passengers. Of course, unless the craft is fully automated, "unsuspended" crew members will be needed for ongoing maintenance and communication duties and their resource requirements will be as usual.

The main challenge in the cryopreservation (freezing) scenario is that, except for the now commonly used case of early-stage cryopreservation of human embryos for in-vitro fertilization (IVF) procedures (which has application to "embryo" or "seed" ships, discussed below), there is currently no freezing process that has demonstrated the safe long-term suspension and then recovery of consciousness in humans beyond this very early stage of development. The type of cryopreservation that would be used would probably involve the whole body and would typically involve preservation at temperatures below $-140°C$ and sometimes below $-80°C$ (Baust et al. 2009). These temperatures are significantly warmer than liquid nitrogen ($-195°C$) and would be applied while cryoprotectants and ice blockers are circulated through the body to reduce the risk of fracturing of cells.

Another significant advantage of the sleeper ship/generation ship model is that it would deliver to a prospective settlement "world" a population that already possesses the culture and embedded knowledge of (at least a subset of) the population of Earth, that is, an awareness of the historical, scientific, and technical education, language acquisition, and (importantly) an understanding of the original reason(s) for the mission.

The main challenges with this method are to provide a safe and reliable process for reanimating the suspended humans prior to reaching the target platform or upon landing, and for dealing with any atrophication or other physical/psychological issues arising from the long period of inaction. The recovery process will require the development of much more advanced medical (and perhaps nanotechnology) techniques at the molecular and cellular level to reverse any damage caused by the preservation process itself, a capability that is currently unavailable.

Hibernation (also known as "torpor") may not have the same negative aspects on humans as full-blown cryopreservation or freezing. Recent research by the ESA (Choukér et al. 2021) has suggested that hibernating would reduce the need for air, food, and water intake (although not as much as freezing) and would also reduce/prevent boredom on years-long missions in a tiny spacecraft by taking the crew out of the environment for significant periods of the journey. Furthermore, studies of hibernating animals indicate that their bodies do not waste away to the same extent as experienced by awake humans in microgravity (or in extended periods of bed rest), and that after awaking they are able to resume normal activities almost immediately, with unimpaired cognitive abilities. Upon arrival, human hibernators might therefore be fit and ready to commence their settlement establishment and exploration duties almost straight away after regaining consciousness.

This is in stark contrast to the observation that, even after the routine use (a minimum of two hours per day) of high-tech fitness machines on the ISS, astronauts lose up to 20 percent of their muscle mass in a month. Images of crews returning to Earth after long flights sometimes show them being carried from their spacecraft in wheelchairs and stretchers by medical personnel. Precautionary or not, such assistance will not be available during the first landing on Mars!

Choukér et al. (2021) also report studies that show the slowed-down cells of a hibernating animal body are not damaged by radiation to the same extent as awake individuals which, if translatable to humans, might reduce one of the most challenging of the health concerns of long spaceflight beyond low Earth orbit.

Regardless of the specific state of suspended animation used (cryopreservation or torpor) a major challenge is the need for very a large spacecraft and fuel to transport the potentially hundreds of "sleeping" humans (to provide the initial settlement population) and their associated life-support systems, including the food and water required for the sleepers after they are extracted from suspended animation. In addition, unless already delivered by previous robotic missions, the ship needs to accommodate all the equipment and other material to establish the initial habitat infrastructure after arrival at the target platform. Subsequent missions to the target will be encumbered with additional settlers and perhaps the materials and equipment for commencing the centuries-long process of terraforming the planet to allow long-term human occupancy (see chapters 6 and 13).

In both sleeper ship models the problems involved in maintaining the passengers in an animate or frozen condition (for example, there certainly must not be any loss of power that allows the passengers to "defrost") and protecting their tissues from radiation damage to cells and DNA, etc., for up to thousands of years is likely to be problematic, since the accumulated dose would be orders of magnitude higher than for a single lifetime of a generation-ship passenger.

Embryo Ships

An embryo, or seed, ship (a variant of the sleeper ship case) is a theoretical interstellar space settlement model that transports donated frozen, early-stage human embryos in a fully robotic craft to a habitable platform and then gestates them in artificial wombs to be birthed just before, or soon after, arrival (Lucas 2004; Crowl 2004). In contrast to sleeper ships, it can use existing, proven frozen embryo technology and does not require the more uncertain and technically challenging freezing and thawing of fully developed humans. Another variation of this scenario involves carrying samples of donated sperm and egg cells that are used to create embryos at or near the destination.

In all cases, significantly advanced robots traveling with the embryos or sperm/eggs would be required to initiate the birthing process at the predetermined time and raise the children to maturity, including educating them with uploaded human history and other knowledge. Alternatively, the infants and children could be raised to adults in a totally virtual world as the target platform is approached, one that provides them with something approaching a "normal" childhood experience. Upon maturation, the adults could be transferred from the virtual world to the real one in their new environment.

There are many difficulties in implementing the embryo/seed ship concept, including several unsolved technical and biological challenges:

1. The need to develop autonomous robots with the (artificial) intelligence required to build the first settlement on the target planet and to successfully raise human children as healthy, culturally aware indi-

viduals while never having had contact with human beings other than their transported cohort. The development of such robots and support systems might be faster than expected because there are already strong incentives to develop such robots on Earth, unrelated to the requirements of space settlement.

2. The current technological (and some legal) limits on the development of artificial wombs for full-term gestation of humans. Human embryos have only been successfully grown in artificial uteri for two weeks or so, although longer terms have been achieved for other mammals.

3. Like mature human passengers, unless adequately protected, the frozen embryos or cells will be exposed to massive, accumulated doses of cosmic rays over the centuries-long journey from Earth that are likely to severely damage their DNA and/or render it unviable at some point during the trip. Immunological functioning of suspended-animation humans in generation ships may also be hampered by this exposure.

4. As is the case for all other transportation models, the development of computer hardware and software that can function reliably over very long periods of time (potentially for many thousands of years) when exposed to damaging cosmic rays. *Voyager 1* has operated successfully and has communicated data back to Earth for more than 45 years, largely due to the ability of controllers on Earth to reprogram the systems when failures occurred, but travel to even the closest exoplanet will entail much longer journeys with very elongated response times and the debilitating potential for the communication "event horizon" blackouts, discussed previously.

On the positive side, it may not be too far in the future that the artificial intelligence of the onboard robot "managers" of the mission will have developed the capability to self-diagnose and fix any hardware or software issues that arise en route and at the destination.

Another even more futuristic variant of the seed ship model is one where the would-be passengers upload the content of their brains into the ship's computers and they are downloaded into cloned bodies that are somehow generated from the transported "seed bank" when the ship reaches its destination. The technology to achieve this scenario is unlikely to be developed for some time but is discussed at length in chapter 12.

Ethical questions are likely to arise in any or all of the above transportation models involving subpopulations of humans, including (1) on what basis will the decisions be made about whose DNA should be the foundation of the space settlement, (2) which cultural values will be learned by the settlers once gestated, and (3) whether it is morally acceptable to deliberately create children who will grow up without (traditional model) parents. One can anticipate a potential minefield arising on all these issues, as demonstrated by some of the dilemmas depicted in the book *Contact* (Sagan 1985) and the 1997 film of the same name.

Hollowed-Out Natural Structures

Levitt and Cole (1963) and Cole and Cox (1964) proposed a very large "interstellar ark" for human habitation based on a hollowed-out ellipsoidal asteroid about 30 kilometers long and rotating about its major axis to provide an outward acceleration force equivalent to gravity. Sunlight inside would be provided with mirrors, and a "pastoral setting" would be created on its inner surface. The shell of the asteroid would provide essential micrometeorite and radiation shielding, along with thermal insulation and thermal inertia, and the "waste" material from hollowing out the asteroid would provide ample building material for constructing the internal accommodation and life-support structures. The idea was reprised by Arthur C. Clarke in his book *Rendezvous with Rama* (1973) as a 50-kilometer-long starship that enters the Solar System and is intercepted by a human crew, and it was also presented in an early episode of *Star Trek*.

While superficially attractive, the overall concept is fraught with many challenges. First among these is the improbability of hollowing out an object of this size in a reasonable time frame and with the necessary gargantuan energy input, followed by the uncertain structural integrity of its constituent minerals and metals, leading to the possibility of collapse. The massive amount of energy required to accelerate this huge object to an acceptable speed for the interstellar journey would also be seriously problematic. Others have suggested that comets might be better suited to this sort of application, as their constituent icy materials may be more easily converted to much-needed fuel, with the contained deuterium isotope of hydrogen having the potential to support a nuclear rocket (see chapter 11); this concept has been labeled as an Einzmann Starship (Crowl et al. 2012).

Stepping Stones

In the stepping-stone, or "island hopping," scenario the mission does not proceed with a single very long flight but, rather, is broken up into shorter stages, like the progressive migration of early humans out of Africa and their subsequent expansion throughout Europe, Asia, the Americas, and, more recently, the islands of the Pacific Ocean. In the case of space travel, it is akin to the proposed use of an engineered platform in orbit around the Earth (or the Moon), located on the surface of the Moon itself, or stationed at a Lagrangian point as a staging post for a trip to Mars. For longer missions to the outer Solar System, a stopover might be provided on Mars, Ceres, or any other asteroid with material resources (water, hydrogen, metals, etc.) that could be used to top up fuel or raw materials for other necessities.

Ideally, these locations will have been stocked in advance with food and other consumables by earlier crewed or robotic missions, thereby reducing significantly the payloads that would otherwise have been required for the prime mission. Sojourns on these intermediate platforms would also serve to break up the monotony of the longer stages of flight, to change over the crew in order to minimize the physical and psychological impact on individuals, and to provide a location for repairs or the construction of spaceships, spaceship components, or other apparatus necessary for the next leg(s) of the journey.

For even longer interstellar missions, the stepping stones might be a series of more distant objects such as dwarf planets in the Kuiper Belt and/or similar bodies in the Ort Cloud that might have been established as settlements in their own right in previous missions. Since the Solar System's Ort Cloud extends out to around halfway to the nearest stars, the latter half of the trip could utilize stopovers on similar bodies in the corresponding outer regions that are expected to surround these star systems.

Of course, the target(s) and itinerary of these missions would need to have been identified and planned in detail well in advance through feedback from robotic "scouting" missions.

Summary and Conclusions from Chapter 10

Each of the current and proposed spacecraft options for the accommodation, support, and well-being of humans during transport have very significant and multiple technical, physical, and social hurdles to overcome before they can be realized in a practical settlement mission, even when it involves relatively close target platforms like Mars and other bodies in the outer reaches of the Solar System.

For missions beyond our local environment, the currently available means of accelerating and decelerating even a small number of humans and their associated paraphernalia severely restrict the speed of the spacecraft to a minute fraction of the speed of light, with all the models then requiring journeys of multiple centuries, at the very least. Thus, without dramatic improvements in propulsion technology, it is very unlikely that humans (in their currently evolved and very fragile form) will become an interstellar species—that is, unless humanity evolves in the meantime to be "represented" by a population of advanced inorganic machines or robots that preserves our civilization in all but its organic nature. More on this prospective scenario in chapter 13.

The following chapter provides a summary of the current status of spaceship propulsion technology, along with some likely and not so likely candidates for the future that might serve to decrease the travel time to acceptable levels for humans.

CHAPTER ELEVEN

Options for Spacecraft Propulsion Technology

Once you get into space, you can really unleash a lot of creativity, but the launch itself? I have been through all of the creative ways, and believe me, chemical rockets are the best.
—Jeff Bezos, founder of Blue Origin, 2019

The revolutionary breakthrough will come with rockets that are fully and rapidly reusable. We will never conquer Mars unless we do that. It'll be too expensive. The American colonies would never have been pioneered if the ships that crossed the ocean hadn't been reusable.
—Elon Musk, founder and CEO of SpaceX, 2012

As enunciated by Jeff Bezos[1] in the first quote above, currently there is only one means of providing the propulsion required to lift a spacecraft into orbit from the gravity well of Earth or another massive celestial body and later on to decelerate it to land safely on or orbit around its target, namely, a rocket based on a liquid or solid chemical fuel. In contrast, there are several proposed and some already tested non-chemical-rocket options for accelerating, cruising, and decelerating a spacecraft during its transit between launch and landing (figure 11.1). Most of the post-launch technologies are theoretical (the white ellipses), two are in the early stages of development and/or testing (mid-gray ellipses), and four

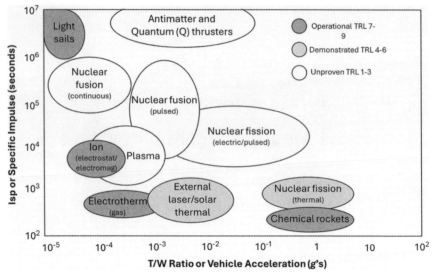

Figure 11.1. General domains of spacecraft propulsion system performance plotted in terms of their specific impulse (I_{sp}) and thrust-to-weight ratio (T/W). The domains are coded as currently operational (dark gray ellipses), in the development phase (mid-gray), and potential future but unproven (white) technologies. The key in the upper right of the figure delineates these three stages of technology in terms of their Technology Level of Readiness (TLR). (*Modified from a NASA graphic reproduced in Matloff and Gerrish [2023] using additional data from Genta [2001] and Kaku [2018].*)

have reached the demonstration and/or operational stage (dark gray ellipses). The relative state of the development of these technologies is indicated by their TLR, or Technology Level of Readiness,[2] shown in the key in the upper right of the figure.

Each technology has different advantages, disadvantages, and performance characteristics. Primary among these performance factors are specific impulse, I_{sp}, and thrust-to-weight ratio, T/W. I_{sp} is directly related to how efficiently the propulsion system produces thrust from the propellant mass leaving the vehicle; the higher the I_{sp}, the less propellant mass is needed to produce the same thrust and/or the higher is the exhaust velocity. The values of I_{sp} (in units of seconds) range from around 10^2 for chemical rockets to 10^7 for light sails (see below for a discussion of some of these specific technologies).

T/W for a spacecraft is calculated by dividing the thrust (in newtons) by the gravitational force acting on its weight, obtained by multiplying the mass in kilograms (kg) by the acceleration due to gravity (m/s^2). It is a measure of spacecraft acceleration (g) and varies (in almost direct contrast to I_{sp}) from values of 10^{-5} for light sails to around 10^1 for chemical rockets.

Table 11.1 provides a compilation of the approximate numerical values of I_{sp} and T/W and the current development status for different propulsion technologies.

Table 11.1. Approximate specific impulse (I_{sp}), thrust-to-weight ratio (T/W), and corresponding development status for various spacecraft propulsion systems.

Propulsion Type	I_{sp} (seconds)	T/W (g)	Status
Chemical (solid)	~250	0.2–5.0	In very common use (mostly for launches)
Chemical (liquid)	~450		
Electrothermal (gas)	150–700	10^{-4}–10^{-3}	In use
Nuclear fission (thermal)	800–1,000	0.1–4	Study/development
Nuclear fission (electric)	~5,000	10^{-3}	Study/development
Nuclear fission (pulsed)	~10,000	10^{-3}–10^0	Studied
Ion thrusters (electrostatic and electromagnetic)	1,500–10,000	10^{-4}	In very common use (mostly for maneuvering)
Plasma (quasi-neutral)	1,000–20,000	10^{-4}	Studied
Nuclear fusion (pulsed)	Up to 10^6	10^{-3}	Laboratory research
Antimatter	Perhaps around 10^7	10^{-1}–10^{-4}	Theoretical research
Quantum (or Q) thruster			
Light sails	10^6–10^7	10^{-5}	In use/development

Source: Adapted in part from Genta (2001) and Kaku (2018)

Figure 11.1 and table 11.1 show that light sails, at one end of the scale, have high I_{sp} (i.e., high propulsion efficiency) but produce very low acceleration, whereas chemical rockets, at the other end, have among the lowest I_{sp} values but have the highest acceleration. High acceleration (T/W ratio) is necessary for and suited to the short (time-wise) launch and landing stages of a flight (i.e., out of and into large gravity wells), whereas high I_{sp} (efficiency) is best suited for much-longer-duration interstellar travel.

The ideal spacecraft would have both high I_{sp} and high T/W, but these parameters seem to be mutually exclusive for options other than the theoretical propulsion technologies in the upper middle of figure 11.1. Thus, the flight may have to be separated into at least two distinct phases, namely, the launch to orbit and then the stage from orbit to target, as in current practice, through the use of different chemical rocket "stages," scaled for the purpose at hand. In future long-term missions, the craft may have to carry two entirely different propulsion technologies on board: a chemical rocket as usual for the launch and one of the other technologies in figure 11.1 for the other legs of the journey and landing.

Current spacecraft propulsion is almost exclusively provided by rockets using a chemical fuel, mostly composed of liquid hydrogen or a hydrocarbon like methane, with liquid oxygen as the oxidant. This system has a low I_{sp}, so its fuel does not last long, but it has a high T/W, so its acceleration is high. Its use is limited to the early stage of the flight concerned with getting the spacecraft out of the deepest part of Earth's gravity well.

Practical options for use in the longer journeys to the far reaches of the Solar System and beyond are lights sails and electrostatic/electromagnetic ion propulsion, considered separately below. These have much lower acceleration but can provide power for much longer than a chemical rocket (figure 11.1). The ion propulsion systems are used routinely on hundreds of current spacecraft for in-course maneuvers, orientation, and rotation control, etc. Other propulsion options with reasonably high I_{sp} in figure 11.1 lie somewhere along the line between the two current technologies of chemical rockets and light sails; they are largely theoretical, untested, and/or have serious issues around safety.

CHEMICAL ROCKET

The limits on rockets are dictated by the so-called rocket equation (equation 4):

$$\Delta v = v_e \ln(m_i/m_f) \qquad \text{Eqn (4)}$$

where Δv is the change in velocity, v_e is the effective exhaust velocity, and m_i and m_f are the initial mass (including propellant) and final mass (minus the propellant used) of the rocket. It is credited to the Russian scientist Konstantin Tsiolkovsky, who derived and published it in 1903 (Blagonravov 1954), although it had been independently derived earlier by the British mathematician William Moore (Moore 1810), the American Robert Goddard in 1912, and German Hermann Oberth about 1920.

The equation describes the motion of a vehicle that can apply acceleration to itself by expelling part of its mass with high velocity and can thereby move in the opposing direction due to the law of conservation of momentum. The change in velocity produced by the engine is directly proportional to the velocity of the gases expelled at the rear of the rocket.

Clearly, carrying more fuel to allow the engine to fire for longer (to achieve a higher velocity) increases the overall mass of the rocket, making it more challenging to escape Earth's gravitational pull and then to accelerate to the desired (cruising) speed. For this reason, as foreshadowed above, launches into Earth orbit are divided into two or three rocket stages, with the spent-stage equipment being jettisoned to reduce the weight that needs to be accelerated by the later stage(s). For a trip to the Moon (and more remote destinations), the spacecraft must leave Earth orbit with a further rocket burn (called trans-lunar injection, or TLI) to achieve a speed approaching the Earth's escape velocity of 11.2 km/s.

The first space probe to achieve TLI was the Soviet Union's *Luna 1* on January 2, 1959, although it missed hitting the Moon. The United States' first lunar impactor attempt, *Ranger 3*, on January 26, 1962, also failed to reach the Moon, but both nations had successes with a total of nearly 50 lunar missions between 1959 and 1976. The first human-crewed mission to perform TLI was *Apollo 8* on December 21, 1968, making its crew the first humans to leave low Earth orbit.

For the Apollo lunar missions, TLI was performed by the J-2 engine in the third stage of the Saturn V rocket; this engine was able to restart once after shutdown. The *Apollo 8* TLI burn lasted around 6 minutes, providing a velocity change of 3.05 to 3.25 km/s, at which point the spacecraft was traveling at approximately 10.4 km/s relative to the Earth.

Insertion into lunar orbit for *Apollo 8* was achieved with an "orbital insertion burn" (OIB) of the Command Service Module AJ10-137 engine which used Aerozine 50 (a 50:50 mix by weight of hydrazine and unsymmetrical dimethylhydrazine) as fuel and nitrogen tetroxide (N_2O_4) as oxidizer to produce 91 kN of thrust. This engine could be fired multiple times, for use in midcourse corrections between the Earth and Moon and to place the spacecraft into and out of lunar orbit. It also served as a retro-rocket to perform the deorbit burn for Earth orbital flights.

For comparison, to escape the Solar System, the rocket (or any other propulsion mechanism that is used) needs to accelerate the spacecraft to 42 km/s, or 0.014c. Any craft sent on such a journey (e.g., *Pioneer 10* and *11*, *Voyager 1* and *2*, and *New Horizons*) is continually being slowed by the gravitational pull of the Sun, but if it starts out traveling more than the relevant escape velocity, it will be able to leave the Solar System and coast into interstellar space. To escape the Milky Way, the velocity needs to be a mammoth 320 km/s, or 0.107c.

Gravity Assist—Propulsion from Other Celestial Bodies

Following exit from Earth orbit into trajectories to other locations in the Solar System or beyond, most spacecraft then remain in cruise mode, using speed generated by the chemical rocket that started them on that course, possibly supplemented by small corrections provided by other small chemical rockets using a monopropellant such as hydrazine (N_2H_4). This chemical spontaneously decomposes in a highly exothermic reaction when passed over a catalyst such as granular alumina coated with iridium. The reaction produces a 1,000°C mixture of nitrogen, hydrogen, and ammonia to provide an I_{sp} of around 340 s. However, these hydrazine thrusters are not designed to significantly change the speed of the spacecraft, which has been set by the burn of the major engine that was used to exit Earth orbit.

To increase cruise speed, interplanetary spacecraft often use the "slingshot," or "gravity assist," effect to provide additional acceleration without requiring rocket thrusters to fire. This process works by transferring orbital momentum from another large body, such as a planet, to the spacecraft. As the spacecraft approaches the planet from behind (in its orbit), its gravity pulls the much lighter spacecraft so that it catches up with the planet. The closer it can fly to the planet, the more momentum it receives, and the faster it flies away from the planet. It was by using close flybys of Jupiter and Saturn that the *Voyager 1* spacecraft was able to reduce its overall travel time to the outer planets by several years.[3]

Another approach to get around the fundamental limitation of the rocket equation and allow greater payload capacity without sacrificing velocity is to use propulsion systems that do not rely solely on onboard fuels, such as solar sails or some varieties of ion propulsion (see below). Alternatively, as discussed above, the mission could invoke in-space refueling, where the spacecraft docks with pre-organized refueling space stations, or lands on "stepping-stone" bodies to appropriate/utilize resources such as water ice on the Moon, asteroids or comets, and other bodies in the Kuiper Belt or the Ort Cloud. Other methods (see the Bussard ramjet below) propose to harvest hydrogen atoms in the interstellar medium in-journey to serve as a continuous supply of fuel for the required propulsion.

Nuclear Power

Proposals for nuclear-powered spacecraft come in two fundamentally different types: those that use the direct energy of multiple nuclear explosions to propel the spacecraft, and those that use heat produced by a nuclear reaction to propel material in a rocket-based process.

Nuclear Pulse(s) Propulsion

The idea of nuclear pulse propulsion was proposed in 1946, and preliminary calculations were made at Los Alamos in 1947 (Everett and Ulam 1955). Initial versions of the vehicle were planned to take off from the ground, but later versions were proposed for use only in space to avoid some of the obvious safety and environmental issues. They are designed to be directly propelled by a series of atomic explosions detonated behind the spacecraft, although an earlier version of the pulse concept proposed containment of the blasts in an internal pressure structure.

The propulsion system was called Orion, in which nuclear explosives would be thrown behind a pusher plate mounted on the bottom of a spacecraft and exploded (Dyson 2002). The shock wave and radiation from the detonation would impact against the underside of the pusher plate, transferring its momentum to the spacecraft. The pusher plate would be mounted on large two-stage shock absorbers that would smoothly transmit acceleration to the rest of the spacecraft.

Non-nuclear tests were conducted with models of this system, but the project was abandoned in deference to the 1963 Partial Test Ban Treaty, which forbade nuclear explosions in space. Other major challenges were/are: the required structural bulk of the ship and its cargo, along with the mass of the bombs and propellant; adequate shielding for the crew; ablation of the pusher plate; and concerns over nuclear fallout back to the surface of the Earth. It is no surprise that this propulsion model has not been adopted to date.

Nuclear Thermal Rocket

A nuclear thermal rocket (NTR) is a type of rocket where the heat from a nuclear reaction, currently restricted to nuclear fission, replaces the chemical energy of the propellants in a chemical rocket. In this design, a working fluid, usually hydrogen, is heated to a high temperature by the energy of a nuclear reactor and then expands through a rocket nozzle to create thrust. In principle, the external nuclear heat source allows a higher effective exhaust velocity (I_{sp}) and could double or triple payload capacity compared to chemical propellants that store energy internally (table 11.1).

Also, the ability to run the nuclear engine for longer periods, rather than just during acceleration from Earth orbit and then during deceleration as the target is approached, would potentially reduce the long periods of coasting during the bulk of the journey and increase the average speed. This could speed up journeys to Mars significantly, thereby reducing both the amount of time the crew spends exposed to the environment of deep space and the mass and volume requirements for life-support services.

The earliest NTR ground tests were in 1955, and the United States maintained a development program through to 1973 when it was shut down, primarily to focus on development of the Space Shuttle. To date, no nuclear thermal rocket has flown, but NASA has plans to test such a propulsion system as part of its strategy to get to Mars. It has recently awarded Ultra Safe Nuclear Corporation a contract to manufacture and test fuel and develop the design of a nuclear thermal propulsion engine for near-term missions (World Nuclear News 2023). In another development, Space Nuclear Power Corporation has partnered with Lockheed Martin Corporation and BWX Technologies for the US Space Force/Air Force's JETSON nuclear electric propulsion demonstration project.

This type of rocket propulsion, like the nuclear pulse model, is designed to run in space but is not suitable for launch from the Earth.

Nuclear Photonic Rocket

A nuclear photon rocket is one that uses thrust from the momentum of photons emitted from the nuclear reaction for its propulsion. It has potential for interstellar flight since, in theory, it can reach speeds at least 10 percent of the speed of light, but the rate of acceleration is very low.

Challenges relate to the inability to use/focus all the photons generated, and the requirement for a large amount of onboard nuclear fuel. For example, if a photon rocket begins its journey in low Earth orbit, one year of thrusting may be required to achieve the Earth escape velocity of 11.2 km/s, and much longer to reach the higher escape speed for the Solar System. Like the other forms of nuclear propulsion, this type of rocket is designed to run only in space.

Light/Solar Sail

Light/solar, or photon, sails are a method of spacecraft propulsion using radiation pressure exerted on large surfaces by photons from sunlight or powerful lasers. The first recognition of the potential for sunlight to exert a force on objects was made in the early 17th century, when Johannes Kepler observed that comet tails point away from the Sun and suggested that the Sun caused the effect (Kepler 1604).

In 1861–1864, James Clerk Maxwell published his theory of electromagnetic fields and radiation, which demonstrated that light has momentum and thus can exert pressure on objects. Maxwell's equations provide the theoretical foundation for propulsion of spacecraft with light pressure, a mechanism that has now been confirmed many times by its influence on the trajectory of spacecraft on long-distance missions. The effect is of such magnitude that in-course corrections must be made for the craft to remain on target.[4]

Light sails are a low-thrust propulsion system considered to be appropriate for relatively small spacecraft and when slow rates of acceleration are acceptable. They require no onboard propellant, and the force (and thus acceleration) is exerted continuously, although it decreases with the square of the distance from the light (photon) source. Other variations of the mechanism include diffractive, electric, and magnetic designs, and still others use the evaporation of chemicals added to the sail surface to supplement the photon pressure.

The sail itself is a large reflective surface, generally constructed from a thin layer of polymer sheet coated with aluminum, such as aluminized 2 μm Kapton film. Sunlight is assumed to be the main source of power in most applications, which means that the targets are preferentially located in a direction away from the Sun in the outer regions of the Solar System.

However, the Breakthrough Starshot project announced on April 12, 2016, aims to develop a fleet of light sail nanocraft carrying miniature cameras, propelled (at least initially) by ground-based, very powerful directed lasers (rather than sunlight), and send them to Alpha Centauri at 20 percent the speed of light (Jessica F 2016). The trip is proposed to take just 20 years, to fit within a human lifetime, but significant technical challenges remain before this can be achieved, not least of which is the number, stable focus, and power of the lasers required for the acceleration phase away from Earth.

The Planetary Society's *LightSail 2* was a crowdfunded project that demonstrated solar sailing as a viable means of propulsion for cubesats (see chapter 7). It was launched on June 25, 2019, and reentered Earth's atmosphere on November 17, 2022, after using sunlight alone to change its orbit.

Most direct applications of light sails proposed so far intend to use them as inexpensive modes of trajectory adjustment and satellite deorbiting, rather than as a means of transporting humans, because of the very low T/W ratio and thus severe limits imposed on the mass of the spacecraft. However, if the spacecraft size could be kept very small, on interstellar missions there might be enough travel time for the acceleration to build to a useful fraction of light speed.

Ion Drive

Ion propulsion engines use beams of ions (electrically charged atoms or molecules generated from onboard reserves) accelerated by electrostatic forces (viz., Coulomb, acting on just the positive ions) or electromagnetic forces (viz., Loretz, acting on a plasma of free electrons as well as positive and negative ions) to create thrust in accordance with the law of conservation of momentum. All designs take advantage of the high charge/mass ratio of the ions, meaning that relatively small potential differences can create high exhaust velocities using lower amounts of propellant relative to chemical rockets. However, like solar sails, they can only provide very low accelerations (figure 11.1, table 11.1) and are therefore unsuitable for launching spacecraft into orbit, but they may be effective for in-space top-up propulsion over longer periods of time, provided enough fuel can be carried. The electric power for ion thrusters is usually proposed to be provided by solar panels or nuclear reactors.

NASA's three-year *Deep Space 1* mission (1998–2001) was a very successful test flight for several new technologies, including proving the effectiveness of ion engines for long-duration spaceflight and advancing the field of spacecraft navigation (Rayman et al. 1999). The ion thruster in this mission changed the spacecraft velocity by 4.3 km/s while consuming less than 74 kg of xenon over a total operating period of 16,265 hours. NASA's *Dawn* spacecraft (2007–2018) broke this record, with a velocity change of 11.5 km/s, though it was only half as efficient, requiring 425 kg of xenon consumed over an operating period of 51,385 hours. Its solar-electric propulsion produced a top speed of around 11 km/s, nearly equal to the velocity provided by *Dawn*'s Delta launch vehicle.

Ion thrust engines are generally practical only in the vacuum of space, as the engine's minuscule thrust and acceleration cannot overcome any significant air resistance without radical design changes and cannot achieve liftoff from any celestial body with significant surface gravity. Current applications include control of the orientation and position of orbiting satellites (some satellites have dozens of low-power ion thrusters), their use as a main propulsion engine for low-mass robotic space vehicles (such as *Deep Space 1* and *Dawn*), and serving as propulsion thrusters for crewed spacecraft and space stations (like *Tiangong*).

Plasma Drive

Plasma drive engines are like ion drives (see above) but do not typically use high-voltage grids or anodes/cathodes to accelerate the charged particles in the plasma. Instead, they utilize currents and potentials that are generated internally to accelerate the ions, resulting in a lower exhaust velocity due to the low accelerating voltages. Advantages over ion drives are that grid ion erosion is reduced due the lower voltages, and there is no need for an electron gun since the plasma exhaust is "quasi-neutral" (positive ions and electrons are present in equal number), which allows simple ion-electron recombination in the exhaust to neutralize the exhaust plume.

Plasma engines have a specific impulse (I_{sp}) value of up to more than 12,000 s, a significant improvement over the bipropellant fuels of conventional chemical rockets, which have an I_{sp} of around 450 s. With this high impulse, plasma thrusters could perhaps reach speeds of 55 km/s over extended periods of acceleration, which would significantly reduce the flight time to Mars. On the other hand, the energy requirement for a plasma engine to produce just 5 newtons (N) of thrust[5] is around 200 kW electrical power, which could be provided by an onboard small nuclear reactor, but then the reactor mass would likely prove to be prohibitive.

Other Futuristic Propulsion Mechanisms
Bussard Ramjet
The Bussard ramjet (Bussard 1960) is a theoretical, but highly popular, method of spacecraft propulsion proposed for interstellar travel in which a fast-moving craft scoops up hydrogen from the interstellar medium

using a huge funnel-shaped magnetic field with a diameter ranging up to thousands of kilometers. The hydrogen is compressed until thermonuclear fusion occurs, which provides thrust to counter the drag created by the funnel and energy to power the magnetic field.

Studies have suggested that various versions of the ramjet would be unable to accelerate even into the solar wind (Andrews and Zubrin 1988) and that, while feasible in principle, the practical construction of a useful Bussard ramjet would be beyond even a civilization of Kardashev type II (Schattschneider and Jackson 2022).

Q-Thruster

In the world of quantum mechanics, virtual particles constantly appear (even in a total vacuum) and then disappear. A quantum vacuum plasma thruster, or Q-thruster, is a form of electric propulsion that uses electric and magnetic fields to push these quantum particles (electrons/positrons) in one direction, while the spacecraft recoils to conserve momentum. The electric fields are created with an onboard nuclear reactor, and like the Bussard ramjet concept, "fuel" is harvested continuously from space.

Laboratory results at NASA's Johnson Space Center (Joosten and White 2014) suggest that continuous specific thrust levels of 0.4–4.0 N for each kW of electricity produced are achievable with essentially no onboard fuel consumption (other than the decay of the nuclear fuel in the reactor). Although this technology is in its infancy, it holds promise as a (distant) future propulsion option.

Antigravity

This is a hypothetical concept, much exploited in science fiction, of creating a place or object that is free from, or reverses, the force of gravity. It has been likened to "dark energy"[6] or the "energy of nothingness" (Kaku 2018) that is pushing the universe apart. It does not refer to the lack of weight experienced in free fall or orbit, or to balancing the force of gravity with some other force, such as electromagnetism or aerodynamic lift.

Antigravity, if possible, would significantly reduce the energy required for spacecraft launches and landings, but it would not provide enough energy vital for the acceleration and deceleration of the spacecraft to and from its cruise speed outside the deep gravity wells at the beginning and end of the flight. In any event, antigravity is not accepted as possible under the currently understood laws of physics.

Antimatter

Antimatter rockets rely on the huge amount of energy released when matter and antimatter[7] are allowed to interact and a large fraction of the rest mass of the mixture is converted to energy under Einstein's famous equation $E=mc^2$.[8] Antimatter rockets would have a far higher energy density and specific impulse than any other proposed class of rocket (Schmidt 2012). They have been proposed in three forms: direct use of the momentum of the products of antimatter annihilation for propulsion, heating a store of onboard material which is then used for propulsion, and heating an intermediate material to generate electricity for some form of ion or plasma drive system.

The major practical impediment to this method of propulsion is that matter and antimatter have the troublesome predilection to destroy anything with which they come in contact, so their creation, separation, and storage is going to be very tricky.

Faster-Than-Light Travel ("Warp Drives")

There are two ways in which faster-than-light (FTL) travel (and thus communication of information) may be possible, namely, to "bend" space and/or to "bend" time. Application of either of these possibilities might be the best hope we have for exploring the universe outside the Solar System in reasonable time frames. It is important to note in this context that in the immediate aftermath of the Big Bang, 13.8 billion years ago, the universe expanded faster than the speed of light (c) but this did not violate the relativity limit because space itself was expanding and nothing within that space moved faster than c.

Particles that exceed c ("tachyons") have been hypothesized but, as explained previously, their existence may violate "causality" in that the associated time traveler then has the potential to change events in the past. Unless, of course, the time travel journey itself has taken the particle into another, parallel universe where this causality condition is no longer relevant (Kaku 2018) since the outcome in that universe can be different from that in which the time-travel journey commenced.

In the first example of potential FTL travel, distorted (warped) regions of space-time (a "shortcut" in the same universe) might permit matter to reach distant locations in less time than light could in normal undistorted space-time (Visser et al. 2000). For example, negative matter or energy might serve to bunch up space like a warped rug, shortening the distance to the destination.

In Wohlforth and Hendrix (2016) it is propositioned that FTL travel is possible inside such a bubble of warped space-time, with an observer on the outside of the bubble observing those inside to be traveling faster than c, while those inside are really traveling at a speed less than c, "something akin to people walking on an airport conveyor belt." Science fiction literature and cinema contain numerous examples of so-called warp drive engines that take full advantage of these concepts by creating a bubble of space around a spaceship and contracting the space in front of it while expanding the space behind it, effectively shortening the distance to its destination and giving the appearance of FTL travel.

The concept of a warp drive was first proposed by Mexican physicist Miguel Alcubierre (1994) and has since been refined by other researchers (White 2013). In White et al. (2021) it is claimed that a way had been found to create a nanoscale warp bubble using negative energy densities generated by "Casimir cavities," the tiny spaces between metal plates in a vacuum.[9] However, this is still far from being a practical solution to FLT, as it would require enormous amounts of energy and exotic forms of matter that we don't know yet how to produce or control.

The second option for achieving apparent FTL travel is based on "Krasnikov tubes" (Krasnikov 1995) in which space-time is warped into permanent tunnels analogous to wormholes or an immobile "Alcubierre drive" (also requiring the existence of exotic matter with negative energy density) with the endpoints displaced in time as well as space.

As of the current state of 21st-century scientific theories, matter is required to travel at slower-than-light (STL, or subluminal) speed with respect to the locally distorted space-time region, and any apparent FTL physical plausibility is currently speculative.

While not FTL travel, the idea of time dilation, discussed in chapter 8, in which time passes differently for different observers depending on their relative speed, can significantly decrease the time for a journey if the speed is high enough. At spaceship speeds close to c, time would slow down significantly for the passengers inside, while it would remain normal for the rest of the universe. In this way, they could travel for hundreds of years as measured by those outside the travelers' frame of reference, while they only aged a few years in the frame on board the spacecraft. This is also not an easy option, as it requires extremely powerful engines and shielding to protect the spaceship from damage from cosmic radiation and tiny (let alone larger) particles of interstellar dust impacting its surfaces at speeds near c.

We don't have the technology or the resources to achieve any of these "solutions" to FTL travel now, but one day we might.

Summary and Conclusions from Chapter 11

The propulsion technology that has been used for all spaceflights to date, other than a few relatively short experimental tests carried out in low Earth orbit for small adjustments to trajectory and orientation and during the "cruise" phases of flights to the outer planets, is based on rockets with a liquid or solid chemical fuel. This is because chemical rockets are the only ones able to provide the required acceleration and deceleration to enable a craft to exit (launch from) and enter (land on) high "gravity wells" like those around the Earth and the Moon. So far, it is only in the intervals between these high acceleration stages of a spaceflight—that is, during the cruise phase between planets in the Solar System and in interstellar space—that other propulsion systems may be able to come into play.

For flights within the Solar System, increased cruise speed can be achieved without consuming onboard fuel by using "gravity assist" or "slingshot" protocols that allow the transfer of momentum from a planet to the smaller spacecraft during a close flyby. Alternatively, high surface area "light sails" constructed from thin layers of polymer and aluminum can be deployed after takeoff to utilize the radiation pressure from photons for acceleration. The source of the photons for acceleration may be the Sun or very powerful Earth-based lasers on the ground or in orbit, but

Options for Spacecraft Propulsion Technology

this means that the method can only be deployed in flights headed away from the Sun or Earth. Another downside with systems involving photon pressure is that the acceleration is very, very small, and their application is therefore likely to be limited to very small spacecraft.

All other long-distance propulsion systems require: (1) the carriage of large quantities of fuel of one sort or another (in the case of ion and plasma drives), (2) a massive commitment of onboard infrastructure (a nuclear fission or fusion reactor), or (3) the ability to collect interstellar hydrogen or other particles in flight as fuel (Bussard ramjet). None of these systems have reached practical implementation as yet. All other futuristic proposals (antimatter, antigravity, and quantum thrusters) and faster-than-light travel (warp drives and "wormholes") are almost exclusively the matter of science fiction. Relativistic time dilation to reduce travel time is certainly possible, but the shortened travel time (for those onboard) only becomes meaningful when the spacecraft is accelerated to a speed close to the speed of light, and this will never be achieved with current propulsion technology.

For all of these reasons, and many more discussed in the following chapters, long-distance space travel by humans—that is, beyond Mars—is likely not to be practical for a century or longer.

Conclusions to Part III

Humanity's hard-won skills and innovation in mathematics, physics, and engineering have led to the achievement of major milestones in the spacefaring history of our species over the past century. These include the first artificial Earth satellite, the first human in orbit, the first transfer of humans to a domain where the Earth was not the dominant gravitational influence, the first landing and occupation of humans on another celestial body, and, through very sophisticated robotic machines, the acquisition of a much greater knowledge and understanding of several bodies in the Solar System and the universe around us. These achievements spawned a sense of excitement and even euphoria about the options that were emerging for our species to "spread its wings" to other worlds, but, over the half-century since humans first landed on the Moon, the emphasis waned significantly.

However, enthusiasm for human spaceflight beyond low Earth orbit has increased in recent years, driven by the emergence of a strong collaboration between government and private enterprise. This interest is tempered by the reality of the enormous difficulties in taking the next logical step beyond leaving a few footprints and scientific equipment on the Moon. Human travel to our nearest celestial neighbor was tricky enough 50 years ago, but we are discovering that expanding the adventure to more-remote locations within the Solar System, let alone to other star systems, carries with it a plethora of much more serious, if not (we may indeed find) ultimately insurmountable, challenges.

The limitations of our current spacecraft propulsion systems mean that the journey times to the outer Solar System are measured in years, and targets beyond this will require centuries or millennia. Furthermore, humans are very fragile creatures when removed from their evolutionary platform, especially in terms of their physical, mental, and social well-being. Add to this the massive amounts of energy and material resources required to construct and accelerate a spacecraft large enough to accommodate larger numbers of humans and their cumbersome life-support systems, even to a small fraction of the speed of light, and the magnitude of the task becomes clear.

Not least among these challenges is obtaining the social license from Earth's population to spend the eye-watering sums of money required to continue our spacefaring destiny (?) beyond returning to the Moon when there are so many other calls on these finances that might improve/sustain life on Earth for billions of us. In the absence of political imperatives like the Cold War with the former Soviet Union, the case for this huge expenditure is harder to make, and, perhaps of equal importance, the appetite for risk-taking in major ventures of any sort is significantly lower than it was in times past.

It is clear that there is a great deal that is still not known and/or is uncertain about human space travel, despite what we have learned from our spacefaring successes and failures so far. We do not yet understand the full effects on the human body and mind of long-term zero or near-zero gravity, cosmic radiation, extreme isolation, and confinement; how to generate enough energy, oxygen, water, and food to sustain us (whether active or in suspended animation) over the entire journey; the psychological, social, and political issues that might/will emerge; or the equipment needed at the target platform to build shelters and to extract and process the necessary resources from the local environment for food, clothing, and energy.

Unless our species becomes an example of the "Great Filter," in which we end up destroying ourselves and our technology by one means or another in the next few years or decades, we may still have enough time to deal with all or most of these issues before any of the other "natural" extinction disasters befalls us. Hopefully, from our experience so far, we now understand what most of the obvious challenges are and can work toward their solution prior to departure to another place by learning through feedback from the next generation of advanced autonomous robotic scouting missions and progressively longer, carefully planned crewed missions.

Part IV examines some of the other big issues of space settlement, including who makes the journey and what to do on arrival.

PART IV

WHAT NEXT?

Humanity looks to me like a magnificent beginning but not the last word.
—FREEMAN DYSON, QUOTED IN REGIS (1990)

Part IV discusses the challenges of transporting humans off Earth in attempts to forestall the complete demise of humanity due to natural events and/or anthropogenic activities (and our lack of action to ameliorate or eliminate them). Humans are too fragile right now to endure the rigors of long-term space travel with current spacecraft technologies. One response is to excuse them from the journey, at least in the short to medium term, until more-advanced propulsion and life-support technologies have been developed that allow humans to adapt to and/or mitigate the threats to their well-being. At the same time, we can continue the exploration of the Solar System and beyond, together with the research and testing of the protocols that are necessary to establish and operate settlements at scale, through the agency of much more robust inorganic machines/robots.

As it stands, machines are becoming ever more powerful, with capabilities that are converging on those of humans at an alarming pace. They are infiltering every corner of human society with competence levels that many pundits predict will soon outstrip those of humans. If this happens, the game is likely to be over for *Homo sapiens* and machines will become the dominant intelligence on Earth.

In the following two chapters, the implications of this eventuality are discussed in respect of the preservation of human civilization and the settlement of other worlds.

Chapter Twelve

A Summary of the Challenges for a Human Settlement Off-Earth

There was no need to burn our ships on arrival, like the explorers of old, to ensure the crew's dedication to the mission. Instead, we'd burned our home.
—Tom B. Night, in *Mind Painter*, 2020

When humans first struck out from their place of origin in East Africa around 45,000 years ago, it is very unlikely that anyone used a formal and deliberate process to decide what route to take, and who would stay and who would go. The departure probably was driven by (1) the natural tendency of a hunter-gatherer society to seek new domains that offered more food and/or fewer competitors and predators, (2) a change in short-term or long-term climatic conditions that made staying where they were unsustainable, and/or (3) a natural curiosity about what was on the other side of the hill.

For these intrepid adventurers, there would have been an expectation that wherever they ended up, there would be water for drinking, food all around them "for the picking or hunting," the ready availability of a surfeit of materials with which to construct shelter from the elements, and an endless supply of air to breathe. It would have been very unlikely that they would encounter other humans to cause disputes over territory, at least in the early stages of their migration, and their major issue would have been to watch out for predators and ensure that they did not get injured or fall sick.

The situation on Earth is now markedly different from 45,000 years ago. Since then, we have occupied, in large numbers, essentially every corner of the planet except the deserts, the polar regions, and the deep oceans. As discussed in chapter 2, humanity now faces many existential challenges of its own making, primarily through the demands of a massive increase in population and its unprecedented impact on the environment, especially since the Industrial Revolution, resulting in an increase in the separation between the "haves" and "have-nots," battles for territory and essential food resources, and rising animosity between religious and cultural groups. These challenges will have to be met, along with the ever-present shadows of a range of globally significant natural calamities that are not of our making. Any or all of them could spell the end, or a severe setback, of civilization as we know it and, probably, our ultimate extinction.

This may not feel so ominous if we knew that there were plenty of other advanced life-forms and civilizations living happily on other life-friendly platforms elsewhere in the galaxy that had survived the same or similar threats to their existence. However, no reliable evidence has been found for life's existence anywhere else in the Solar System or the galaxy, whether simple or complex, and this despite decades of intensive searching for any and all direct and indirect signs. Indeed, if advanced life is or has been present elsewhere throughout the long history of the universe, logic dictates that we should have become aware of it many times by now—Fermi's paradox.

Thus, at the moment, our presence appears to be unique, and if Earth is extinguished, we and our civilization may as well have never existed. Despite this, we continue to exacerbate the negative aspects of the current situation that are contributing to our demise—overcrowding, overuse and inefficient use of natural resources, massive waste accumulation, climate change, and associated political unrest. We seem unable to put our differences aside and give up (or at least share more equitably) the advantages enjoyed by subsets of the population that resulted almost entirely from the fortuitous, or otherwise, time and place of their birth. We fail miserably to marshal our efforts to reverse the degradation of the environment and the increase in our numbers to make our planet more livable, for longer.

A Summary of the Challenges for a Human Settlement Off-Earth

The discussion in this book began by suggesting that because of our unwillingness or inability to address these anthropogenic and existential threats, the imperative lay in finding an alternative home for humanity off-Earth that would ensure the survival of our species and its civilization if the Earth was rendered uninhabitable or destroyed. The urgency of this need to relocate has been disputed at one level or another by some (e.g., Weinersmith and Weinersmith 2023) but strongly endorsed by others (e.g., Zubrin 2019).

After a half-century of inaction since the Apollo Moon landings, private and public initiatives to develop the technology and know-how to embark on the exploration and settlement of other bodies have accelerated, not necessarily to avoid the extinction imperative but mostly in response to the promise of resource riches beyond our imagination or as a natural expression of humanity's "inherent" desire to "boldly go where no man has gone before" (the five-year mission of the *Star Trek* science fiction franchise by Gene Roddenberry from 1966 to 1969).

In parts I and II, the origin of life on Earth and the possibility that it exists elsewhere in our Solar System and/or in other stellar systems was examined as a guide to the search for the kind of off-Earth platform that might sponsor life as we know it. The nooks and crannies of our Solar System and exoplanets in other stellar systems were reviewed for their potential as platforms for human settlement in case Earth becomes unsuitable for occupation. The physical, health, and technological challenges uncovered during hundreds of spaceflights in low Earth orbit (some for periods of more than a year) and for short periods of human occupation of the lunar surface were discussed, highlighting the fragility of human biology under conditions of zero or low gravity and deadly radiation, the potential sociological impacts of long-term confinement and isolation, and many other conditions for which we have not evolved adequate resilience.

We do not have any empirical evidence that these risks are magnified, decreased, or adapted to when we are required to survive (1) for the necessary tens if not hundreds of thousands of years of time (with current propulsion technologies) required for the journey to the far reaches of our Solar System and beyond, and (2) the conditions extant on the target settlement platform which, unless we are very fortunate, will be manifestly different from those on Earth.

Moreover, the life-support requirements for the human travelers over these extended times, along with the massive quantities of additional equipment and materials required to set up and maintain camp at the destination, would make the size of any space transportation platform well outside the capability of any propulsion technology that is, or may become, feasible in the short- to medium-term future. Thus, any attempt to reduce these transit times by accelerating a vehicle to speeds that are even a small proportion (say, 10 percent) of the speed of light is well beyond our current reach.

To make the situation even more complicated (and difficult), technologies for the local paraterraformation, let alone global transformation, of the environments of the target platforms that are very likely to be necessary to allow large numbers of humans to survive and thrive there are also far beyond current capabilities and would require hundreds if not thousands of years to be achieved.

We are not even certain that humans can reproduce themselves in space or that the resultant offspring will be fertile and develop normally (physically and emotionally) when they are conceived, gestated, and raised under extended periods of low or zero gravity and exposure to radiation. There is also no certainty that the off-Earth settlement will be immune from the same social, political, nationalistic, and "tragedy of the commons" flaws that have dogged our existence on Earth for millennia and that have contributed to the very environmental, territorial, and overcrowding issues that caused us to relocate in the first place, in shades of the notorious Great Filter.

The situation becomes even more indeterminate and challenging if the settlement aims to have complete resource and economic independence from Earth—that is, to be able to feed, nurture, and replenish its inhabitants; accommodate differences in individual competence and capability without threatening the survival of the settlement; and deal with technical, medical, governance, and other emergencies as they occur. For this level of true independence to be delivered, the minimum population for long-term economic and genetic viability may need to be of the order of several hundred thousand individuals (Weinersmith and Weinersmith 2023, and references therein). Establishing a population of this magnitude early in the settlement is not going to be easy; it would require a thousand spacecraft to be launched from Earth at short intervals, each carrying a few hundred passengers.

What has not frequently been discussed in the literature so far is how to select the initial population for these settlements. To date, all the space farers (other than a small number of recent "space tourists") have been selected solely because of their superior physical and intellectual skills and mental resilience, rather than to provide a diverse set of specialist skills required by a full-blown self-sufficient community. Thus, the members of the pioneering settlement crews, in total, must possess the skills to deal with a whole raft of engineering design, construction, resource processing, food production, medical services, education, social/psychological, management, labor, and legal, governance, and political matters.

For smaller populations of settlers, especially, we need to consider the selection of a genetically "appropriate" human crew or bank of embryos to occupy the spaceship if it is desired to provide a "representative" cross section of humanity, not just for the preservation of the diversity of the human genetic code, but also to prevent its otherwise inevitable decay through inbreeding. The initial settlement population would be required to reflect the full spectrum of racial, cultural, religious, and political characteristics of the population on Earth if the raison d'être for the settlement is to preserve human civilization in its full blossoming.

These considerations involve a multitude of moral and ethical dilemmas. Would it not be advantageous from a community survival perspective for the individuals in the community to have the highest possible IQs and EQs, to be in the healthiest and most robust physical condition, to have a disease-free ancestry, and to possess the most favorable, resilient, and nonviolent behavior traits? What does this say about the worth of those that are not represented? What are the ethics of sending frozen embryos and the resulting artificially nurtured humans to an unknown future, with no chance of return?

Mental and emotional issues of isolation and confinement, with no possibility of escape, will have to be managed, along with the poorly understood psychological phenomenon of "Earth-out-of-view," in which (at least) the first generation of settlers will have to deal with the disconnection resulting from seeing the Earth as just another "star" in the sky (if on Mars), or it being completely invisible if viewing from farther afield. For later generations, of course, the settlement will be their only "home," so this may not be an issue, perhaps.

On top of these challenges, current space treaties have not yet had to deal with actual cases of competing claims for off-Earth real estate of any sort, let alone when it involves access to highly sought-after and limited resources or areas of high prospectivity. This could become especially contentious when, for example, the claims relate to the small domains in the polar regions of the Moon that may have large reserves of water ice in the perpetually dark floors of deep craters, or other places nearby that are sufficiently elevated to allow almost continuous solar power generation. Treaties have worked successfully on Earth to regulate access to Antarctica and the exploitation of the seabed, but the existing space treaties are written in broad and ambiguous terms that allow a multitude of interpretations. At the moment, it appears that real estate and resources on the Moon (for example) will be allocated on a first-come, first-served basis, as there is no higher authority to police the actions of claimants.

While these "big picture" humanity-centric issues are prominent in some of the discussion on the viability of future off-Earth communities (Weinersmith and Weinersmith 2022), rarely are they accompanied by a discussion of the plethora of more mundane "operational" and practical factors that significantly complicate off-Earth journeys and settlements. This neglect is very apparent in most descriptions of space travel both in current reality and especially in science fiction and includes the following examples:

- Spacecraft docking is a very slow (hours-long) and dangerous process, even when undertaken by small, bespoke vehicles using well-refined protocols. The maneuverability of the much larger craft required for settlement-scale missions will be even slower and more ponderous, making the achievement of accurate orbital trajectories and especially surface landing particularly complex and difficult. In practice it will make more sense to leave the massive spaceship in

A Summary of the Challenges for a Human Settlement Off-Earth

orbit around the settlement target body, with transport to and from the surface undertaken by much smaller vehicles like the Zodiacs used for this purpose to ferry passengers to and from cruise ships. Even so, these small-ship transfers will still be slow and dangerous.

- Launches into space from Earth and their corresponding landings are loaded with risk. Uncrewed missions have a woeful record, with even small satellite missions to Earth orbit having a success rate somewhere between 40 and 70 percent. A total of 15 astronauts and 4 cosmonauts have died in five separate accidents of crewed missions so far. This includes the death of a cosmonaut in 1967 when his *Soyuz 1* reentry parachute failed to open; an astronaut who lost control of his *X-15* suborbital flight in 1967; 3 *Soyuz 11* cosmonauts in 1971 through unplanned depressurization in the only accident recorded above the Kármán line (i.e., while in orbit); 7 astronauts in the launch of the *Challenger* shuttle in 1986; and another 7 lost during the reentry of the *Columbia* shuttle in 2003. A further 11 individuals have been killed during training or tests (e.g., the 3 *Apollo 1* astronauts in a fire on the launch pad in January 1967), and there have been a host of "near misses."

- Landings off-Earth are also far from perfect. The experience with "controlled/soft" landings on the Moon demonstrates the magnitude of the difficulties, with around half of all lunar landing attempts failing, including recent attempts by India, Israel, and Russia in the important and difficult landscape of the polar regions. The surface of the Moon is made more complicated for landings not only because of the roughness of the surface, but also because parachutes cannot be deployed because of the absence of an atmosphere (unlike the case of a landing on the return to Earth). Atmospheric braking of this kind is also not possible on Mars, where the approach speed is high and the margin for error is tiny. For these (and other) reasons, a high proportion of Mars missions (25 of 55 up to 2021) have failed. The challenge will be even more difficult for landings on bodies without atmospheres in the outer reaches of the Solar System (such as asteroids and the smaller moons of planets) and on remote exoplanets where much less is known currently about the specific conditions in play.

- Spacecraft structures are very complicated, costly, and energy hungry and have materials-dense engineering. With $150 billion invested in its construction and operation so far, the International Space Station is the most expensive structure built by humans, yet it can only accommodate up to seven crew and is unable to provide anywhere near a self-sufficient habitat. Its ecosystem is not fully recycling and requires routine, regular replenishment of water, food, and replacement materials and equipment from Earth. The situation will be much more difficult to manage for settlement-size craft carrying hundreds of passengers.

- In the absence of human-friendly atmospheric composition, pressure, and temperature, exiting the spacecraft for exploration (let alone for serious construction and repair work) will be a very slow and dangerous process (to wit, extravehicular activity on the ISS and the lunar surface), and the return to the spacecraft may be fraught with issues of contamination and incursion of toxic and/or corrosive surface dust (as on Mars and the Moon). It will be nothing at all like the virtually instantaneous "Beam me up, Scotty" protocols that are the mainstay of *Star Trek* and other popular science fiction books and films.

- The extraction of essential life-support resources (food, gases, fuel, construction materials, etc.) from the local environment will not be achieved easily. Unlike for our forebears' expansion into unoccupied parts of the Earth tens of thousands of years ago, there will be no food for the picking on local plants, no animals for killing for protein, and no trees for felling to use in the construction of shelters. Oxygen may be able to be recovered from in situ water (ice or liquid) with appropriate electrolysis equipment (and energy), but it will almost certainly not be pure, requiring careful and complete removal of toxins by industrial-scale processes that will have to be constructed by and in the settlement. Similarly, the winning of metals and oxides from the local soil or rocks and their refinement to appropriate purity standards for their use in shelter construction, energy production, manufacturing processes, etc., will require (like on Earth) massive infrastructure for anything larger than bench-scale processes. This will be complex and will require huge amounts of energy and appropriate waste disposal or recycling procedures.

None of the above challenges are trivial, and many will likely require decades to be researched comprehensively and solutions proposed and implemented before the initial settlement craft will be cleared to depart Earth. They raise the fundamental question of whether off-Earth, fully self-sufficient human settlements are possible within the inner Solar System in the foreseeable future, let alone their establishment in more-remote locations in the outer regions and in other stellar systems. It also raises the question of whether the massive financial and resource commitments required to establish these settlements would be better applied to addressing the manifold problems emerging on Earth in order to ensure the survival of humans there.

But there may be another way.

Chapter Thirteen

Humans or Machines?

> *A time will come when science will transform [our bodies] by means which we cannot conjecture. . . . And then, the earth being small, mankind will migrate into space, and will cross the airless Saharas which separate planet from planet, and sun from sun. The earth will become a Holy Land which will be visited by pilgrims from all quarters of the universe.*
> —Winwood Reade, in *The Martyrdom of Man*, 1872

A clear distinction (and choice) can and should be made between missions in which the primary goals are short-term (i.e., weeks to a few years) exploration and scientific research and ones that aim to establish a permanent, self-sustaining community of millions of humans as a "refuge" against extinction scenarios on Earth.

The technological/inorganic and human/biological requirements of each of these two mission types are very, very different. For a start, over the past 50 years, short-term missions with science and exploration goals have been and are being achieved very well by robotic spacecraft, orbiters, and rovers operating throughout the Solar System and, in five cases, in the interstellar neighborhood. As we have seen in part III, as soon as humans are involved, especially in long-term missions and even more so when the goal is to establish a full-blown human community off-Earth, things get far more complex and the technological, social, and political issues are much more uncertain.

The discussion below explores some of these scenarios in more detail.

Another Path to Survival

Almost everything about living off-Earth is uncompromisingly deadly to humans. This is because *Homo sapiens* is the product of millions of years of exquisite Darwinian evolution that has enabled us to deal with a multitude of threats to our existence. These include very complex things like, inter alia, the ability to think our way through problems, immunity to diseases, and recovery from cell damage. However, it has not had to deal with (i.e., protect us from) cosmic radiation, gravitational forces lower (or higher) than Earth's, toxic atmospheres, temperatures outside the range of a few tens of degrees either side of zero centigrade, or unrelenting confinement in small spaces.

Unfortunately, as described in previous chapters, these are precisely the conditions that confront human spaceflight, short or long. For the longer missions aimed at settlement, we need to construct, launch, and fly for hundreds or thousands of years a fleet of spacecraft large and robust enough to support generations of live or suspended-animation humans and/or their seeds/embryos, with all of the associated paraphernalia and a huge bank of DNA material to create the plants and other life-forms necessary to establish, feed, and clothe an entire community.

But there is another way to establish a sustainable settlement of "individuals" that is reflective of human civilization on Earth and deserving of serious consideration (Canning and Hill 2008). This option is ethically and conceptually challenging, but (in the medium-term future, at least) technically more feasible, and perhaps more socially and ethically acceptable, than the current model.

The "other" way is to remove biological humans with their inherent fragility from the picture altogether and, instead, send our consciousness and our civilization embedded within the artificial intelligence software of advanced inorganic machines.[1] Putting religion and spirituality aside for the moment, humans are nothing more and nothing less than the sum total of the electrochemical processes and molecular forms and functions that have evolved from simple component molecules in the primordial "soup," through single-celled organisms and ultimately to *Homo sapiens* over the previous 4 billion years or so of life on Earth.

It can be argued that all that differentiates humans from other lifeforms on Earth is our advanced level of self-awareness or consciousness. Is it solely this extraordinary capability of our central processor/brain that defines what it is to be human? If so, would not an equivalent inorganic processor be capable of reproducing the organic electrochemical processes that are the foundation and essence of our personalities and civilization, including thought, intelligence, consciousness, joy, sadness, humor, selflessness, altruism, and morality? Is it not possible, even likely, that since these characteristics of "humanity" have evolved in a biological framework, there is a possibility that they could also be reproduced in an inorganic platform?

If we are content to accept that a robot or other inorganic form with advanced artificial intelligence can provide the means for the survival and/or preservation of "humanity" (in its generic, non-spiritual sense), then there is no need to send fragile and frail humans on these journeys beyond Earth (Canning and Hill 2008; Kaku 2018; Goldsmith and Rees 2022, and references therein). We could then engineer robots with artificial consciousness that could survive in any physical environment that allows machines to function, human friendly or not.

The boundary conditions in this machine-friendly world already have been shown to be much more relaxed than is permissible for humans, namely, the robotic exploration of the high-pressure deep oceans of Earth, the weightlessness of low Earth orbit, the airless surface of the Moon, the extreme temperature swings and global dust of Mars, the very low temperature and toxic dense atmosphere of Titan, and the exotic conditions on a few comets and asteroids.

Furthermore, so far, the operational software and hardware on the *Voyager*, *Pioneer*, and *New Horizons* spacecraft have not been significantly affected by unfiltered cosmic radiation damage and temperatures near absolute zero, despite being exposed to these harsh conditions for up to 60 years. There is little doubt that space-based computers will need to be "hardened" against the intense radiation flux when they are close to stellar systems, just as satellites in Earth orbit require extra protection against solar flares and coronal mass ejections, but not so much when they are in the relatively benign expanses of interstellar space.

Machines also have the advantage that they are much less expensive to maintain/sustain than humans. Goldsmith and Rees (2022) estimate that astronaut missions to Mars will cost roughly 50 times more than rovers and other inanimate devices, and suggest that robots will not be subject to the "Earth-out-of-view" effect discussed in chapter 12 or the sense of isolation and monotony that might impact humans on long spaceflights. Whether robots "couldn't care less" is certainly true of current models, but it may not apply to the more advanced and sophisticated machines which might be developed soon and which will be expected (and designed) to possess consciousness of their environment and their relationship to it. In this case, the machines would need to be programmed to have this resilience.

Inorganic robots have their limitations, even if these limitations are far more relaxed than for organic matter. Specifically, the temperature must not be such that any of the hardware softens or melts, embrittles, or seizes up through a loss of lubricant function and/or dust, and, most of all, there must always be a supply of energy, the machine's equivalent of food. Unlike humans', the machine's "food" does not need to be grown in a carefully curated medium (i.e., soil) under specific environmental conditions and patiently and expertly tended for weeks or months in specially constructed "glass houses" (consuming other valuable resources like settlement energy, surface area, and water). The food of robots would be energy derived from insolation (if the settlement is close enough to the local star) and/or nuclear reactors.

On the downside, the energy generated by these putative reactors depletes continuously as radioactive half-life after half-life passes by—as is happening to the power sources on all the current nuclear-powered craft and rovers, wherever they are—and none of them have a reserve supply of nuclear fuel. While this is of little concern for relatively short-lived missions of up to a few tens of years, it is a very important residual issue for longer operating periods outside the inner Solar System on interstellar missions or on settlement platforms where solar energy is too weak to be harnessed effectively.

If the extraction and refinement of nuclear fuel from local resources is not possible (by no means a simple task, even if the raw materials are available for mining!), an alternative source of energy for the robots, such as fuel/galvanic cells, would need to be manufactured from less-esoteric local mineral resources. The same long-term energy supply issue would apply for the overall energy source required to operate equipment, lights, pumps, etc., in a corresponding human settlement, including the energy for the production of plant-based food/energy for the humans themselves.

Another limitation of inorganic machines is that the technology for full self-replication by a von Neumann–style machine has not yet been developed, although this capability will no doubt be forthcoming when the machines develop superior general intelligence. When this happens, it is likely that the machines will also be able to construct infrastructure for the mining and/or synthesis of fuel for energy production. As a case in point, both Zubrin (2019) and Wohlford and Hendricks (2016) have described the utilization of CO_2 from the atmosphere on Mars and complex hydrocarbons present in the atmosphere of Titan, respectively, to generate fuel for all manner of purposes in settlements on those bodies.

All problems solved, except those that we know nothing about!

This brings us to a discussion of the seemingly unstoppable march of artificial intelligence.

Artificial Intelligence

Computers, automation, machine learning, and AI are already approaching or surpassing humanity's current level of competence and superiority in specific areas such as game-playing (e.g., chess, *Jeopardy*, and Go) and are starting to encroach on many other areas of human activity, including autonomous vehicles (driverless cars), warfare (autonomous drones and missiles), medical diagnosis and surgery, remote sensing, and farming (GPS-directed tractors, fenceless paddocks, etc). Robots are even assuming roles as companions and personal assistants for humans, namely, Apple's digital assistant Siri, Google's Gemini, Microsoft's Copilot, and Amazon's cloud-based voice service Alexa.

It is foreshadowed that the "digital cloud" will soon be able to be connected directly to human brains, making access to information essentially boundless and instantaneous. By accessing the knowledge and experiences of millions of other humans through the cloud, we will be able to expand massively the data that is currently obtained only from our personal experience of social human interactions and information from internet search engines and recent AI developments like ChatGPT. But will the concept of the "individual" survive this extreme connectivity, or is it inevitable that this future expanded, essentially limitless connectable intelligence renders useless the concept of individual thought and individual consciousness?

Harari (2015) discusses the rise of the importance of data, leading to the emergence of dataism, the "worship of data," in human existence. He posits to the humanists (i.e., those whose philosophy attaches prime importance to human rather than divine or supernatural matters) that while God may be a product of human imagination, dataism suggests that the imagination capability is itself merely a product of biochemical algorithms.

Just as we did not upgrade horses when the car replaced the horse-drawn carriage, Harari condemns *Homo sapiens* to "join the mammoths and the Chinese river dolphins in oblivion" and suggests that, through the increasingly ubiquitous role of data, human consciousness is decoupling from intelligence to produce a new species, "*Homo deus*," possessing the "divine powers of creation and destruction." Bohan (2022) says a similar thing in a different way when she points out that we are "innovating our way to our own obsolescence and we are fading away as we become something new."

Such access to the entire information database on the internet is a potential resolution to the oft-mentioned objection (Kaku 2018) raised against the replication of human self-awareness in machines, namely, that

> *the robot would have to understand millions of rules of common sense—the simple laws of physics, biology, and human behaviour that we take for granted. Moreover, it would have to understand causality and anticipate the consequences of certain actions.*

While acknowledging that human infants and children absorb this enormous capability through their exposure to billions of interactions and shared experiences with other humans over the first few decades of life, Kaku does not focus on the fact that even current crude "machine learning" draws upon a database of such information. Its extension to include access to the entire internet of data and billions of human stories (as is beginning to be embodied, for example, within software such as ChatGPT—see the next section) would provide just such a (albeit vastly accelerated and powerful) learning process for these inorganic "humans."

Passing Through the "Singularity"

The more or less inevitable emergence and potentially dystopian impact of a second dominant (artificial) intelligence on Earth is discussed in elegant detail in Bostrom (2014), Tegmark (2017), and Kaku (2018). A similar but more optimistic theme is presented by Bohan (2022) through the concept of a more forgiving "integration" of humans with advanced robotics and artificial intelligence to create a "future superhuman."

As profound and potentially prophetic as are these elegant books and many others on the future of humanity, we are already becoming more and more aware of the positive and negative impacts of AI on our lives. The pace of change in AI is rapidly ushering in an age when AI "machines" of many kinds have superior skills and capabilities to humans in certain specialized areas, and we already have access to the almost ubiquitous machine learning, facial and voice recognition, melanoma diagnosis, internet search engines, and marketing software that seems to know us better than we do. Perhaps the best example of this coming tsunami of AI was the release in November 2022 by the OpenAI company of the large language model, generative software ChatGPT. This software is now available to more than a billion people, and it seems that there will be no stopping its further incursion into our daily lives and for the enhancement of machine capability.

Green (2024) has pointed out that the T in GPT is the "transformer neural network" architecture invented at Google in 2017, itself built on the "deep learning" revolution of Geoffrey Hilton and others over the last decade, the "back propagation algorithm" from the 1960s, and Gottfried Leibniz's "chain rule" from 17th-century mathematics. It has been developing for a decade or more but is now rapidly becoming an integral part of the way we do business, solve problems, undertake research, create art, innovate, and entertain ourselves.

Opinion is divided on its impact: Harari has stated that GPT will be "the best or worst thing that ever happened to humanity," while Alphabet CEO Sundar Pichai said it will be "better than fire, electricity or the internet" (Green 2024). Others worry about its impact on education and the creation of deep fakes that can manipulate and/or change the way we think about the events around us and confuse our ability to discriminate between truth and lies. Concern has also been expressed about the creation of unemployment, social disruption, magnification of the economic divide within and between countries, and threats to intellectual property (i.e., fair use of the raw data).

There will be no stopping this progressive enhancement of generic machine capability, and it seems inevitable that there will come a time in the not-too-distant future when machines will cross over from servant to master. Some predict that this transformational event, or "singularity," will occur within the next 20 years, although it should be recalled that corresponding statements that the achievement of net positive energy from nuclear fusion "is only 20 years away" have been made for at least the past 60 years.

Nevertheless, from the singularity on, the acceleration in machine capability will rise exponentially since they will be able (by definition) to enhance themselves more rapidly and effectively than we can, and their power will be beyond our control. Bostrom (2014) and other pundits have noted that the fate of humans would then be totally in the superintelligence's hands, just as the fate, for example, of gorillas is currently in ours. At this point, our only hope might be that we serve a purpose for them.

To skip forward in the discussion momentarily, the passage of these machines through the AI singularity, if it occurs, would have been a conscious, if not willing, outcome of our own making and could be viewed as the generation of a new species of *Homo*, dubbed *Homo deus* by Harari (2015) and *Homo machinus* by others.

The development of machines with significantly advanced, let alone superhuman, intelligence also has profound implications for space exploration and travel because it could then elevate uncrewed orbiters, landers, and rovers from their current roles as (admittedly very capable) semiautonomous, arm's length extensions/proxies of humans to the status of a full and complete replacement for biological human explorers and settlers. In so doing, it could provide an alternative mechanism for the continuation of the next manifestation of human civilization, whether or not *sapiens* is doomed on Earth. Equally importantly, it releases fragile, organic humans from the debilitating, if not deadly, and preclusive dangers of space travel.

But let's not get ahead of ourselves: machines are not immune from all the dangers of space. Table 13.1 summarizes the relative capabilities and limitations of humans and machines in space.

Both potential populations—organic humans and inorganic machines—will not be able to operate in short- or long-term settlements without a continuous source of energy, generated locally in large quantities. For humans, there is a dual requirement, first for the humans themselves to operate (i.e., the energy generated from food and water) and second for operation of the overall support infrastructure for the settlement (i.e., electricity and heat). For the machines, they need only to tap into the electricity supply that is already in place to power the settlement.

Individuals in both populations will also need comprehensive repair and replacement strategies. For humans, this will involve drugs and medical services for treating ailments and accidents, along with support for the gestation, birth, and development of new generations of infants through to maturity. The machines will require the equivalent repair and maintenance facilities, and processes for their replication and enhancement. Both humans and machines will need equipment for the mining and processing of all construction and engineering materials, and comprehensive workshops to undertake construction and repairs.

On balance, table 13.1 suggests that advanced machines will have fewer operational and safety issues, will last longer, and will be much more productive in space than humans. This is because machines are less sensitive to ambient temperature, pressure, and radiation; require no atmosphere; do not need to sleep or relax, etc.; produce less waste; and their performance can be enhanced through routine software updates. But sustainable "life" in the early decades of the establishment of the settlement will be very difficult for both humans and machines until all the support infrastructure is in place. During this period, regular "care packages" from home will be essential.

Table 13.1. Comparative characteristics and operational requirements of humans and advanced AI machines in space.

Issue of Concern	Humans (Organic)	Advanced AI Machines (Inorganic)
Intelligence	As currently.	Approaching parity with humans now, but will be far superior after the AI "singularity" is reached.
Longevity	120 years, at current best, depending on general state of health, absence of serious debilitating diseases, and access to a comprehensive range of medical services.	With maintenance and replacement parts, probably indefinite, and progressively upgradable with further enhancements of hardware and software.
Home base off-Earth	Require a pressurized structure for living quarters and storage of consumables, batteries, equipment, etc., and protection of occupants from a toxic atmosphere, extreme temperature, radiation, dust, and micrometeorites, that is, all the dangers to organic beings arising from the external environment.	Require a structure only for storage of consumables, batteries, and equipment; space for repairs; and a manufacturing workshop with protection from wind, dust, corrosive chemicals (if present), micrometeorites, etc.

(*continued*)

Table 13.1. *Continued*

Issue of Concern	Humans (Organic)	Advanced AI Machines (Inorganic)
Operating conditions	Require a biologically compatible, breathable atmosphere and pressure for all activities. Temperature around 20°C. Gravity not too far from Earth's g. Protection from radiation. Continuous supply of food and water, generated locally to provide functional energy and to maintain health. Activities outside the shelter require spacesuits to reproduce human survival conditions and to be impenetrable to micrometeorites, etc.	No atmosphere required, but if present it needs to be noncorrosive to the machine materials. A wide range of pressures is acceptable. Wide temperature range allowed but must be lower than the lowest softening/melting point of component materials. Can survive zero g, but there will be an upper limit ($>g$) dependent on machine component strength. Radiation limits uncertain (for software resilience) but will be less onerous than for humans. Onboard energy source and effective lubrication for moving parts are required for all functions.
Dexterity	Excellent, although encumbered by bulky spacesuits when operating external to the base.	Probably less dextrous than humans inside the shelter, but probably more dextrous when outside.
Productivity	Limited by mental state (morale and social compatibility with other individuals) and the need for sleep cycles, meals, relaxation, entertainment, camaraderie, etc.	Can work around the clock, except for periodic maintenance and fuel top-up. Likely programmed not to be limited by motivation, morale, boredom, or violence.

Issue of Concern	Humans (Organic)	Advanced AI Machines (Inorganic)
Maintenance and repair	Facilitated by good nutrition and a wide range of specialized medical supplies and services, along with frequent, lengthy rest and sleep periods.	Require periodic repair, maintenance, and replacement of lubricants and fixed components due to wear and tear. Otherwise, able to operate continuously.
Waste	Liquid and solid waste from bodily functions (and death) must be fully recycled and leftover solids destroyed.	The only waste produced is in the form of heat. Defunct parts and machines and used lubricants can be recycled.
Reproduction	Untested in space, but likely to be some version of normal conception and birth, adjusted for local conditions. Requires a dedicated human to be assigned as a parent and educator for one to two decades. DNA genetic diversity must be maintained.	Require a workshop to manufacture materials and parts for the generation of replicas and/or the development of more advanced versions of the machines. New and improved specialized functions, capabilities, intelligence, etc., can be generated during production of new individuals.
Energy for individuals to operate	Derived from ingested food and water, which needs to be generated continuously from local resources.	Require a portable onboard power source, probably an electric battery, for all functioning. This source will need periodic recharging from the overall power supply in the settlement.

(*continued*)

Table 13.1. *Continued*

Issue of Concern	Humans (Organic)	Advanced AI Machines (Inorganic)
Energy for settlement operation (air-conditioning, lights, pumps, transport, fuel/food production, etc.)	Solar (unlikely to be sufficient) or nuclear. Need backup source for nuclear fuel when initial source is depleted. This could be chemical in nature (galvanic/fuel cell) if the consumed raw materials are available locally for their construction and the replenishing of consumed chemicals.	As for humans.
Education/ development	Performed by parents and/or formal educators in the society.	Self-education from massive information data banks, continuously updated.
Relationships/ mental health	Uncertain and fragile.	Unlikely to be a problem.

MERGING WITH MACHINES

In another, more hopeful than the singularity, take on the future of humanity during the rest of the 21st century, Bohan (2022) argues that we will increasingly merge with various forms of artificial intelligence, reengineer our biology to remove many of our fragilities and cruel vestiges of our evolution, and lengthen our lives to become something that she defines as "posthuman" or "future human." In *Homo Deus*, Harari (2015) also proposes a gradual merger between the organic and the inorganic that offers a gentler (at least short-term) future than our extinction or complete subjugation to machines.

Homo sapiens's bodies have already responded to major changes in our living habits and physical form over the past 2 million years, namely, inter alia, the shortening of our jaws through the "invention" of fire and thence the ability to soften food through cooking instead of chewing, increases in skin pigmentation to cope with insolation differences at diverse latitudes, and, more recently, the dramatic increase in our height due to better nutrition. Given enough time for space settlements to become a reality and to be populated for significant periods, we may further evolve by natural selection to better deal with the quite different gravity and radiation environments of the relevant platform. Whether this evolution can/should be accelerated by conscious selection of progeny that are able to cope better and survive in the new environment (i.e., eugenics) is another matter that has massive and very obvious moral and ethical dimensions.

In many cases, we have adapted to changes in climate and the strictures of our local environment not though physical changes to our bodies, but through the development of coping strategies and technologies that mitigate the environmental impacts. These include congregation into villages and cities; the domestication of animals and the development of machines to do our work and transport our goods; the establishment of intricate food, water, and materials supply chains; advanced medical treatments; and even air-conditioning. It is therefore reasonable to assume that there will be further adjustments and accommodations as technology advances and lifestyles change on Earth and/or in the settlements.

Humans are already well on the road to becoming cyborgs (viz., an integrated human-machine system) using both inorganic augmentations or replacements and/or electronically enhanced body parts. These range from superficial cosmetic changes to function and appearance to highly complex replacements for failed and/or damaged body parts and the extension of lifetimes, including:

- tooth fillings and false teeth
- wigs and hair extensions
- false fingernails and eyelashes
- eyeglasses (prescription, normal and sun)
- artificial corneas, lenses, and retinas
- hearing aids and implants

- speech synthesis
- titanium and ceramic replacements for knees, hips, and shoulder joints
- static and increasingly controllable limb prostheses
- stents, valves, and pacemakers for the heart
- heart and lung machines
- kidney dialysis
- in vitro fertilization and early attempts at artificial wombs

In this sense, we have already begun to "merge" or "morph" with inorganic materials and machines, and chances are that we will continue to do so at an increasing pace. As discussed by Bostrom (2014), Harari (2015), Tegmark (2017), Kaku (2018), Bohan (2022), and a host of other authors, recent advances in the integration of machine learning, neural networks, artificial intelligence, and electronic social networks into human lives are set to revolutionize our lifestyle, relationships, longevity, capability, and functionality. With further massive increases in computing power and ever more capable and lifelike robots, the distinction between the organic and inorganic domains will become increasingly blurred. This form could represent something between an evolutionary extension of the species *Homo sapiens*, along the lines of the "future superhuman," considered inevitable and essential for survival in the latter part of the 21st century by Bohan (2022), and the new all-knowing, all-powerful species dubbed *Homo deus* proposed by Harari (2015).

As disturbing as this prospect might seem, given humanity's recent propensity for self-destruction through nuclear or biological war and our inability to reduce our contributions to climate change, a cyborg-like merger with inorganic machines may be the only chance that we have of maintaining our position as the dominant species on Earth. Maybe by then, we might have more resilience to the rigors of space travel and be better positioned (in whatever form we have become) to find and settle an alternative space to live off-Earth.

Is the creation of this next version of ourselves—be it "future" biological, solely machine-based, or a mixture of both—inevitable? Will these future creations, with their inherent dataism connection to the "internet of all things," continue to reflect our values and aspirations and respect and value our history? Would the machine version have a "soul," and is this characteristic even necessary to define what it is to be human? In the face of a global extinction calamity, would any of these future humans consider that there is anything worth preserving of human civilization?

The following section describes how an inorganic machine might develop and/or be provided with capability that extends beyond mere access to an essentially unlimited bank of data and human experiences to achieve a real sense of feeling and consciousness.

Simulation or Copying of "Consciousness"

At least some neuroscientists (e.g., Koch and Tononi 2008) believe that the learning, memory, and consciousness functions of the human mind arise from purely physical and electrochemical natural processes in the brain and that consciousness "does not arise from some magical or otherworldly quality." This means that it should, in theory, be possible to simulate the thinking processes and consciousness of the human brain in the software of a machine (Llinás 2001). Thus, it should be possible to scan, map, copy, and transfer the information, processes, and neural networks of a human mind to the control system/computer of a machine/robot to simulate the brain of a human (Bostrom 2014). In so doing, the memory, capability, data content, and lifespan of that brain could, in theory, be preserved and extended indefinitely (Martin 1971). This is referred to as "mind uploading" or "whole brain emulation" (Sandberg and Bostrom 2008) and has been proposed as a mechanism to achieve "digital immortality."

Absent the gradual, evolutionary development of human-superior AI, a "quick fix" of mind uploading to robots might be humanity's best option for preserving the nonphysical (personality, thought processes, and consciousness) characteristics of *Homo sapiens* and, simultaneously, enabling interstellar space travel without risking human crews. It would provide an alternative to cryopreservation of the complete human body or of embryos, as discussed in chapter 10. When implemented/embodied on a nonbiological scaffold that has already been established on another world, it could also provide humans on Earth with the experience of a virtual presence on that remote world without the need to travel there themselves.

On the downside, the full uploading or simulation of a human brain requires far more computing power than is currently available due to the enormous number of neurons in the brain, the huge complexity of each neuron and the uncertain role of associated proteins, and other yet-unknown contributing factors. There are also serious ethical issues related to the creation of artificial consciousness and animal/primate welfare arising from the need for animal experimentation (Sandberg 2014). There are likely to be legal implications relating to whether the simulated brains ("emulations") should be assigned the same rights as biological humans, and impacts on religious and spiritual perceptions of immortality and divinity.

SUMMARY AND CONCLUSIONS FROM CHAPTER 13
The answer to the simple question raised in the title of this chapter is that it is not a binary choice between the end-member options of humans or machines. Our exploration of the Solar System and beyond to date already encompasses a combination of semiautonomous machines and human crewed spaceflights, and it is inevitable that this trend will continue as technology advances.

In the case of our exploration of the Moon, the first human presence there was preceded by a huge number of exploration or "scouting" missions undertaken by machines to suss out the characteristics of the target and to put landing and departure technologies to the test. Human survival strategies were assessed first in laboratories on Earth and then in the riskier environment of low Earth orbit. A similar experimental situation is unfolding now as various government and private agencies seek to examine new rocket propulsion technologies, spacecraft designs, and methods for reducing launch costs prior to establishing temporary settlements around and on the Moon, and to investigate and test options for similar settlements on Mars.

However, this mix of missions by machines and human crews is taking place in the context of the relatively well-known Earth-Moon system, which we have been playing in for more than 60 years. The risks to human health and well-being are significantly higher as travel times lengthen and the distance from Earth increases. For these more remote targets, the role of machine-only missions is sure to be exclusive for quite some time yet, especially as the machines are becoming more and more capable of undertaking a wider range of formerly human-only tasks. Tellingly, there will be no opportunity, unlike the case of the *Apollo 13* mission to the Moon, to execute an immediate return-to-Earth escape strategy for the onboard crew to Mars when something goes wrong. Furthermore, such crewed missions are unlikely to receive a social "license" when we still know so little about the impact on humans of long-term low or zero gravity, radiation damage, and a host of other risks to our fragile bodies.

Humans may return to lunar orbit and the Moon's surface in the next decade or so, but the momentum of efforts to take the next step to Mars and beyond is likely to be reconsidered in light of the huge increase in risk and materials and financial costs, and in the context of a rising expectation that we should focus more on urgent social and environmental matters of various sorts on Earth. We have already seen this very situation play out in the decline in public support and eventual abandonment of the *Apollo 18–20* missions planned for the early 1970s due to increased social unrest in the United States and the ongoing massive cost of the Vietnam War.

Thus, based on both cost and risk factors, it is likely that machines will dominate exploratory and scientific research-based missions to Mars and targets beyond for the foreseeable future. Whether machines can, or should, be the sole means for the preservation of human civilization by replacing humans in permanent settlement missions is another matter.

More on this conundrum in the wrap-up provided in the epilogue.

Epilogue

Life, for ever dying to be born afresh, for ever young and eager, will presently stand upon this earth as upon a footstool, and stretch out its realm amidst the stars.
—H. G. Wells, in *The Outline of History*, 1920

Humanity is living in the shadow of several existential risks:

- The number of humans on Earth is likely to peak at around 10 billion people by the end of the 21st century through a combination of the limit to the carrying capacity of the planet and a declining reproduction rate.
- The resource impact of the population increase, improvements in the overall standard of living, and the need for more and more energy to support this lifestyle has produced a continuing rise in Earth's average temperature and sea level, and associated increased incidence of extreme weather events that, acting together, will endanger the lives and livelihoods of large swathes of the population.
- In parallel, competition for land, food, water, and other essential resources, together with increased tension between competing political and religious groups, is emerging with a ferocity that threatens global peace and the rule of law.
- The specter of the continued emergence of robots and other machines with massively powerful artificial intelligence and instantaneous global connectivity hovers over human lives like never

before, changing everything about our way of life and threatening to replace us as the most powerful species on the planet.
- If this wasn't enough, nonanthropogenic risks such as the ever-present prospect of, inter alia, a collision with a massive asteroid or comet in the near term and, in the longer term, inevitable engulfment by the Sun in a few billion years will guarantee human extinction.

All but the last set of these risks are of our own making and may constitute humans on Earth playing out their own version of the "Great Filter" that has been suggested as precipitating the end of other technologically advanced civilizations that may have emerged throughout the galaxy, and the reason why we appear to be alone in the cosmos.

Many would consider the impactor risk, alone, is a good enough reason to begin the process of establishing a settlement/refuge for humans on another planet or moon, but the urgency of preparing for this eventuality is up for debate, given the stochastic character of asteroid impacts and other ex-Earth catastrophes. Whether the anthropogenic (Great Filter) risks likewise warrant urgent action is also open to debate, but if life as we know it is reasonably likely to expire or to be significantly degraded over the next few decades, it would seem like a good idea to start preparing right away to get out of the frying pan (literally) and start afresh somewhere else. There is, of course, no guarantee that we will not take all our undesirable, self-destructive "baggage" with us to the new platform and simply repeat on a regular time frame of a few thousand years the whole depressing process over, and over, again.

Regardless of these existential risks, it seems to be beyond argument that we humans have evolved an inherent and incessant desire to "see what is over the next hill," that we seek new challenges, and that we require answers to some of the fundamental questions of the why, how, and where life came into being. Having largely explored and occupied most of the corners of our planet, it is not surprising, then, that we want to go to other places that have been beyond our reach until recently. It is also possible that we are arrogant enough to want to preserve our civilization, especially, but not only, because it may be the only example of such complex life in the galaxy.

The United States of America sent 12 humans outside the gravitational dominance of the Earth to spend time on the surface of the Moon in 1969–1972, but this was under very unusual circumstances. The landings may not have been attempted, except that it was at the height of the Cold War "space race," where the demonstration of political and technical superiority over the then-rival Soviet Union was deemed to be more important than almost anything else. The momentum to expand our presence on the Moon and beyond after this spectacular technical achievement was lost, for economic, political, and "social license" reasons, and now, in more than 50 years since then, we have not sent humans anywhere but in circles around the Earth. However, the impetus for a return to the Moon has recently reemerged, this time in a potential space race with the People's Republic of China.

The stated primary intention of this renewed initiative is to use the Moon as "practice" for an expedition to Mars, not specifically for this planet to become a platform for the preservation of human civilization, but "because it is there" and someone wants to be first to plant their flag and footprints there. The trouble is that getting to the Moon was difficult enough, but going to Mars is much, much trickier. We have discovered after 60 years of spaceflight that humans are not designed to survive space and that it will be very difficult just to get to Mars and back safely, let alone trying to establish a viable, long-term, self-sustaining, independent-from-Earth community there. We don't know enough about the full range of responses of the human body and mind to long-term exposure to zero or low gravity or radiation; or the isolation, boredom, and confinement of spacecraft; or reproductive success; or how we might extract food, water, energy, and materials from the local environment (even if we knew for certain that it is there in useful quantities).

We also don't know how to handle ownership, access, or governance issues on the surface of another celestial body, or the moral and ethical grounds for selection of those individuals who will populate the settlement if they are to "represent" and "preserve" humanity as a whole. Once we get there, even if it is for a stay of only a few months until the next time the orbits of Earth and Mars are optimally aligned for the journey back, interest and acceptance of the inevitable push for longer and longer visits will likely fade soon after the initial euphoria of the first human presence subsides—just as it did for the first landing on the Moon in 1969.

Mars is very tricky, but going to locations farther afield is a whole lot more challenging again. A few exploratory visits on and around some of the moons of Jupiter and Saturn, and a few orbits of and landings on asteroids, have revealed some promising signs for the potential existence of simple life-forms on or under the surface, but conditions on all of these bodies are way outside the comfort zone for humans.

We know even/much less about the conditions on planets in other stellar systems, and studying them in detail will require a long and slow process of exploration by telescopes and robotic missions that, in the latter case, will require decades, if not centuries, to execute. We need to know about the conditions on and around those exoplanets, whether there is life there already (not to mention how we will deal with it, if it is), and whether it is possible to terraform them to be more suitable for the establishment of a settlement there.

What we do know already is that it will require much, much better technology for spacecraft propulsion if we are to reduce the massive travel times (even to the outer Solar System) and to deal with the much greater mass that is required to transport us and all the equipment, food, and energy that we need for the journey and to sustain us for the first few months or years after our arrival. Methods for maintaining us either fully alive or in some form of suspended animation or as embryos during transport have also yet to be worked out. What will be the response of those left behind? Will they provide the social license for the huge cost of the preparations and the journey itself?

With all this uncertainty, the establishment of settlements off-Earth anytime soon seems clearly to be out of the question. Short-term "scouting" and research expedition trips with crews of a small number of humans and the minimum two-year turnaround time to Mars might be achieved in the next 25 years or so. However, we will need to do a lot of homework to even understand the full scope of our ignorance and incapacity for trips of these lengths, let alone to provide solutions to the multifarious challenges that will need to be confronted. With a lot of effort, money, and luck, we might be ready to embark on a bona fide long-term settlement mission on Mars by the end of the 21st century. Maybe.

In this writer's view, the most effective and lowest-risk approach to achieving the knowledge and understanding required for a trip to Mars and beyond is to remove ourselves completely from the firing line and to give robotic machines the responsibility for doing the "dirty" work of exploration and research. Later, they can also be utilized for the establishment of short-term settlements, namely, the construction of dwellings, the planting and managing of food production systems, the mining and processing of essential local materials, the creation of manufacturing facilities for equipment, and the production of energy.

Why risk human lives and massive expenditure at perhaps 50 times the cost when even current-generation robots are already providing us with extremely valuable scientific information about the surfaces and atmospheres of a host of Solar System bodies, despite the decidedly human-unfriendly local conditions? The next generation(s) of robots will surely be even more competent and increasingly autonomous. In perhaps a few decades, machines with very advanced AI are likely to have progressed to the point where they will provide the most practical (or indeed, perhaps the only) means by which a new version of "humanity" can explore and assess for habitability through direct witness another platform beyond Earth. Humans may then be able to follow up on this background work with crewed expeditions or, instead, we might decide to use virtual reality constructed by these robotic missions to substitute for our physical presence altogether.

In any event, at the current rate of artificial intelligence development, by then these machines may have developed a superior intellect to us, and we may no longer be the dominant intelligence on our home base. We may then have, in effect, transported our "consciousness" to an inorganic, much more resilient version of ourselves in a bid to create the next chapter in the record of our civilization, along with the continuation of its existence.

Before long, and on the other side of this singularity, we may have been able to eliminate many of *Homo sapiens*'s physical, emotional, moral, and ethical deficiencies, driven by a desire for collective (rather than familial or tribal) preservation. We may have then morphed into a new species, dubbed by some as *Homo machinus*, with a genuine chance of surviving the Great Filter.

We might even dare to hope that *Homo machinus* might, at some point in this long journey, have turned its attention to saving the Earth itself, at least until the expanding Sun inevitably has its way and consumes us, if we are still there and have not found another space in which to live.

Notes

Chapter 1: Are We Alone? The Search for Extraterrestrial Life

1. For a broad summary of the history of life on Earth, see https://en.wikipedia.org/wiki/History_of_life.

2. Snowball Earth is a hypothesis that proposes the planet's surface became completely or nearly completely frozen (hence "Slushball Earth") during one or more of Earth's global glaciation periods. Support for the hypothesis comes from the observation of sedimentary deposits that are believed to be of glacial origin at tropical palaeolatitudes. The period of such is believed to have occurred sometime before 650 million years ago, before the Cambrian explosion. The most recent snowball episode may have triggered the evolution of multicellularity.

3. A virus is a small collection of genetic code, either DNA (deoxyribonucleic acid) or RNA (ribonucleic acid), surrounded by a protein coat. A virus cannot replicate alone and must infect cells and use components of the host cell to make copies of itself. A virion is an entire virus particle consisting of an outer protein shell called a capsid and an inner core of RNA or DNA. The core confers infectivity, and the capsid provides specificity to the virus.

4. A nova occurs in a binary star system composed of a white dwarf and another star that are close enough to each other for the white dwarf to accrete material from its companion. The material, mostly hydrogen, remains on the surface of the white dwarf until there is enough to kick-start nuclear fusion. This then triggers a runaway thermonuclear reaction, ejecting the material and emitting massive amounts of energy. Because the white dwarf remains intact after blowing away this excess, a stellar system can experience multiple classical novae.

A supernova occurs when the hydrogen fuel at the center of a star with at least five times the mass of our Sun starts to run out and the outward pressure from radiation is no longer able to balance the inward pressure from gravity. The star suddenly collapses, resulting in enormous shock waves that cause the outer part of the star to explode. Usually, a very dense core is left behind, which, if the star is more than about 10 times the size of our Sun, may leave behind a object so dense that it becomes a black hole.

5. Plate tectonics is a mechanism dating from the 1960s that relates major geological events such as mountain building, volcanoes, and earthquakes to global movements of large rocky plates in the Earth's outermost layer, or lithosphere (its crust and upper mantle), above a partially molten layer below. Due to the convection in the subterranean layers, the plates move relative to each other at different rates, from 2 to 15 centimeters per year, on

Notes

the one hand (in mid-oceanic regions) generating new areas of crust, and on the other, burying or "subducting" old crust back into the mantle.

6. In the greenhouse effect, sunlight with a wavelength of ~0.6 microns (600 nm) is not absorbed efficiently by the atmosphere, so it scatters its way through the clouds and eventually makes it to the surface. The surface warms and then radiates this energy away in the longer wavelength infrared part of the spectrum. CO_2 in the atmosphere absorbs some of this infrared radiation and so not all of it escapes back into space, remaining trapped in the surface and the atmosphere. In the case of Venus, a "runaway" greenhouse mechanism heated the surface to ~460°C, whereupon the energy was able to radiate away at the same rate it was absorbed by the surface, but the damage to life (had it existed) would have already been done.

7. A "trace fossil," or "ichnofossil," is a record of biological activity, but not the preserved remains of the plant or animal itself. On the other hand, a "body fossil" is the fossilized remains of parts of an organism's body, commonly altered or preserved by later chemical activity or mineralization.

8. Wells's book has been interpreted as a commentary on the theory of evolution, imperialism, and Victorian era fears, superstitions, and prejudices, especially in relation to invasion and colonization. Wells later indicated that the inspiration for the book's plot was the effect of European colonization on the Aboriginal Tasmanians. Some historians have argued that Wells wrote the story to encourage his readership to question the morality of imperialism (Schwartz 2015).

9. The frequency of the spectral lines of the neutral hydrogen atom was selected as a target for radio emissions detection because this element is by far the most abundant (74 percent) in the universe and is considered the most likely to be selected by any other civilization as the most obvious and sensible wavelength for communication and detection.

10. A light year is the distance traveled by light in one year, at a speed of around 300,000 kilometers per second. This corresponds to 9.5×10^{12}, or 9.5 trillion kilometers.

11. Von Neumann's self-reproducing schema proposed that open-ended evolution requires inherited information to be copied and passed to offspring separately from the self-replicating machine, an insight that preceded the discovery of the structure of the DNA molecule by Watson and Crick in 1953 and how it enables the translation and replication of life's architecture in cells to provide heritable information about the organism.

12. Soviet astronomer Nikolai Kardashev (1932–2019) proposed in 1964 a hypothetical ranking of a civilization's level of technological advancement based on its energy consumption requirements (Kardashev 1964). A Type I civilization accesses all the available energy on its host planet (around 10^{16} watts on Earth). Type II directly harvests all the energy of its host star (around 10^{26} watts), and Type III can capture all the energy emitted by its galaxy (around 10^{37} watts). Currently, humanity consumes around 10^{13} watts, so it is still well below the level suggested for a fully fledged Type I civilization.

13. An astronomical unit (AU) is a unit of distance measurement equal to 149.6 million kilometers, the mean distance from the center of the Earth to the center of the Sun. It is a convenient way of expressing large distances when the usual "light year" is too large, and it also represents a time unit for a distance of about 8 light minutes, the time taken for light to travel from the Sun to the Earth.

14. The heliosphere is the magnetosphere, astrosphere, and outermost atmospheric layer of the Sun. It has the shape of a vast, bubble-like region of space that constitutes the spherical volume of influence by the Sun in the surrounding interstellar medium. The heliosphere extends far beyond the orbit of Pluto and is dominated by plasma originating from the Sun, known as the solar wind. Outside the heliosphere, this solar plasma gives way to the interstellar plasma permeating regions between stars in the Milky Way.

15. The heliopause is the outer edge of the heliosphere and marks the boundary between matter originating from the Sun and matter originating from the rest of the galaxy. Spacecraft that depart the heliosphere (such as the two *Voyagers*) are in "interstellar space."

16. In orbital mechanics, a "gravity assist," "gravitational slingshot," or "swing-by" maneuver of a spacecraft close to a planet or other astronomical object takes advantage of that object's relative movement (e.g., orbital trajectory) and gravity to alter the path and speed of the spacecraft, with the general aim of saving propellant.

17. Bacteria are a domain/kingdom of single-celled microorganisms that were among the first life-forms to appear on Earth and are present in most of its habitats. Bacteria inhabit soil, water, acidic hot springs, radioactive waste, and the deep portions of Earth's crust, and they are also known to have flourished in and on spacecraft. Bacteria are vital in recycling nutrients, with many of the stages in nutrient cycles dependent on these organisms.

18. Archaea are a domain/kingdom of single-celled microorganisms that have no cell nucleus or any other membrane-bound organelles in their cells. Archaea were viewed as extremophiles living in harsh environments, but they have been found in a broad range of habitats and are a major part of Earth's life.

19. Eukaryota are all animals, plants, fungi, and many unicellular organisms with cells that have a defined membrane-bound nucleus. They constitute a major group of life-forms alongside the two groups of prokaryotes: the bacteria and the archaea. They represent a small minority of the total number of organisms, but since they generally have a much larger size, their collective global biomass is much larger than that of bacteria and archaea (prokaryotes). Eukaryotes may be either unicellular or multicellular, but prokaryotes are typically unicellular.

20. Carl Sagan convinced NASA to take spectroscopic observations of the presence of a range of gases in Earth's atmosphere during the spacecraft *Galileo*'s gravity-assisted flybys of Earth in 1990 and 1992 on its way to Jupiter. These measurements showed that the relative abundance of O_2 and CH_4 were out of equilibrium by many orders of magnitude, indicating the presence of life processes on Earth and demonstrating that similar measurements could be used to search for the presence of life on exoplanets. At the same time, *Galileo* detected radio signals generated by humans, indicating that measurements of this kind could provide a definitive signature of *intelligent* life.

21. An object or a system is "chiral" if it is distinguishable from its mirror image (i.e., its enantiomorph); that is, it cannot be superimposed onto it (for example, human right and left hands). Many specific organisms, individual compounds, organs, or behaviors are observed to possess the same enantiomorphic form ("left" or "right") and so this property is used as a characteristic of life as we know it on Earth.

22. Theia is a hypothetical planet with a diameter of around 6,000 kilometers, about the size of Mars, that was formed at about the same time as the Earth around 4.5 billion years

ago. According to the "giant-impact hypothesis" for the origin of the Moon, Theia was formed in and occupied either the L4 or the L5 Lagrangian point in the Sun-Earth system, eventually colliding with the proto-Earth after its position was destabilized due to the gravitational influence of Jupiter and/or Venus. Some of the collision debris was incorporated into the Earth, while the remainder coalesced in orbit around the Earth to form the Moon. This model explains why the Earth's core is larger than expected for a body its size and why the Moon has a relatively small core, since most of Theia's core mixed with Earth, while its mantle mostly went into orbit to form the Moon. Another theory suggests that Theia might have formed in the volatiles-rich outer Solar System before moving into the vicinity of Earth, and that much of Earth's water originated from the subsequent collision.

23. The sources of the detailed information about the missions provided in this section are listed in the "Additional Internet Sources," chapter 1, section at the end of the book.

24. The Doppler effect/shift is the apparent change in frequency of a wave in relation to an observer moving relative to the wave source. It is named after the Austrian physicist Christian Doppler, who described the phenomenon in 1842. An increase in frequency (decrease in wavelength) arises when the source of the waves is moving toward the observer because each successive wave crest is emitted from a position closer to the observer than the crest of the previous wave. The reverse applies when the source is moving away from the observer.

25. Tidal locking between two co-orbiting celestial bodies arises when at least one of them presents the same face to the other over the course of its orbit, either all the time (like the Moon relative to the Earth) or at certain integral multiples of its rotation period (like Mercury's 3:2 rotation and revolution period resonance relative to the Sun). All of the 20 moons in the Solar System large enough to be spherical are tidally locked to their host planet. Pluto and its largest moon, Charon, are tidally locked to each other.

26. A "black body" is an idealized opaque, nonreflective physical body that absorbs all incident electromagnetic radiation, regardless of frequency or angle of incidence. The radiation emitted by a black body in thermal equilibrium with its environment is called "black-body radiation." The body has a specific, continuous spectrum of wavelengths, inversely related to intensity, that depends only on its temperature.

27. The mediocrity principle is the philosophical notion that "if an item is drawn at random from one of several sets or categories, it's more likely to come from the most numerous category than from any one of the less numerous categories" (Bates 1972). The principle has been taken by some to suggest that there is nothing very unusual about the evolution of the Solar System, Earth's history, the evolution of biological complexity, or human evolution.

28. Some philosophers and theologians have argued that "fine-tuning" implies that there is a "fine tuner," namely, that the universe could only be as it is because of (1) the intervention of a deity, (2) an alien life form operating the universe as a gigantic simulation experiment, or (3) another unknown presence.

Chapter 2: Scenarios for the Demise of Humanity

1. A red giant is a luminous giant star of low or intermediate mass (around 0.3 to 8 solar masses) in a late phase of stellar evolution. The outer atmosphere is inflated and tenuous, with a radius ten to hundreds of times larger than that of the Sun and a surface temperature

NOTES

of around 4,700°C or lower. By this stage, the star has exhausted the supply of hydrogen in its core and has begun thermonuclear fusion of hydrogen in a shell surrounding the core. When this energy supply is no longer available, the core will contract and start fusing helium, and when this energy source is no longer available, the star will throw off its outer regions to create a "planetary nebula" and resume the rest of its life as a white dwarf with its luminosity coming solely from the emission of residual thermal energy.

2. The Earth is around 4.5 billion years old. Geological time over this period is divided into time sections or units delineated by events that took place within and between each of these units that served to change the nature and/or the content of the relevant geological strata, thereby distinguishing them from other such units. In order of decreasing time span, these units are eon, era, sub-era, system/period, series/epoch, and stage/age.

3. The Permian system/period occurred between (approximately) 299 and 251 million years ago, and the Triassic period ran from about 252 to 201 million years ago.

4. Gamma-ray bursts are extremely energetic explosions that have been observed in distant galaxies. They are the most energetic and luminous electromagnetic events in the universe and can last from 10 milliseconds to several hours. The intense radiation is thought to be released during a supernova or super-luminous supernova as a high-mass star implodes to form a neutron star or a black hole, or as a result of the merger of binary neutron stars.

5. Nuclear winter (sometimes referred to as a "nuclear twilight/autumn") is a severe and prolonged global climatic cooling effect that is hypothesized to occur due to major firestorms following a large-scale nuclear war, blanket conventional bombing, or widespread wildfires. The thesis is that such fires can inject soot and other particulates into the stratosphere, where it can block direct sunlight from reaching the surface of the Earth. The resulting cooling would lead to widespread crop failure and famine for months or years, depending on the magnitude of the nuclear conflagration.

6. It is noted in respect to figure 2.2 that the most severe of the nuclear threats occurred during the Cuban missile crisis of 1962, but this threat emerged and was resolved before it could be incorporated into the relevant version of the Doomsday Clock.

7. In my opinion, this definition of genocide should be moderated to allow exceptions that relate to deaths that eventuate as part of the legitimate defense of the members of any of those groups following an unjustified attack. Examples include the bombings of civilian targets in Germany and Japan during the Second World War and the killing of members of an acknowledged terrorist group.

8. In a historical context, communities established by "foreign" parties far away from their places of origin are more commonly referred to as "colonies," this term carrying with it all the undesirable and destructive behaviors and impacts normally implied by its use. In the discussion on attempts to establish an alternative "home" for humanity, provided later in this book, the less evocative or pejorative descriptor of these communities as a "settlement" is preferred because of the absence of the usual historical baggage.

9. Nanobots are machines or robots whose components are at or near the scale of a nanometer (10^{-9} meters). The idea of these machines was first proposed for medical applications in drug and protein delivery.

10. The Human Development Index is a statistical composite index of life expectancy, education, and per capita income indicators, used to rank countries into four tiers of human development.

Notes

11. Carbon dioxide usually behaves as a gas at standard temperature and pressure, or as a solid called "dry ice" when cooled and/or pressurized sufficiently. If the temperature and pressure are both increased to be at or above its critical temperature (31°C) and critical pressure (73 atm), it behaves as a "supercritical fluid" with properties intermediate between those of a liquid and a gas.

12. In this discussion a distinction can be made between the clearly disastrous effect of a runaway increase in temperature due to climate change that occurred on Venus and the many significant increases in temperature that have been experienced throughout the history of the Earth which have remained in relatively stable balance with competing natural forces (figure 2.5). Thus, some authors (Ridley 2010, Rosling 2018, Plimer 2021, and Shellenberger 2023, to name just a few) are much more positive about the future of the world than are the climate change "activists." The more-optimistic authors point to the massive improvements in almost all aspects of human life that have been produced by ready access to cheap and reliable energy throughout modern history, mostly by the oxidation of fossil fuels, and from which huge benefits continue even now. They suggest that human intervention to mitigate and/or adapt to further temperature rise through technological innovation and adjustments to social behavior can prevent any runaway from occurring. However, this optimistic view requires that the increases in global emissions that are produced while this innovation takes place do not push us past a trigger point for runaway warming.

13. Trinitrotoluene, commonly known as TNT, is a chemical compound with the formula $C_6H_2(NO_2)_3CH_3$ that is best known as an explosive material. The explosive yield of a ton of TNT is the standard comparative convention for bombs and asteroid impacts.

14. All the planets in the Solar System *revolve* in an anticlockwise, or prograde, orbit around the Sun, as viewed from the north pole of the ecliptic. A clockwise orbital direction is called "retrograde." With respect to *rotation* direction, all planets except Venus, Uranus, and the minor planet Pluto move in an anticlockwise, "prograde" direction.

15. The Late Heavy Bombardment is an event thought to have occurred approximately 4.1 to 3.8 billion years ago (Bottke and Norman 2017; Mann 2018). During this period, a disproportionately large number of asteroids and comets are presumed to have collided with the early terrestrial planets in the inner Solar System, including Mercury, Venus, Earth, and Mars, arising from both post-accretion and planetary instability.

16. Tephra are rock fragments, ash, and other particles ejected by a volcanic eruption.

17. The Local Cluster is a group of more than 30 galaxies that includes 2 large spiral galaxies—the Milky Way and Andromeda—as well as numerous smaller galaxies, many of which are called "dwarf" galaxies. The Local Cluster has a diameter of approximately 10 million light years, with its gravitational center located between our galaxy and the Andromeda galaxy. It is part of a larger grouping known as the Virgo Supercluster.

18. It is possible that life on the planet may have also used the energy arising from radioactive decay of an abundance of radioactive elements in its crust, so the departure of that planet from its host star may not necessarily signal the extinguishing of this life-form.

19. The Oort Cloud is a spherical cloud of predominantly icy (i.e., water, ammonia, and methane) planetesimals 1 to 3 light years from the Sun that were left over from the formation of the Solar System. This places the outer edge of the cloud at around half the distance to Proxima Centauri, the nearest star to the Sun. The Kuiper Belt is another

reservoir of trans-Neptunian objects that forms a flat disk in the plane of the Solar System, at less than one-thousandth of the Oort Cloud's distance. The outer limit of the Oort Cloud defines the cosmographical boundary of the Solar System and the end of the Sun's gravitational dominance.

20. A main sequence star is one that lies on a roughly diagonal band on a plot of stellar brightness and color, the so-called Hertzsprung-Russell diagram. The band runs from high brightness and blue color to low brightness and red color. After the hydrogen fuel in the core of the star is exhausted, the core collapses, leading to increased nuclear reaction rates and thus a massive increase in luminosity and size. The star is then in its red giant phase of evolution.

Chapter 3: How Urgent Is the Imperative to Search for Other Locations for Humans to Live?

1. Supernovae are key sources of elements heavier than iron in the Periodic Table. Elements from He up to ^{56}iron are produced by nuclear fusion in a main sequence star, while elements heavier than ^{56}iron are created only by nucleosynthesis during a supernova explosion of such a star. These ejected elements enrich the molecular clouds that are the sites of star formation. Thus, each stellar generation has a slightly different composition, going from an almost-pure mixture of hydrogen and helium to a more "metal-rich" composition. The different abundances of elements in the material that forms a star have important influences on the star's life and the possibility of having planets orbiting it, and of those planets then having the necessary elemental composition to enable the evolution and support of life.

2. The amount of CO_2 emitted by the cement industry is nearly 900 kilograms of CO_2 for every 1,000 kilograms of cement produced.

3. Stony, or S-type, asteroids are moderately bright and have a spectral type indicative of a siliceous (i.e., "stony") mineralogical composition. They have relatively high density (around 3.0 g/cm^3) and represent about 17 percent of all asteroids, second only to the carbonaceous, or C-type. They dominate the inner part of the asteroid belt within 2.2 AU, are common in the central belt within about 3 AU, but become rare farther out.

4. See https://en.wikipedia.org/wiki/Near-Earth_object for references to these frequency estimates.

5. Apollo asteroids are near-Earth asteroids orbiting the Sun with highly elliptical orbits ranging from inside the Earth's orbit out to the main asteroid belt between Mars and Jupiter. As of November 2023, the number of known Apollo asteroids is close to 19,000, making them the largest class of near-Earth objects. Their sizes are less than 10 kilometers (1866 Sisyphus is the largest discovered so far, with a diameter of 7 km) and they form most of the population of Earth-crossing and potentially hazardous asteroids.

Conclusions to Part I

1. The maximum speed of the *Apollo 11* spacecraft was about 41,000 km/h, but the speed of the craft gradually slowed to a value of around 3,600 km/h at the gravitational midpoint between Earth and the Moon (about 38,000 kilometers from the Moon), after which the spacecraft sped up again until it reached lunar orbit. Based on the total time elapsed between launch and lunar orbit (three days), the average speed of *Apollo 11* was around 5,400 km/h,

which would have produced a transit time to Proxima Centauri of about 840,000 years. On longer spaceflights, when the launched-from and target bodies are much farther away from each other than the Earth and the Moon, much more of the transit time is taken up "coasting" in deep space with little reduction in the starting speed. Furthermore, when the target body has a similar or higher mass than the launched-from body, the influence of the acceleration imparted by the target body manifests itself sooner, and is stronger, than that of the Earth-Moon system, thereby increasing the average speed of the craft.

Chapter 4: The Record of Space Travel So Far

1. Two Space Shuttle flights ended in failure, with the deaths of the entire seven-member crews: *Challenger* in 1986 and *Columbia* in 2003. In addition, the three crew members of the *Apollo 1* mission died on the launch pad during a test in 1967, but the craft never flew.

2. The CapCom for NASA spaceflights was an astronaut assigned to be the sole communicator with the mission astronauts during any particular flight.

3. A Hohmann transfer orbit (HTO) is an elliptical trajectory between orbiting celestial bodies (planets, moons etc.) that is tangential to the initial and target orbits. An HTO requires the starting and destination points be at particular locations/alignments in their orbits relative to each other. For a flight between Earth and Mars, these alignments, or launch windows, occur every 26 months and the travel time is about 9 months. If the HTO is performed between orbits close to celestial bodies with significant gravitation, much less onboard fuel is required. This process takes advantage of the Oberth effect, wherein the rocket engine is fired during the higher speeds generated close to the body (ideally at "periapsis," the closest point in the orbit to that body), thereby providing a greater change in kinetic energy than a firing of the same rockets at lower speeds. The application of an HTO in the case of the Earth-Moon system is called a "lunar transfer orbit" and has been used for all spacecraft flights to the Moon.

The Oberth effect is not to be confused with a "gravitational slingshot," "gravity assist maneuver," or "swing-by," which requires no rocket firing and uses only the relative orbital movement and gravity of a planet or other astronomical object to alter the path and speed of a spacecraft. The "assist" is provided by the motion of the gravitating body as it pulls on the spacecraft. Any gain or loss of kinetic energy and velocity by a passing spacecraft is correspondingly lost or gained by the gravitational body. The gravity assist maneuver was first used in 1959 when the Soviet probe *Luna 3* photographed the far side of Earth's Moon, and it has been used by interplanetary probes from *Mariner 10* onward, including the flybys of Jupiter, Saturn, and Uranus by one or the other or both of the two *Voyager* probes. Using this procedure, the *Voyager 2* spacecraft was able to shorten its journey to Neptune from a "normal" 30 years down to just 12 years without using any additional fuel.

Chapter 5: Options for Establishing a Permanent Settlement in Earth's Immediate Environment

1. A low Earth orbit (LEO) is defined as an orbit around Earth with a period of 128 minutes or less and an eccentricity less than 0.25. Most of the artificial objects in outer space are in LEO, the region below an altitude of 2,000 kilometers, or about one-third of Earth's radius. Objects in orbits that pass through this zone, even if they have an apogee (their

farthest distance from the Earth) farther out or are suborbital in nature, are carefully tracked since they present a collision risk to the many LEO satellites. All crewed space stations to date have been within LEO. From 1968 to 1972, the Apollo program's lunar missions sent humans beyond LEO, but since then, no human spaceflights have ventured beyond LEO.

2. Lagrange points are positions in space where an object under the gravitational influence of two much larger bodies can remain in a semi-stable orbital state around that point due to the approximate equality of its centripetal force and the gravitational pull of the other two. They can be used by spacecraft as "parking spots" in space since they can be maintained there with minimal fuel requirements. They are sometimes called "libration points" because an object occupying one of these positions in space appears to oscillate, jiggle, or librate around the mean position when viewed from another location. They are named after the mathematician Joseph-Louis Lagrange, who first described their mathematical characteristics in 1772. Further details are provided in later sections of this chapter.

3. A magnetotail is the region of the magnetosphere of a celestial body (such as a planet) that is swept back by the solar wind in the direction away from the Sun, akin to the plasma trail of a comet.

4. The cosmic microwave radiation (CMB) is the residual radiation left over from the Big Bang, the time when the universe began. It was discovered in 1965 by Arno Penzias and Robert Wilson from Bell Telephone Laboratories, who were awarded the 1978 Nobel Prize in Physics for this work. Studies of the magnitude and fluctuations of the CMB help scientists learn how the early universe was formed and understand what caused its overall structure.

Chapter 6: Options for Settlement Beyond the Earth-Moon Setting but Within the Solar System

1. The compact audio-visual discs were created by the Jet Propulsion Laboratory with their contents selected by a committee chaired by Carl Sagan of Cornell University. They carry photographs of the Earth and its life-forms, a range of scientific information, spoken greetings from such people as the secretary-general of the United Nations and the president of the United States, and a collection of the "sounds of Earth," including whales, a baby crying, waves breaking on a shore, and a collection of musical works by Mozart, Blind Willie Johnson, Chuck Berry ("Johnny B. Goode"), and Valya Balkanska. Other Eastern and Western classics are included, as well as various performances of indigenous music from around the world. The record also contains greetings in 50 different languages.

2. This leads to a surface pressure that is well below the so-called Armstrong limit of 6.2 percent of Earth's surface pressure, below which water boils at the normal temperature of the human body and is therefore fatal.

3. For more information on Ceres, see http://www.universetoday.com/26587/life-on-ceres-could-the-dwarf-planet-be-the-root-of-panspermia/#ixzz33jXLHGDn.

Chapter 7: Options for Settlement Outside Our Solar System

1. The Rare Earth hypothesis maintains that the origin of life and biological complexity (i.e., multicellular organisms, sexual reproduction, human intelligence) requires an improb-

able combination of astrophysical and geological events and circumstances (see chapter 1 for details). As a result, it is concluded that complex extraterrestrial life is improbable and thus likely to be rare throughout the universe.

2. Stellar spectral types reflect the classification code assigned to stars based on their electromagnetic radiation spectral characteristics. This code primarily summarizes the ionization state of the star's photosphere and thus its temperature. Most stars are currently classified under the Morgan-Keenan (MK) system using the letters O, B, A, F, G, K, and M, a series ranging from the hottest (O type) to the coolest (M type). Each letter class is then subdivided using a numeric digit with 0 being hottest and 9 being the coolest. The sequence has been expanded with classes for other stars and starlike objects that do not fit in the classical system, such as class D for white dwarfs and classes S and C for carbon stars, and also designations that indicate luminosity.

3. Breakthrough Starshot is part of a multi-initiative, science-based program founded in 2016 by Yuri Milner, Stephen Hawking, and Mark Zuckerberg and funded by Julia and Yuri Milner to search for extraterrestrial intelligence. Breakthrough Listen will search for artificial radio or laser signals, Breakthrough Message will create a message "representative of humanity and planet Earth," and Breakthrough Watch will identify and characterize Earth-size, rocky planets around stars within 20 light years of Earth.

4. The Planetary Society is a non-government, nonprofit organization founded in 1980 by Carl Sagan, Louis Friedman, and Bruce Murray to provide members of the general public with an active role in advancing space exploration. Now led by CEO Bill Nye, it is the world's largest and most influential private space global community with more than 2 million members.

5. A cubesat is a class of miniaturized satellite with a size generally in multiples of 10-centimeter cubes and a mass of less than 2 kilograms per unit, often using commercial off-the-shelf components for their electronics. In the case of *LightSail 2*, the satellite had extendable "sails" measuring 5.6 meters on each edge to provide the surface used to harness photon pressure for propulsion.

6. Red dwarfs are the smallest, coolest, and most common (perhaps up to 75 percent) kind of star in the Milky Way, at least in the neighborhood of the Sun. They are very-low-mass stars with a low nuclear fusion rate, a low temperature, and very long lives, sometimes far longer than the present age of the universe.

7. A flare star is one that can undergo unpredictable large increases in brightness during events that last for a few minutes. The brightness increase is across the spectrum, from X-rays to radio waves, and if sufficiently large, can be a major danger/inconvenience to life on nearby planets (as do the relatively benign flares from our own Sun). Most flare stars are red dwarfs.

8. For example, red dwarfs (spectral class M) are the coolest main-sequence stars, similar to the Sun (spectral class G), but with a surface temperature of about 1,770 to 3,230°C, compared to the Sun's temperature of around 5,600°C.

Chapter 8: The Tyranny of Distance and Time

1. While average life expectancy has been increasing significantly over the past 200 or so years due to advances in medical technology, better nutrition, etc., the maximum lifes-

pan (i.e., the maximum recorded age at death) has increased only a little, if at all, over that period (Weon and Je 2009).

2. An event horizon is usually associated with the outer boundary of a black hole, loosely considered to be its "surface." It is the point, according to NASA, that the gravitational influence of the black hole becomes so great that not even light can escape it. It is also used to describe a "point of no return," as in the case of communication described here.

3. Causality is influence by which one event, process, state, or object (a cause) contributes to the production of another event, process, state, or object (an effect). Thus, a common objection to the idea of traveling back in time is the so-called grandfather paradox in which if one could go back in time and the time traveler were to change anything (such as to kill their grandfather), there would arise a contradiction in that the past would become inconsistent with the present (namely, the birth of the time traveler). Some have sought to answer this paradox by arguing that backward time travel could be possible if natural laws prevented the traveler from changing anything in the past such as to cause this paradox. It might also be possible for the time traveler to kill their grandfather if this action triggered a move into a parallel quantum universe in which the grandfather was indeed killed.

Chapter 9: Physical, Social, and Psychological Challenges of Space Travel

1. A cyborg can be described as a living being whose powers are enhanced by inorganic mechanical body parts and/or computer software implants.

2. Metal shielding is commonly used for protection against solar radiation, but it provides little defense against cosmic rays. To deal with these much more penetrating rays, hydrogen-rich materials like water, special plastics, and hydrogen fuel tanks can be used on spacecraft, or if the humans are on land, underground locations are also suitable.

3. Indeed, Cook met his end in Hawaii in February 1779, during his third and final voyage of discovery in the Pacific Ocean. He had returned to Hawaii for repairs after failing to find the much-vaunted Northwest Passage, believed to link the Atlantic and Pacific Oceans. Having worn out his welcome during a previous stay on the island, he was stabbed to death by the local inhabitants during an altercation associated with their theft of one of the ship's boats.

4. The Coriolis effect is a force (the Coriolis force) experienced by a mass moving in a rotating system. It acts perpendicular to the direction of motion and to the axis of rotation and is most commonly associated with differences in wind deflection in the northern and southern hemispheres of the Earth, and discussions about the swirl direction of water exiting a bathtub.

Chapter 10: Spacecraft Options for Human Transportation

1. Assuming that a temporary or permanent off-Earth platform had been established to accommodate the bulk of the "refugees," similar challenges would present themselves in selecting a potentially small number of humans to remain on Earth in a heavily protected (paraterraformation) environment for the lengthy period required to ride out the impact and/or to find a remedy for the impact of climate change, lethal virus, nuclear holocaust, or whatever other disaster has caused the evacuation.

Notes

Chapter 11: Options for Spacecraft Propulsion Technology

1. Jeff Bezos is the founder, executive chairman, and former president and CEO of Amazon, the world's largest e-commerce and cloud computing company. In 2000 he founded the aerospace manufacturer and suborbital spaceflight services company Blue Origin.

2. TLR in the key for figure 11.1 is the abbreviation for Technology Level of Readiness, referenced on a scale of 1 to 9, where 1 is a completely theoretical concept and 9 is full commercialization.

3. NASA's *Parker Solo Probe* spacecraft is the fastest man-made object ever recorded with a maximum speed of 692,000 km/h (191 km/s) in its very close flyby of the Sun in 2024. The maximum speed for missions into the outer Solar System was achieved by the *Juno* spacecraft in 2016, at roughly 365,000 km/h (101 km/s) as it approached Jupiter. The fastest launch velocity belongs to *New Horizons*, which reached 58,000 km/h (16.1 km/s).

4. For example, a typical spacecraft going to Mars experiences a trajectory displacement of thousands of kilometers from solar pressure, and its similar effect on orientation must be included in spacecraft design.

5. One newton (N) is the force needed to accelerate 1 kilogram of mass at the rate (acceleration) of 1 meter per second per second. As a reference point, the three Space Shuttle main engines could produce a thrust of 1.8M N.

6. Dark energy has been estimated to represent around 69 percent of all "matter" in the universe (noting the equivalence of matter and energy in Einstein's theory of relativity), whereas "dark matter" makes up about 26 percent, and the ("baryon") matter of which humans and everything tangible around them are made constitutes around 5 percent.

7. Antimatter is matter that is composed of the antiparticles (or "partners") of the corresponding particles in "ordinary" matter. The two forms of matter have reversed charge, parity, and magnetic moment. Antimatter occurs in cosmic ray collisions, in some types of radioactive decay, and in some particle accelerator experiments.

8. The equation $E=mc^2$ describes the equivalence between mass (m) and energy (E) in a system's rest frame, where the constant (c) is the speed of light. Since c is a large number (around 300,000 kilometers per second), the formula indicates that a small amount of mass corresponds to a very large amount of energy.

9. The Casimir effect (Casimir and Polder 1948) is a physical force acting on the boundaries of a very small, confined space which arises from the different pressure generated inside and outside the cavity by quantum fluctuations (the appearance and disappearance) of virtual particles. The effect falls off rapidly with the increased size of the cavity; at separations of 10 nm—about 100 times the typical size of an atom—the Casimir effect produces the equivalent of about 1 atmosphere of pressure.

Chapter 13: Humans or Machines?

1. The meaning of "advanced machines" is taken here to be inorganic bodies (or robots) capable of undertaking physical and intellectual tasks at levels close to, the same as, or perhaps someday higher than humans.

References

Adhikari, Mohamed. 2021. In *Civilian-Driven Violence and the Genocide of Indigenous Peoples in Settler Societies*, edited by Mohamed Adhikari. New York: Routledge.
Agnew, Matt. 2022. *Dr Matt's Guide to Life in Space*. Sydney: Allen & Unwin.
Alcubierre, Miguel. 1994. "The Warp Drive: Hyper-Fast Travel within General Relativity." *Classical and Quantum Gravity* 11, no. 5: L73–77.
Aldrin, Buzz. 2009. *Magnificent Desolation: The Long Journey Home from the Moon*. London: Bloomsbury.
Alfrey, C. P., M. M. Udden, C. L. Huntoon, and T. Driscoll. 1996. "Destruction of Newly Released Red Blood Cells in Space Flight." *Medicine & Science in Sports & Exercise* 28 (10 Suppl): S42–44.
Altermann, Wladyslaw, and Daniele L. Pinti. 2021. "Apex Chert, Microfossils." In *Encyclopedia of Astrobiology*, edited by Murial Gargaud, et al. Berlin and Heidelberg: Springer. https://link.springer.com/referenceworkentry/10.1007/978-3-642-27833-4_1866-7.
Andrews, Dana G., and Robert M. Zubrin. 1998. "Magnetic Sails and Interstellar Travel." 39th International Astronautical Congress, Bangalore, October 8–15, 1998. Paper IAF-88-533.
Anglada-Escudé, G., P. J. Amado, J. Barnes, et al. 2016. "A Terrestrial Planet Candidate in a Temperate Orbit around Proxima Centauri." *Nature* 536, no. 7617: 437–40.
Armstrong McKay, D. I., A. Staal, J. F. Abrams, R. Winkelmann, B. Sakschewski, S. Loriani, I. Fetzer, et al. 2022. "Exceeding 1.5°C Global Warming Could Trigger Multiple Climate Tipping Points." *Science* 377, no. 6611.
Atkins, S., M. Taylor, M. McAdam, R. Morrison, and J. Feldman. 2022. "Governance in Outer Space: The Case for a New Global Order." Norton Rose Fulbright, November 2022. https://www.nortonrosefulbright.com/en/knowledge/publications/e8862684/governance-in-outer-space-the-case-for-a-new-global-order.
Ayres, Philip. 1999. *Mawson: A Life*. Melbourne: Melbourne University Press.
Bahr, Jeff, 2009. *Amazing and Unusual USA: Hundreds of Extraordinary Sights*. Lincolnwood, IL: Publications International.
Ballesteros, F. J., A. Fernandez-Soto, and V. J. Martinez. 2019. "Diving into Exoplanets: Are Water Seas the Most Common?" *Astrobiology* 19, no. 5: 642–54.
Balter, Michael. 2012. "100,000 Years of Dramatic Population Changes." *Discover Magazine*, October 17, 2012. https://www.discovermagazine.com/the-sciences/100-000-years-of-dramatic-population-changes.

References

Baran, R., M. Wehland, H. Schulz, M. Heer, M. Infanger, and D. Grimm. 2022. "Microgravity-Related Changes in Bone Density and Treatment Options: A Systematic Review." *International Journal of Molecular Sciences* 23, no. 15: 8650.

Barnes, Rory. 2017. "Tidal Locking of Habitable Exoplanets." *Celestial Mechanics and Dynamical Astronomy* 129, no. 4: 509–36.

Basner, M., D. F. Dinges, D. Mollicone, A. Ecker, C. W. Jones, E. C. Hyder, A. Di Antonio, et al. 2013. "Mars 520-d Mission Simulation Reveals Protracted Crew Hypokinesis and Alterations of Sleep Duration and Timing." *Proceedings of the National Academy of Sciences* 110, no. 7: 2635–40.

Bates, D. R. 1972. "Communication with Galactic Civilizations." *Physics Bulletin* 23, no. 1: 26–29.

Baum, Seth D. 2010. "Is Humanity Doomed? Insights from Astrobiology." *Sustainability* 2, no. 2: 591–603.

Baust, John G., Dayong Gao, and John M. Baust. 2009. "Cryopreservation: An Emerging Paradigm Change." *Organogenesis* 5, no. 3: 90–96.

Binzel, Richard P. 2000. "Torino Impact Hazard Scale." *Planetary and Space Science* 48, no. 4: 297–303.

Blagonravov, A. A., ed. 1954. *Collected Works of K. E. Tsiolkovskiy*. Vol. 2, *Reactive Flying Machines*. Moscow: Akademii Nauk SSSR.

Bohan, Elise. 2022. *Future Superhuman: Our Transhuman Lives in a Make-or-Break Century*. Sydney: New South Publishing.

Bolonkin, Alexander. 2011. *Universe, Human Immortality and Future Human Evaluation*. Amsterdam: Elsevier.

Bostrom, Nick. 2008. "Where Are They? Why I Hope the Search for Extraterrestrial Life Finds Nothing." *MIT Technology Review*, May/June 2008: 72–77.

———. 2014. *Superintelligence: Paths Dangers, Strategies*. Oxford: Oxford University Press.

Bottke, William F., R. Jedicke, A. Morbidelli, J. M. Petit, and B. Gladman. 2000. "Understanding the Distribution of Near-Earth Asteroids." *Science* 288, no. 5474: 2190–94.

Bottke, William F., and Marc D. Norman. 2017. "The Late Heavy Bombardment." *Annual Review of Earth and Planetary Sciences* 45, no. 1: 619–47.

Boyd, Robert, and Peter J. Richerson. 2009. "Culture and the Evolution of Human Cooperation." *Philosophical Transactions of the Royal Society B* 364: 3281–88.

Bracewell, R. N. 1960. "Communications from Superior Galactic Communities." *Nature* 186, no. 4726: 670–71.

Breitman, Daniela. 2017. "Wow! Signal Explained After 40 Years?" EarthSky, June 7, 2017. https://earthsky.org/space/wow-signal-explained-comets-antonio-paris.

Brennan, Pat. 2023. "Life on Other Planets: What Is Life and What Does It Need?" NASA, June 20, 2023. https://exoplanets.nasa.gov/news/1762/life-on-other-planets-what-is-life-and-what-does-it-need.

Broad, William J. 1991. "As Biosphere Is Sealed, Its Patron Reflects on Life." *New York Times*, August 24, 1991.

Bruhaug, Gerrit, and William Phillips. 2021. "Nuclear Fuel Resources of the Moon: A Broad Analysis of Future Lunar Nuclear Fuel Utilization." *NSS Space Settlement Journal*, issue 5 (June 2021). https://arc.nss.org/wp-content/uploads/NSS-JOURNAL-Nuclear-Fuel-Resources-of-the-Moon-2021-June.pdf.

References

Bryan, Scott E. 2010. "The Largest Volcanic Eruptions on Earth." *Earth-Science Reviews* 102, no. 3–4: 207–29.

Buchhave, L. A., D. W. Latham, A. Johansen, M. Bizzarro, G. Torres, J. F. Rowe, N. M. Batalha, et al. 2012. "An Abundance of Small Exoplanets around Stars with a Wide Range of Metallicities." *Nature* 486, no. 7403: 375–77.

Buckley, Jay C., Jr. 2006. *Space Physiology*. Oxford: Oxford University Press.

Budyko, M. I. 1969. "The Effect of Solar Radiation Variations on the Climate of the Earth." *Tellus* 21, no. 5: 611–19.

Bussard, Robert W. 1960. "Galactic Matter and Interstellar Flight." *Astronautica Acta* 6: 179–95.

Bussey, D. B. J., K. Fristad, P. Schenk, M. S. Robinson, and P. D. Spudis. 2004. "Constant Sunlight at the Lunar North Pole." *Meteoritics and Planetary Science* 39 (Suppl) A20.

Bussey D. B. J., M. S. Robinson, K. Fristad, and P. D. Spudis. 1999. "Permanent Sunlight at the Lunar North Pole." *Geophysical Research Letters* 26, no. 9: 1187–90.

Canning, W., and R. J. Hill. 2008. "Beyond the Planet." Alfred Deakin Innovation Lectures, Melbourne.

Caroti, Simone. 2011. *The Generation Starship in Science Fiction: A Critical History, 1934–2001*. Jefferson, NC: McFarland.

Casimir, Hendrik B. G., and Dirk Polder. 1948. "The Influence of Retardation on the London–van der Waals Forcesm. *Physical Review* 73, no. 4: 360–72.

Chancellor, J. C., G. B. Scott, and J. P. Sutton. 2014. "Space Radiation: The Number One Risk to Astronaut Health beyond Low Earth Orbit." *Life* (Basel) 4, no. 3: 491–510.

Chesley, S. R., P. W. Chodas, A. Milani, and G. B. Valsecchi. 2002. "Quantifying the Risk Posed by Potential Earth Impacts." *Icarus* 159: 423–32.

Choukér, A., T. J. Ngo-Anh, R. Biesbroek, G. Heldmaier, M. Heppener, and J. Bereiter-Hahn. 2021. "European Space Agency's Hibernation (Torpor) Strategy for Deep Space Missions: Linking Biology to Engineering." *Neuroscience & Biobehavioral Reviews* 131: 618–26.

Clark, Stuart. 2002. "Acidic Clouds of Venus Could Harbour Life." NewScientist, September 26, 2002. https://www.newscientist.com/article/dn2843-acidic-clouds-of-venus-could-harbour-life.

Cleaves, H. J., J. H. Chalmers, A. Lazcano, S. L. Miller, and J. L. Bada. 2008. "A Reassessment of Prebiotic Organic Synthesis in Neutral Planetary Atmospheres." *Origins of Life and Evolution of Biospheres* 38, no. 2: 105–15.

Cockell, C. S., et al. 2016. "Habitability: A Review." *Astrobiology* 16, no. 1: 89–117.

Cohen, Joshua, and Joel Rogers. 1995. *Associations and Democracy: The Real Utopias Project*. Vol. 1. London: Verso Trade.

Cole, Dandridge M., and Donald W. Cox. 1964. *Islands in Space: The Challenge of the Planetoids*. Philadelphia: Chilton Books.

Collins, Michael. 2019. *Carrying the Fire: An Astronaut's Journeys*. New York: Farrar, Straus & Giroux.

Criscuolo, François, Cédric Sueur, and Audrey Bergouignan. 2020. "Human Adaptation to Deep Space." *Frontiers of Public Health* 8, Article 119.

Crowl, Adam. 2012. "Embryo Space Colonisation to Overcome the Interstellar Time Distance Bottleneck." *Journal of the British Interplanetary Society* 65: 283–85.

References

Crowl, Adam, K. F. Long, and R. K. Obousy. 2012. "The Enzmann Starship: History and Engineering Appraisal." *Journal of the British Interplanetary Society* 65: 185–99.

Cumming, A., R. P. Butler, G. W. Marcy, S. S. Vogt, J. T. Wright, and D. A. Fischer. 2008. "The Keck Planet Search: Detectability and the Minimum Mass and Orbital Period Distribution of Extrasolar Planets." *Publications of the Astronomical Society of the Pacific* 120, no. 867: 531–54.

Darimont, C .T., C. H. Fox, H. M. Bryan, and T. E. Reimchen. 2015. "The Unique Ecology of Human Predators." *Science* 349, no. 6250: 858–60.

Dartnell, L. R., T. A. Nordheim, M. R. Patel, and J. P. Mason. 2015. "Constraints on a Potential Aerial Biosphere on Venus: I. Cosmic Rays." *Icarus* 257: 396–405.

Darwin, Charles. 2009. *On the Origin of Species by Means of Natural Selection; or The Preservation of Favoured Races in the Struggle for Life*. London: Penguin Classics. First published 1859 by John Murray (London).

Dembitzer, J., R. Barkai, M. Ben-Dor, and S. Meira. 2022. "Levantine Overkill: 1.5 Million Years of Hunting Down the Body Size Distribution." *Quaternary Science Reviews* 276: 107316.

Des Marais, D. J., J. A. Nuth, L. J. Allamandola, A. P. Boss, J. D. Farmer, T. M. Hoehler, B. M. Jakosky, et al. 2008. "The NASA Astrobiology Roadmap." *Astrobiology* 8: 715–30.

Diamond, Jared. 2005. *Collapse: How Societies Choose to Fail or Succeed*. London: Viking Penguin.

Doherty, Peter. 2013. *Pandemics: What Everyone Needs to Know*. Oxford: Oxford University Press.

Drake, F. D. 1961. "Project Ozma." *Physics Today* 14, no. 4: 40–46.

Drexler, K. Eric. 1986. *Engines of Creation: The Coming Era of Nanotechnology*. New York: Anchor Press/Doubleday.

Dyson, Freeman J. 1960. "Search for Artificial Stellar Sources of Infrared Radiation." *Science* 131, no. 3414: 1667–68.

———. 1979. *Disturbing the Universe*. New York: Harper and Row.

Dyson, George. 2002. *Project Orion: The True Story of the Atomic Spaceship*. New York: Henry Holt and Company.

Einstein, Albert, and Nathan Rosen. 1935. "The Particle Problem in the General Theory of Relativity." *Physical Re*view 48: 73–77.

Everett, C. J., and S. M. Ulam. 1955. "On a Method of Propulsion of Projectiles by Means of External Nuclear Explosions: Part I." Los Alamos Scientific Laboratory of the University of California, August 1955.

Faria, J. P., A. S. Mascareño, P. Figueira, A. M. Silva, M. Damasso, O. Demangeon, F. Pepe, et al. 2022. "A Candidate Short-Period Sub-Earth Orbiting Proxima Centauri." *Astronomy & Astrophysics* 658, A115.

Fennell, Paul, Justin Driver, Christopher Bataille, and Steven J. Davis. 2022. "Cement and Steel—Nine Steps to Net Zero." *Nature* 603: 574–77.

Fields, B. D., A. L. Melott, J. Ellis, A. F. Ertel, B. J. Fry, B. S. Lieberman, Z. Liu, et al. 2020. "Supernova Triggers for End-Devonian Extinctions." *Proceedings of the National Academy of Science* 117, no. 35: 21008–10.

Fogg, Martyn J. 1995. *Terraforming: Engineering Planetary Environments*. Warrendale, PA: Society of Automotive Engineers.

References

Foster, Gavin L., Dana L. Royer, and Daniel J. Lunt. 2017. "Future Climate Forcing Potentially without Precedent in the Last 420 Million Years." *Nature Communications* 8: 14845.

Freitas, Robert A. 1980. "A Self-Reproducing Interstellar Probe." *Journal of the British Interplanetary Society* 33: 251–64.

Freitas, Robert. A., and Ralph C. Merkle. 2004. *Kinematic Self-Replicating Machines*. Georgetown, TX: Landes Bioscience.

Garrett-Bakelman, F. E., M. Darshi, S. J. Green, R. C. Gur, L. Lin, B. R. Macias, M. J. McKenna, et al. 2019. "The NASA Twins Study: A Multidimensional Analysis of a Year-Long Human Spaceflight." *Science* 364, no. 6436: eaau8650.

Ge, Yong, and Xing Gao. 2020. "Understanding the Overestimated Impact of the Toba Volcanic Super-eruption on Global Environments and Ancient Hominins." *Quaternary International: Current Research on Prehistoric Central Asia* 559: 24–33.

Gelling, Cristy. 2013. "Atom & Cosmos: Mars Trip Would Mean Big Radiation Dose: Curiosity Instrument Confirms Expectation of Major Exposures." *Science News* 183, no. 13: 8.

Genta, Giancarlo. 2001. *Propulsion for Interstellar Space Exploration*. COSPAR Colloquia Series 11: 421–30.

Gildenberg, B. D. 2003. "A Roswell Requiem." *Skeptic* 10, no. 1: 60.

Gilster, Paul. 2012. "ESO: Habitable Red Dwarf Planets Abundant." Centauri Dreams, March 29, 2012. https://www.centauri-dreams.org/2012/03/29/eso-habitable-red-dwarf-planets-abundant.

Gleiser, Marcelo. 2021. "The Mediocrity of the Mediocrity Principle (for Life in the Universe)." Big Think, October 6, 2021. https://bigthink.com/13-8/mediocrity-principle-life.

Goldblatt, Colin, and Andrew J. Watson. 2021. "The Runaway Greenhouse: Implications for Future Climate Change, Geoengineering and Planetary Atmospheres." *Philosophical Transactions of the Royal Society A: Mathematical, Physical and Engineering Sciences* 370, no. 1974: 4197–216.

Goldsmith, Donald, and Martin J. Rees. 2022. *The End of Astronauts: Why Robots Are the Future of Exploration*. Cambridge, MA: Belknap Press of Harvard University Press.

Greaves, J. S., A. M. S. Richards, W. Bains, P. B. Rimmer, H. Sagawa, D. L. Clements, S. Seager, et al. 2020. "Phosphine Gas in the Cloud Decks of Venus." *Nature Astronomy* 5, no. 7: 655–64.

Green, T. 2024. "Don't Believe the Doom and Gloom: AI Can be a Force for Good in a World Where You Can No Longer Trust Anything You See or Hear." *The Australian*, January 8, 2024.

Guivarch, C., E. Kriegler, J. Portugal-Pereira, V. Bosetti, J. Edmonds, M. Fischedick, P. Havlík, et al., eds. "IPCC, 2022: Annex III: Scenarios and Modelling Methods." In *IPCC, 2022: Climate Change 2022: Mitigation of Climate Change. Contribution of Working Group III to the Sixth Assessment Report of the Intergovernmental Panel on Climate Change*. Cambridge and New York: Cambridge University Press.

Hadžić, Jasna. 2023. "Homo Sapiens Is #9. Who Were the Eight Other Human Species?" Big Think, August 7, 2023. https://bigthink.com/the-past/other-human-species.

References

Hanson, Robin. 1998. "The Great Filter—Are We Almost Past It?" George Mason University, September 15, 1998. https://mason.gmu.edu/~rhanson/greatfilter.html.

Harari, Yuval N. 2015. *Homo Deus: A Brief History of Tomorrow.* London: Harvill Secker.

Hardin, Garrett. 1968. "The Tragedy of the Commons." *Science* 162, no. 3859: 1243–48.

Harrison, Edward R. 1981. *Cosmology: The Science of the Universe.* Cambridge: Cambridge University Press.

Hawking, Stephen. 2016. "The Best or Worst Thing to Happen to Humanity?" Speech at the launch of the Centre for the Future of Intelligence, Cambridge University, October 19, 2016. https://www.cam.ac.uk/research/news/the-best-or-worst-thing-to-happen-to-humanity-stephen-hawking-launches-centre-for-the-future-of.

Hein, A. M., M. Pak, M., D. Pütz, C. Bühler, and P. Reiss. 2012. "World Ships—Architectures & Feasibility Revisited." *Journal of the British Interplanetary Society* 65, no. 4: 119.

Held, Isaac M., and Brian J. Soden. 2000. "Water Vapor Feedback and Global Warming." *Annual Review of Energy and the Environment* 25: 441–75.

Hewer, Mariko, and Scott Sleek. 2018. "Teams in Space: It Isn't Just Rocket Science." *APS Observer*, October 31, 2018. https://www.psychologicalscience.org/observer/teams-in-space-it-isnt-just-rocket-science.

Hill, Roderick J. 2020. "The Lagrangian Points in Astronomy." *Bulletin of the Astronomical Society of South Australia* 129, no. 11: 7–8.

———. 2022. "Gravitational Clearing of Natural Satellite Orbits." *Publications of the Astronomical Society of South Australia* 39: e006.

Hu, W., Z. Hao, P. Du, F. Di Vincenzo, G. Manzi, J. Cui, Y-X. Fu, et al. 2023. "Genomic Inference of a Severe Human Bottleneck during the Early to Middle Pleistocene Transition." *Science* 381, no. 6661: 979–84.

Jessica F. 2016. "Stephen Hawking, Mark Zuckerberg, Yuri Milner Launch $100M Space Project Called Breakthrough Starshot." Nature World News, April 14, 2016. https://www.natureworldnews.com/articles/20799/20160414/stephen-hawking-mark-zuckerberg-and-russian-millionaire-yuri-milner-launch-100m-space-project-called-breakthrough-starshot.htm.

Joosten, B. Kent, and Harold G. White. "Spacecraft Propulsion and Power Report." Report Number JSC-CN-32129, Johnson Space Center, Houston, TX.

Joy, Bill. 2000. "Why the Future Doesn't Need Us." Wired, April 2000. https://www.wired.com/2000/04/joy-2.

Kaku, Michio. 2018. *The Future of Humanity: Terraforming Mars, Interstellar Travel, Immortality and Our Destiny Beyond Earth.* New York: Penguin Books.

Kanas, Nick. 2023. *Behavioral Health and Human Interactions in Space.* Cham, Switzerland: Springer.

Kanas, Nick, V. Salnitskiy, E. M. Grund, V. Gushin, D. S. Weiss, O. Kozerenko, A. Sled, and C. R. Marmar. 2000. "Interpersonal and Cultural Issues Involving Crews and Ground Personnel during Shuttle/Mir Space Missions." *Aviation, Space and Environmental Medicine* 71, no. 9: A11-6.

Kanas, Nick, V. Salnitskiy, J. E. Boyd, V. I. Gushin, D. S. Weiss, S. A. Saylor, O. P. Kozerenko, and C. R. Marmar. 2007. "Crew Member and Mission Control Personnel

References

Interactions during International Station Missions." *Aviation, Space and Environmental Medicine* 78, no. 6: 601–7.

Kaneda, Toshiko, and Carl Haub. 2022. "How Many People Have Ever Lived on Earth?" Population Reference Bureau, November 15, 2022. https://www.prb.org/articles/how-many-people-have-ever-lived-on-earth.

Kardashev, N. S. 1964. "Transmission of Information by Extraterrestrial Civilizations." *Soviet Astronomy* 8, no. 217: 282–87.

Karow, Armand M., Jr., and Watts R. Webb. 1965. "Tissue Freezing: A Theory for Injury and Survival." *Cryobiology* 2, no. 3: 99–108.

Kasting, J. F. 1993. "Earth's Early Atmosphere." *Science* 259, no. 5097: 920–26.

———. 2014. "Modeling the Archean Atmosphere and Climate." In *Treatise on Geochemistry*, 2nd ed., edited by Heinrich D. Holland and Karl K. Turekian, 157–75. Amsterdam: Elsevier.

Kawaguchi, Y., M. Shibuya, I. Kinoshita, J. Yatabe, I. Narumi, H. Shibata, R. Hayashi, et al. 2020. "DNA Damage and Survival Time Course of Deinococcal Cell Pellets During 3 Years of Exposure to Outer Space," *Frontiers in Microbiology* 11: 2050.

Keller, Gerta. 2014. "Deccan Volcanism, the Chicxulub Impact, and the End-Cretaceous Mass Extinction: Coincidence? Cause and Effect?" In *Special Paper 505: Volcanism, Impacts, and Mass Extinctions: Causes and Effects*, edited by Gerta Keller and Andrew C. Kerr, 57–89. Boulder, CO: Geological Society of America.

Kelly, Scott. 2017. *Endurance: A Year in Space, a Lifetime of Discovery*. New York: Alfred A. Knopf.

Kennedy, Andrew. 2006. "Interstellar Travel—The Wait Calculation and the Incentive Trap of Progress." *Journal of the British Interplanetary Society* 59: 239–46.

Kepler, Johannes. 1604. "De cometis libelli tres." In *Ad vitellionem paralipomena, quibus astronomiae pars optica traditur . . . de modo visionis, & humorum oculi usu, contra opticos & anatomicos*. Frankfurt: C. Marnius & Heirs of J. Aubrius, 1604.

Kerr, Richard A. 1978. "Climate Control: How Large a Role for Orbital Variations?" *Science* 201, no. 4351: 144–46.

Koch, Christof, and Giulio Tononi. 2008. "Can Machines be Conscious?" *IEEE Spectrum* 45, no. 6: 55–59.

Koonin, E. V. 2007. "The Biological Big Bang Model for the Major Transitions in Evolution." *Biology Direct* 2: 21.

Krasnikov, Serguei. 1995. "Hyperfast Interstellar Travel in General Relativity." *Physical. Review D* 57: 4760–66.

Krueger, David. 2023. "Why Do some AI Researchers Dismiss the Potential Risks to Humanity?" *New Scientist*, April 19, 2023.

Large, Ross, and John Long. 2015. "Plate Tectonics May Have Driven the Evolution of Life on Earth." The Conversation, July 15, 2015. https://theconversation.com/plate-tectonics-may-have-driven-the-evolution-of-life-on-earth-44571.

Lawton, John H., and Robert M. May. 1995. *Extinction Rates*. Oxford: Oxford University Press.

Levitt, I. M., and Dandridge M. Cole. 1963. *Exploring the Secrets of Space: Astronautics for the Layman*. Englewood Cliffs, NJ: Prentice Hall.

References

Lewis, Geraint F., and Luke A. Barnes. 2016. *A Fortunate Universe: Life in a Finely Tuned Cosmos*. Cambridge: Cambridge University Press.

Lewis, Robert E., ed. 2023. "Human System Risk Board." NASA, March 16, 2023. https://www.nasa.gov/directorates/esdmd/hhp/human-system-risk-board.

Lin, David, Leopold Wambersie, and Mathis Wackernagel. 2021. "Estimating the Date of Earth Overshoot Day 2021." Nowcasting the World's Footprint & Biocapacity for 2021, Global Footprint Network, May 2021. https://www.overshootday.org/content/uploads/2021/06/Earth-Overshoot-Day-2021-Nowcast-Report.pdf.

Llinás, Rodolfo R. 2001. *I of the Vortex: From Neurons to Self*. Cambridge, MA: MIT Press.

Loeb, Abraham. 2021. *Extraterrestrial: The First Sign of Intelligent Life beyond Earth*. London: John Murray Publishers.

Logsdon, John M. 2024. "Space Exploration: Major Milestones." *Encyclopedia Britannica*, modified August 2, 2024. https://www.britannica.com/science/space-exploration/Major-milestones.

Lorenz, Ralph. 2020. *Saturn's Moon Titan: From 4.5 Billion Years Ago to the Present*. Sparkford, UK: Haynes.

Lorenz, Ralph, and Jacqueline Mitton. 2010. *Titan Unveiled: Saturn's Mysterious Moon Explored*. Princeton, NJ: Princeton University Press.

Lucas, Paul. 2004. "Cruising the Infinite: Strategies for Human Interstellar Travel." Strange Horizons, June 21, 2004. https://web.archive.org/web/20061114040746/http://www.strangehorizons.com/2004/20040621/travel.shtml.

Mann, Adam. 2018. "Bashing Holes in the Tale of Earth's Troubled Youth." *Nature* 553, no. 7689: 393–95.

Margari, V., D. A. Hodell, S. A. Parfitt, N. M. Ashton, J. O. Grimalt, H. H. Kim, K-S. Yun, et al. 2023. "Extreme Glacial Cooling Likely Led to Hominin Depopulation of Europe in the Early Pleistocene." *Science* 381, no. 6658: 693–99.

Marino, B. D. V., and Howard T. Odum. 1999. *Biosphere 2: Research Past and Present*. Amsterdam: Elsevier Science.

Martin, Brian. 1988. "Nuclear Winter: Science and Politics." *Science and Public Policy* 15, no. 5: 321–34.

Martin, G. M. 1971. "Brief Proposal on Immortality: An Interim Solution." *Perspectives in Biology and Medicine* 14, no. 2: 339–40.

Matloff, Greg, and Harold Gerrish. 2023. "The Scale of the Problem: Interstellar Distances, Time, and Energy Considerations." In *Interstellar Travel: Purpose and Motivations*, edited by Les Johnson and Kenneth Ray, 51–82. Amsterdam: Elsevier.

Matthews, Naomi E., Jorge A. Vazquez, and Andrew T. Calvert. 2015. "Age of the Lava Creek Supereruption and Magma Chamber Assembly at Yellowstone Based on $^{40}Ar/^{39}Ar$ and U-Pb Dating of Sanidine and Zircon Crystals." *Geochemistry, Geophysics, Geosystems* 16, no. 8: 2508–28.

Mawson, D. 1915. *The Home of the Blizzard*. Kent Town, Australia: Wakefield Press. First published in two volumes by William Heinemann (London).

Mazouffre, Stéphane. 2016. "Electric Propulsion for Satellites and Spacecraft: Established Technologies and Novel Approaches." *Plasma Sources Science and Technology* 25, no. 3: 033002.

References

McCallum, Malcolm L. 2015. "Vertebrate Biodiversity Losses Point to a Sixth Mass Extinction." *Biodiversity and Conservation* 24, no. 10: 2497–519.

McDowell, Jonathan C. 2018. "The Edge of Space: Revisiting the Kármán Line." *Acta Astronautica* 151: 668–77.

Melott, Adrian L. and Brian C. Thomas. 2009. "Late Ordovician Geographic Patterns of Extinction Compared with Simulations of Astrophysical Ionizing Radiation Damage." *Paleobiology* 35, no. 3: 311–20.

Miller, Stanley L., and Harold C. Urey. 1959. "Organic Compound Synthesis on the Primitive Earth." *Science* 130, no. 3370: 245–51.

Mitchell, Jere H., Benjamin D. Levine, and Darren K. McGuire. 2019. "The Dallas Bed Rest and Training Study Revisited After 50 Years." *AHA Circulation* 140, no. 16: 1293–95.

Moore, William. 1810. "On the Motion of Rockets Both in Nonresisting and Resisting Mediums." *Journal of Natural Philosophy, Chemistry & the Arts* 27: 276–85.

NASA. 2024. "About CHAPEA." Updated February 14, 2024. https://www.nasa.gov/humans-in-space/chapea/about-chapea.

———. n.d. "Kepler-1606 b." https://science.nasa.gov/exoplanet-catalog/kepler-1606-b.

Nelson, Mark. 2018. *Pushing Our Limits: Insights from Biosphere 2*. Tucson: University of Arizona Press.

Niiler, Eric. 2018. "The Apollo Mission That Nearly Ended with a Mutiny in Space." History.com, October 11, 2018. https://www.history.com/news/apollo-7-near-mutiny-ground-control-astronauts.

O'Neill, Ian. 2008. "Bad News: Interstellar Travel May Remain in Science Fiction." Universe Today, August 19, 2008. https://www.universetoday.com/17044/bad-news-insterstellar-travel-may-remain-in-science-fiction.

Ozasa, Kotaro, Harry M. Cullings, Waka Ohishi, Hida Ayumi, and Eric J. Grant. 2019. "Epidemiological Studies of Atomic Bomb Radiation at the Radiation Effects Research Foundation." *International Journal of Radiation Biology* 95, no. 7: 879–91.

Paez, Yosbelkys M., Lucy I. Mudie, and Prem S. Subramanian. 2020. "Spaceflight Associated Neuro-Ocular Syndrome (SANS): A Systematic Review and Future Directions." *Eye and Brain* 12: 105–17.

Pagnini, F., D. Manzey, E. Rosnet, D. Ferravante, O. White, and N. Smith. 2023. "Human Behavior and Performance in Deep Space Exploration: Next Challenges and Research Gaps." *NPJ Microgravity* 9, no. 1: 27.

Pariset, E., S. Malkani, E. Cekanaviciute, and S. V. Costes. 2021. "Ionizing Radiation-Induced Risks to the Central Nervous System and Countermeasures in Cellular and Rodent Models." *International Journal of Radiation Biology* 97 (Sup 1): S132–50.

Patel, Z. S., T. J. Brunstetter, W. J. Tarver, A. M. Whitmire, S. R. Zwat, S. M. Smith, and J. L. Huff. 2020. "Red Risks for a Journey to the Red Planet: The Highest Priority Human Health Risks for a Mission to Mars." *NPJ Microgravity* 6, no. 33: 1–13.

Petigura, Erik A., Andrew W. Howard, and Geoffrey W. Marcy. 2013. "Prevalence of Earth-Size Planets Orbiting Sun-Like Stars." *Proceedings of the National Academy of Sciences* 110, no. 48: 19273–78.

Planet Habitability Laboratory. n.d. "Habitable Worlds Catalog." Updated March 21, 2024. https://phl.upr.edu/hwc.

References

Plimer, Ian R. 2021. *Green Murder: A Life Sentence of Net Zero with No Parole*. Redland Bay, Australia: Connor Court Publishing.

Podsiadlowski, P., P. A. Mazzali, K. Nomoto, D. Lazzati, and E. Cappellaro. 2004. "The Rates of Hypernovae and Gamma-Ray Bursts: Implications for Their Progenitors." *Astrophysics Journal Letters* 607, no. 1: L17–20.

Prosser, J., L. Hink, C. Gubry-Ranjin, and G. W. Nicol. 2020. "Nitrous Oxide Production by Ammonia Oxidizers: Physiological Diversity, Niche Differentiation and Potential Mitigation Strategies." *Global Change Biology* 26, no. 1: 103–18.

Rampino, Michael R., and Stanley H. Ambrose. 2000. "Volcanic Winter in the Garden of Eden: The Toba Supereruption and the late Pleistocene Human Population Crash." In *Special Paper 345: Volcanic Hazards and Disasters in Human Antiquity*, edited by Floyd W. McCoy and Grant Heiken, 71–82. Boulder, CO: Geological Society of America.

Raup, D. M., and J. J. Sepkoski. 1982. "Mass Extinctions in the Marine Fossil Record." *Science* 215, no. 4539: 1501–3.

Rayman, M. D., P. A. Chadbourne, J. S. Culwell, and S. N. Williams. 1999. "Mission Design for Deep Space 1: A Low-thrust Technology Validation Mission." *Acta Astronautica* 45, no. 4–9: 381–88.

Regis, Edward. 1990. *Great Mambo Chicken and the Transhuman Condition*. Reading, MA: Addison-Wesley.

Ridley, Matt. 2010. *The Rational Optimist*. New York: Harper Collins.

Riley, Maree. 2021. "Life in the Australian Antarctic Program: Psychological Considerations in Isolated, Confined and Extreme Environments." *InPsych, Bulletin of the Australian Psychological Society* 43, no. 1: 8–15.

Ritchie, Hannah. 2022. "There Have Been Five Mass Extinctions in Earth's History." Our World in Data, November 30, 2022. https://ourworldindata.org/mass-extinctions.

Ritchie, Raymond J., Anthony W. D. Larkum, and Ignasi Ribas. 2018. "Could Photosynthesis Function on Proxima Centauri b?" *International Journal of Astrobiology* 17, no. 2: 147–76.

Rogers, Alan R. 1995. "Genetic Evidence for a Pleistocene Population Explosion." *Evolution* 49, no. 4: 608–15.

Roser, Max, and Hannah Ritchie. 2023. "Two Centuries of Rapid Global Population Growth Will Come to an End." Our World in Data, March 18, 2023. https://ourworldindata.org/world-population-growth-past-future.

Rosling, Hans. 2018. *Factfulness. Ten Reasons Why We're Wrong About the World—and Why Things Are Better Than You Think*. London: Sceptre.

Rumpf, Clemens M., Hugh G. Lewis, and Peter M. Atkinson. 2017. "Asteroid Impact Effects and Their Immediate Hazards for Human Populations." *Geophysical Resource Letters* 44, no. 8: 3433–40.

Sagan, Carl. 1985. *Contact*. New York: Simon and Schuster.

Sagan, Carl. 1994. *Pale Blue Dot: A Vision of the Human Future in Space*. New York: Random House.

Sagan, Carl, and George Mullen. 1972. "Earth and Mars: Evolution of Atmospheres and Surface Temperatures." *Science* 177, no. 4043: 52–56.

Sandberg, Anders. 2014. "Ethics of Brain Emulations." *Journal of Experimental & Theoretical Artificial Intelligence* 26, no. 3: 439–57.

References

Sandberg, Anders, and Nick Boström, N. 2008. *Whole Brain Emulation: A Roadmap.* Technical Report 2008-3, Future of Humanity Institute, Oxford University.

Sandberg, Anders, Eric Drexler, and Toby Ord. 2018. "Dissolving the Fermi Paradox." Submitted to *Proceedings of the Royal Society A*, June 6, 2018. https://arxiv.org/abs/1806.02404.

Sauro, F., R. Riccardo Pozzobon, M. Massironi, P. De Berardinis, T. Santagata, and J. De Waele. 2020. "Lava Tubes on Earth, Moon and Mars: A Review on Their Size and Morphology Revealed by Comparative Planetology." *Earth-Science Reviews* 209): 103288.

Sawyer, Kathy. 2006. *The Rock from Mars.* New York: Random House.

Sawyer, Robert J. 2009. *Calculating God: A Novel.* New York: Tor Books.

Schattschneider, Peter, and Albert A. Jackson. 2022. "The Fishback Ramjet Revisited." *Acta Astronautica* 191: 227–34.

Schechner, Sam, and Deepa Seetharaman. 2023. "How Worried Should We be About AI's Threat to Humanity? Even Tech Leaders Can't Agree." *Wall Street Journal*, September 4, 2023. https://www.wsj.com/tech/ai/how-worried-should-we-be-about-ais-threat-to-humanity-even-tech-leaders-cant-agree-46c664b6.

Schirrmeister, B. E., J. M. de Vos, A. Antonelli, and H. C. Bagheri. 2013. "Evolution of Multicellularity Coincided with Increased Diversification of Cyanobacteria and the Great Oxidation Event." *Proceedings of the National Academy of Sciences of the United States of America* 110, no. 5: 1791–96.

Schmidt, George. 2012. "Nuclear Systems for Space Power and Production." In *62nd International Astronautical Congress 2011: (IAC 2011): Cape Town, South Africa, 3–7 October 2011*, 6792–812. Paris: International Astronautical Federation.

Schneider, Peter. 2010. *Extragalactic Astronomy and Cosmology: An Introduction.* Berlin: Springer.

Schulze-Makuch, Dirk, René Heller, and Edward Guinan. 2020. "In Search for a Planet Better than Earth: Top Contenders for a Superhabitable World." *Astrobiology* 20, no. 12: 1394–404.

Schwartz, A. Brad. 2015. "The Infamous 'War of the Worlds' Radio Broadcast Was a Magnificent Fluke." *Smithsonian Magazine*, May 6, 2015.

Shellenberger, Michael. 2023. "A Pro-Human Environmental Policy: Don't Fall for the Malthusian Trap." Alliance for Responsible Citizenship Conference, London, October 30, 2023.

Shukman, David. 2016. "Hawking: Humans at Risk of Lethal 'Own Goal.'" BBC, January 19, 2016. https://www.bbc.com/news/science-environment-35344664.

Siddiqui, Tabassum. 2023. "Risks of Artificial Intelligence Must be Considered as the Technology Evolves: Geoffrey Hinton." U of T News, June 29, 2023. https://utoronto.ca/news/risks-artificial-intelligence-must-be-considered-technology-evolves-geoffrey-hinton.

Siegel, Ethan. 2017. "Voyager's 'Cosmic Map' of Earth's Location Is Hopelessly Wrong." *Forbes*, August 17, 2017. https://www.forbes.com/sites/startswithabang/2017/08/17/voyagers-cosmic-map-of-earths-location-is-hopelessly-wrong/.

Simko, Thomas, and Matthew Gray. 2014. "Lunar Helium-3 Fuel for Nuclear Fusion: Technology, Economics, and Resources." *World Futures Review* 6: 158–71.

References

Smil, Vaclav. 2022. *How the World Really Works: A Scientist's Guide to Our Past, Present and Future.* New York: Penguin Books.

Smith, S. M., S. R. Zwart, G. Block, B. L. Rice, and J. E. Davis-Street. 2005. "The Nutritional Status of Astronauts Is Altered after Long-Term Space Flight Aboard the International Space Station." *Journal of Nutrition* 135: 437–43.

Specktor, Brandon. 2023. "'Planet Killer' Asteroids Are Hiding in the Sun's Glare. Can We Stop Them in Time?" LiveScience, November 13, 2023. https://www.livescience.com/space/asteroids/the-sun-is-blinding-us-to-thousands-of-potentially-lethal-asteroids-can-scientists-spot-them-before-its-too-late.

Speyerer, Emerson J., and Mark S. Robinson. 2013. "Persistently Illuminated Regions at the Lunar Poles: Ideal Sites for Future Exploration." *Icarus* 222, no. 1: 122–36.

Stapledon, Olaf. 1937. *Star Maker.* London: Methuen & Co.

Stern, Robert J. 2016. "Is Plate Tectonics Needed to Evolve Technological Species on Exoplanets?" *Geoscience Frontiers* 7, no. 4: 573–80.

Stothers, Richard B. 1984. "The Great Tambora Eruption in 1815 and Its Aftermath." *Science* 224, no. 4654: 1191–98.

Stringer, Chris. 2012. "What Makes a Modern Human?" *Nature* 485, no. 7396: 33–35.

Tang, K. L., N. P. Caffrey, D. B. Nóbrega, S. C. Cork, P. E. Ronksley, H. W. Barkema, et al. 2017. "Restricting the Use of Antibiotics in Food-Producing Animals and Its Associations with Antibiotic Resistance in Food-Producing Animals and Human Beings: A Systematic Review and Meta-Analysis." *The Lancet: Planetary Health* 1, no. 8: 316–27.

Tegmark, Max. 2017. *Life 3.0: Being Human in the Age of Artificial Intelligence.* New York: Alfred A. Knopf.

Tim the Yowie Man. 2020. "Saucer Serial Hysteria: The Case of the Tully Crop Circle." *Australian Geographic*, September 21, 2020. https://www.australiangeographic.com.au/blogs/tim-the-yowie-man/2020/09/saucer-serial-hysteria-the-case-of-the-tully-crop-circle.

Tischler, M. E., and M. Slentz. 1995. "Impact of Weightlessness on Muscle Function." *Bulletin of the American Society of Gravitational and Space Biology* 8, no. 2: 73–81.

Toon, O. B., A. Robock, and R. P. Turco. 2008. "Environmental Consequences of Nuclear War." *Physics Today* 61, no. 12: 37–42.

Torchinsky, Rina. 2022. "Elon Musk Hints at a Crewed Mission to Mars in 2029." NPR, March 17, 2022. https://www.npr.org/2022/03/17/1087167893/elon-musk-mars-2029.

Turner, Michael. 2013 "Exovivaria as Simulacra for Generation Starship Societies." *Proceedings of 100 Year Starship Symposium*, September 9–22, 2013, 157–73.

United Nations Office for Outer Space Affairs. n.d. "Treaty on Principles Governing the Activities of States in the Exploration and Use of Outer Space, including the Moon and Other Celestial Bodies." https://www.unoosa.org/oosa/en/ourwork/spacelaw/treaties/introouterspacetreaty.html.

van Valkenburgh, Blair. 1999. "Major Patterns in the History of Carnivorous Mammals." *Annual Review of Earth and Planetary Sciences* 27: 463–93.

References

Ventura, Tim. 2019 "Sleeper Ship: Frank Merkle on Space Cryonics & Nanotechnology." Medium, November 26, 2019. https://medium.com/discourse/ralph-merkle-on-space-cryonics-nanotechnology-eb760dc2bbc7.

Vesteg, Matej, and Juraj Krajcovic. 2008. "Origin of Eukaryotic Cells as a Symbiosis of Parasitic Alpha-proteobacteria in the Periplasm of Two-Membrane-Bounded Sexual Pre-karyotes." *Communicative & Integrative Biology* 1, no. 1: 104–13.

Visser, Matt, B. A. Bassett, and S. Liberati. 2000. "Superluminal Censorship." *Nuclear Physics B: Proceedings Supplements* 88, no. 1–3: 267–70.

von Kármán, Theodore, and Lee Edson. 1967. *The Wind and Beyond: Theodore von Kármán, Pioneer in Aviation and Pathfinder in Space.* Boston: Little, Brown.

Waltham, David. 2017. "Star Masses and Star-Planet Distances for Earth-like Habitability." *Astrobiology* 17, no. 1: 61–77.

Warner, H., J. Anderson, S. Austad, E. Bergamini, D. Bredesen, R. Butler, B. Carnes, et al. 2005. "Science Fact and the SENS Agenda." *EMBO Reports* 6, no. 11: 1006–8.

Weinersmith, Kelly, and Zach Weinersmith. 2023. *A City on Mars: Can We Settle Space, Should We Settle Space, and Have We Really Thought This Through?* London: Particular Books.

Wells, H. G. 1898. *The War of the Worlds.* London: William Heinemann and New York: Harper & Bros.

Weon, Byung Mook, and Jung Ho Je. 2009. "Theoretical Estimation of Human Lifespan." *Biogerontology* 10, no. 1: 65–71.

White, Harold. 2013. "Warp Field Mechanics 101." *Journal of the British Interplanetary Society* 66: 242–47.

White, Harold, J. Vera, A. Han, A. R. Bruccoleri, and J. MacArthur. 2021 "Worldline Numerics Applied to Custom Casimir Geometry Generates Unanticipated Intersection with Alcubierre Warp Metric." *European Physical Journal C* 81, no. 7: 677–86.

Wikipedia. 2024a. "List of Space Travelers by Nationality." https://en.wikipedia.org/wiki/List_of_space_travelers_by_nationality.

———. 2024b. "Timeline of Space Exploration." https://en.wikipedia.org/wiki/Timeline_of_space_exploration.

———. 2024d. "List of Missions to Mars." https://en.wikipedia.org/wiki/List_of_missions_to_Mars.

———. 2024c. "Biosphere 2." https://en.wikipedia.org/wiki/Biosphere_2.

———. 2024e. "List of Exoplanet Extremes." https://en.wikipedia.org/wiki/List_of_exoplanet_extremes.

Wohlforth, Charles, and Amanda R. Hendrix. 2016. *Beyond Earth: Our Path to a New Home in the Planets.* New York: Pantheon Books.

Wolfe, Tom. 1979. *The Right Stuff.* New York: Farrar, Straus and Giroux.

Wolszczan, A., and D. A. Frail. 1992. "A Planetary System around the Millisecond Pulsar PSR1257 + 12." *Nature* 355, no. 6356: 145–47.

Wood, J., L. Schmidt, D. Lugg, J. Ayton, T. Phillips, and M. Shepanek. 2005. "Life, Survival, and Behavioral Health in Small Closed Communities 10 Years of Studying Isolated Antarctic Groups." *Aviation, Space and Environmental Medicine* 76 (6 Suppl): B89–93.

References

World Nuclear News. 2023. "Nuclear Companies Sign Up for Space Technology Missions." October 20, 2023. https://www.world-nuclear-news.org/articles/nuclear-companies-sign-up-for-space-technology-mis.

Wuebbles, D. J., D. W. Fahey, K. A. Hibbard, D. J. Dokken, B. C. Stewart, and T. K. Maycock, eds. 2017. *Climate Science Special Report: Fourth National Climate Assessment.* Vol. 1. Washington, DC: U.S. Global Change Research Program.

Xi-Liu, Yue, and Gao Qing-Xian. 2018. "Contributions of Natural Systems and Human Activity to Greenhouse Gas Emissions." *Advances in Climate Change Research* 9, no. 4: 243–52.

Yeomans, Don, and Paul Chodas. 2013. "Asteroid 2012 DA14 to Pass Very Close to the Earth on February 15, 2013." Center for Near Earth Object Studies, February 1, 2013. https://cneos.jpl.nasa.gov/news/news177.html.

Yin, An. 2012. "Structural Analysis of the Valles Marineris Fault Zone: Possible Evidence for Large-Scale Strike-Slip Faulting on Mars." *Lithosphere* 4, no. 4: 286–330.

Zealley, Ben, and Aubrey D. N. J. de Grey. 2013. "Strategies for Engineered Negligible Senescence." *Gerontology* 59, no. 2: 183–89.

Zubrin, Robert. 2011. *The Case for Mars: The Plan to Settle the Red Planet and Why We Must.* New York: Free Press.

———. 2019. *The Case for Space: How the Revolution in Spaceflight Opens Up a Future of Limitless Possibilities.* Amherst, NY: Prometheus.

Additional Internet Sources

Chapter 1: Are We Alone? The Search for Extraterrestrial Life
https://en.wikipedia.org/wiki/Cambrian_explosion
http://en.wikipedia.org/wiki/Habitable zone
http://en.wikipedia.org/wiki/Fermi_paradox
https://en.wikipedia.org/wiki/Bracewell_probe
https://exoplanets.nasa.gov/alien-worlds/ways-to-find-a-planet
https://www.esa.int/ESA_Multimedia/Images/2019/12/Exoplanet_phase_curve
https://en.wikipedia.org/wiki/List_of_NASA_missions

Chapter 2: Scenarios for the Demise of Humanity
https://en.wikipedia.org/wiki/Homo
https://en.wikipedia.org/wiki/Holcene_extinction
https://www.britannica.com/explore/savingearth/climate-change-throughout-history
https://bigthink.com/the-past/other-human-species
https://en.wikipedia.org/wiki/Human_extinction
https://en.wikipedia.org/wiki/Extinction_event
https://en.wikipedia.org/wiki/List_of_epidemics_and_pandemics
https://www.who.int/news-room/fact-sheets/detail/antibiotic-resistance
https://en.wikipedia.org/wiki/Generative_artificial_intelligence
https://en.wikipedia.org/wiki/Gray_goo
https://www.footprintnetwork.org
https://en.wikipedia.org/wiki/Projections_of_population_growth
https://en.wikipedia.org/wiki/Ice- albedo_feedback
https://en.wikipedia.org/wiki/Plate_tectonics
https://www.ipcc.ch/report/ar6/wg3/downloads/report/IPCC_AR6_WGIII_Annex-III.pdf
https://en.wikipedia.org/wiki/Near-Earth_object

Chapter 3: How Urgent Is the Imperative to Search for Other Locations for Humans to Live?
https://en.wikipedia.org/wiki/List_of_natural_disasters_by_death_toll
https://en.wikipedia.org/wiki/Near-Earth_object

https://en.wikipedia.org/wiki/Tragedy_of_the_commons
http://neo.jpl.nasa.gov/stats/images/web_total.png
https://www.nasa.gov/dartmission

Chapter 4: The Record of Space Travel So Far
https://en.wikipedia.org/wiki/List_of_spaceflight_records
https://www.unoosa.org/oosa/en/ourwork/spacelaw/treaties/introouterspacetreaty.html
https://en.wikipedia.org/wiki/Human_spaceflight
https://en.wikipedia.org/wiki/List_of_space_stations

Chapter 5: Options for Establishing a Permanent Settlement in Earth's Immediate Environment
https://www.sylviaengdahl.com/space/spacequotes.htm
https://www.nasa.gov/international-space-station

Chapter 6: Options for Settlement Beyond the Earth-Moon Setting but Within the Solar System
https://voyager.jpl.nasa.gov/frequently-asked-questions
https://en.wikipedia.org/wiki/Titan_(moon)
http://en.wikipedia.org/wiki/Tragedy_of_the_commons
https://en.wikipedia.org/wiki/List_of_missions_to_the_outer_planets
https://en.wikipedia.org/wiki/Terraforming_of_Mars
https://marspedia.org/Paraterraforming
https://terraforming.fandom.com/wiki/Mercury

Chapter 7: Options for Settlement Outside Our Solar System
https://en.wikipedia.org/wiki/Planet-hosting_star
https://phl.upr.edu/projects/habitable-exoplanets-catalog
https://en.wikipedia.org/wiki/Breakthrough_Initiatives
https://www.planetary.org/sci-tech/lightsail
https://en.wikipedia.org/wiki/List_of_nearest_exoplanets
https://exoplanetarchive.ipac.caltech.edu
https://exoplanets.nasa.gov/exoplanet-catalog/2252/kepler-1606-b
https://en.wikipedia.org/wiki/Rare_Earth_hypothesis
https://en.wikipedia.org/wiki/Mediocrity_principle
https://en.wikipedia.org/wiki/Red_dwarf
https://en.wikipedia.org/wiki/Flare_star

Chapter 8: The Tyranny of Distance and Time
https://en.wikipedia.org/wiki/Interstllar_travel
https://en.wikipedia.org/wiki/Wormhole

ADDITIONAL INTERNET SOURCES

Chapter 9: Physical, Social, and Psychological Challenges of Space Travel
https://en.wikipedia.org/wiki/Interstellar_travel
https://en.wikipedia.org/wiki/Effect_of_spaceflight_on_the_human_body
https://www.nasa.gov/humans-in-space/the-human-body-in-space
https://spacecenter.org/twins-study-research-reveals-effects-of-space
https://en.wikipedia.org/wiki/Artificial_gravity
https://en.wikipedia.org/wiki/MARS-500
https://www.esa.int/Science_Exploration/Human_and_Robotic_Exploration/Mars500/Mars500_quick_facts
https://en.wikipedia.org/wiki/Biosphere_2

Chapter 10: Spacecraft Options for Human Transportation
https://en.wikipedia.org/wiki/Life_extension
https://en.wikipedia.org/wiki/Strategies_for_engineered_negligible_senescence
https://en.wikipedia.org/wiki/Effects_of_ionizing_radiation_in_spaceflight
https://en.wikipedia.org/wiki/Sleeper_ship
https://en.wikipedia.org/wiki/Embryo_space_colonization

Chapter 11: Options for Transport Propulsion Technology
https://en.wikipedia.org/wiki/Spacecraft_propulsion
https://en.wikipedia.org/wiki/Rocketdyne_J-2
https://en.wikipedia.org/wiki/Apollo_command_and_service_module
https://en.wikipedia.org/wiki/Monopropellant_rocket
https://en.wikipedia.org/wiki/Project_Orion_(nuclear_propulsion)
https://en.wikipedia.org/wiki/Nuclear_thermal_rocket
https://www.google.com/search?q=lightsail+propulsion+system+advantages+and+disadvantages
https://en.wikipedia.org/wiki/Breakthrough_Starshot
https://en.wikipedia.org/wiki/Plasma_propulsion_engine
https://solarsystem.nasa.gov/missions/dawn/overview/#otp_major_engineering_achievements
https://en.wikipedia.org/wiki/Plasma_propulsion_engine

Chapter 12: A Summary of the Challenges for a Human Settlement Off-Earth
https://en.wikipedia.org/wiki/List_of_spaceflight-related_accidents_and_incidents

Chapter 13: Humans or Machines?
https://en.wikipedia.org/wiki/List_of_missions_to_the_Moon
https://researchfeatures.com/rethinking-consciousness
https://en.wikipedia.org/wiki/Mind_uploading

Acknowledgments

Warren Canning, for catalyzing my interest in writing more deeply on this subject through his invitation for me to be part of a discussion panel for the 2008 Alfred Deakin Innovation Lecture on "Beyond the Planet: Space Travel Imagined" in Melbourne in 2008.

My late mother-in-law, Mrs. Lisbeth Calwell, for helping me to organize my thoughts into intelligible sentences and illustrations by facilitating my presenting a popular talk on the subject at her retirement village in 2014.

My wife, Monica, for her unflinching support for the project, for her valuable advice on style and comprehension, and for her patient acquiescence to the long hours that I have spent at the computer over the course of the writing process.

My friends, for innumerable helpful, insightful, and humorous conversations on the topic of the future of humanity.

Jonathan Kurtz at Prometheus Books for recognizing the potential of the preliminary manuscript. Meredith Dias and the other editorial staff at Globe Pequot for their very helpful, timely, and invaluable assistance with the final copyediting of the manuscript.

While sources of specific information/data are documented in the "References" section and embedded in the figure captions, other sources of a more general background nature are listed under the relevant individual chapter in the "Additional Internet Sources" section.

Index

AAE. *See* Australian Antarctic Expedition
absorption lines, 62
accretion disc, 18
Advanced Composition Explorer (ACE), 167
advanced generalized intelligence (AGI), 234
Age of Enlightenment, 68
AGI. *See* advanced generalized intelligence
aging processes, 227
Agreement Governing the Activities of States on the Moon and Other Celestial Bodies, 165
Agreement on the Outer Space Treaty: Article 1, 164–65; Article 2, 165; principles of, 164
AI. *See* artificial intelligence
"The Aims of Astronautics" (Tsiolkovsky), 131
albedo extinction scenarios, 104
Alcubierre drive, 285
Aldrin, Buzz, *44*, 142
ALH84001 meteorite, 63; controversy over, 64
Allen, John P., 188, 260
Almaz, 148
American Skylab, 148
amino acids: metabolism-first experiments, 9; synthesizing, 8–9
Anders, Bill, 143
Andromeda Galaxy, 108

antibiotic resistant microorganisms, 78
antigravity propulsion, 283–84
antimatter propulsion, 284, 335n6
Apollo 7, 250–51
Apollo 8, 143
Apollo 10, 223
Apollo 11, 135; speed of, 132, 330n1
Apollo 12, 155
Apollo 13, 250–52
Apollo 17, 156
Apollo asteroids, 330n5
archaea: classification of, 43; habitat of, 326n18
Arecibo Observatory, 32, 54
Arianespace, 139
Armstrong, Neil, 142
Armstrong limit, 332n2
Artemis-P1, 168
Artemis-P2, 168
Artemis program, 159
artificial intelligence (AI), 10, 117; alignment of, 81; Bracewell probe and, 38; brain connection to, 304–5; concerns over, 80–81; dystopian impact of, 306; extinction and, 80–81; Hinton on, 80; humans compared with, *309–12*
asteroids: Apollo, 330n5; average time between impacts, 124; in popular culture, 124; stony, 330n3. *See also* near-earth asteroids

astronomical unit (AU), 325n13
astronomy, 56–57
astrosphere, 326n14
atmosphere: carbon dioxide in, 92–93; Mars, 184, 192; of Moon, 161; radiation absorbed by, 19–20, 85–86; spectral analysis, 62; terraforming, 184; thickness of, 19–20; of Titan, 51, 197; water preserved by, 19–20
AU. *See* astronomical unit
Australian Antarctic Expedition (AAE), 245–47
Australian Apex chert rocks, 10–11
autonomization, 250

Bacillus subtilis, 43
back propagation algorithm, 306
bacteria: overview, 326n17; population of, 43; in vacuum, 10, 43. *See also specific bacteria*
basalt flooding, 103–4, 121
Bass, Ed, 188, 260
Baxter, Stephen, 67
behavioral health, 244–48
Bennu, 46
Bernal, J. D., 176
BETA. *See* Billion-Channel ExtraTerrestrial Assay
Betelgeuse, 101
Bezos, Jeff, 139, 271
Big Ear Radio Observatory, 32
Billion-Channel ExtraTerrestrial Assay (BETA), 31
biological removal, 301–4
biosignatures, 28; agnostic, 45; debate over, 191; extraterrestrial, 44–45; Mars, 191; measuring, 44–45; spectral analysis and, 62; telltale compounds, 45
biosphere, 185–86; of Earth, 10

Biosphere 2, *188*, 260; establishment of, 188; management transfer of, 189; operational problems with, 189; problems with, 261
black body temperature, 85–86, 327n26
Black Death, 76–77, 115
black smokers, 43
Blue Origin, 139
body fossil, 325n7
Borman, Frank, 143
Bracewell probe, 208; artificial intelligence and, 38; extraterrestrial contact and, 37–38; life classification of, 39; as local representative, 39; von Neumann probe compatibility and, 39
Brahe, Tycho, 5
brain cloning, 267–68, 301–4, 315–16
Branson, Richard, 139
Breakthrough Initiative, 41
Breakthrough Listen, 32–33
Breakthrough Starshot, 208, 224, 280; funding for, 333n3
Brera Observatory, 34–35
Bulletin of the Atomic Scientists, 75
Bussard ramjet propulsion, 282–83

Calment, Jeanne, 226–27
Cambrian period: diversity rise during, 12–13; oxygen availability of, 13; ozone increase during, 13; plate tectonics and, 19
Campanian Ignimbrite, 103
carbohydrates, 8
carbon, 7; chemistry, 15
carbon dioxide: in atmosphere, 92–93; from cement industry, 330n2; emissions, 120; forms of, 329n11; regulating, 16
Casimir cavities, 285, 335n9

355

Index

Cassini, 52
causality, 334n3
cell membranes, 8
cellular communication, 227
cellulose, 8
central nervous system, 243
Ceres: destination time to, 224; discovery of, 199; settlement, 199–200
Cernan, Eugene, 141
chain rule, 306
Challenger, 331n1
Chandrayaan-3, 157–58
Chang'e-4, 170
Chang'e-5-T1, 168, 170
CHAPEA. *See* Crew Health and Performance Exploration Analog
#ChasingUFOs, 32
ChatGPT, 80, 117, 306–7
chemical equilibrium, 45
chemical fossils, 44–45
chemical rocket propulsion, 275–76
Chernobyl, 74
China, People's Republic of, Moon missions, 153
chiral compounds, 45; overview, 326n21
Churyumov-Gerasimenko comet, 46
circumstellar habitable zone (CHZ), 16
civilization rankings, 325n12
Clarke, Arthur C., 67, 145, 203; on extraterrestrial life, 7
climate change: adapting to, 90–91; consequences of, 86–90; extinction and, 85–94, 118–21; historical, *87*, 94; impacts of, 90–91; Industrial Revolution and, 89–90; mechanisms of, 86–90; plate tectonics and, 94; population growth and, 88–89, 95; runaway, 91–94, 118–21; Venus and, 91–92; warnings about, 119
CMB. *See* cosmic microwave radiation
Cold War, 74, 114, 320

Collins, Michael, 143, 257
Columbia, 331n1
Comet 266P/Christensen, 32
Comet P/2008 Y2 (Gibbs), 32
comets: extinction scenario, 7, 95–100, 122–25. *See also specific comets*
Commercial Transportation Service (COTS), 139
confined environments: dependencies of, 187; on Earth, 188–89; experiments, 187; life support in, 186–90
Conrad, Pete, *155*
consciousness copying, 267–68, 301–4, 315–16
Contact (Sagan), 31
continental drift, 105
Cook, James, 236–38, 248–49
Copernicus, Nicolaus, 5, 59–60
Coriolis force, 334n4
cosmic microwave radiation (CMB), 332n4
COTS. *See* Commercial Transportation Service
Crew Health and Performance Exploration Analog (CHAPEA), 263
crop circles, 30
cross contamination, 63–64
cryopreservation, 264–66
cubesat, 333n5
Curiosity, 49–50
cyanobacteria, 11

Dallas Bed Rest Study, 239
dark energy, 283–84, 335n6
DART. *See* Double Asteroid Deflection Test
Darwin, Charles: on extinction, 68; on universal common descent, 8
dataism, 305
data storage, 10

Index

Davies, Paul, 5
Dawn, 281—282
Deccan Traps, 103–4
deep learning, 306
Deep Space 1, 281–82
Deep Space Climate Observatory (DSCOVR), 167
Deinococcus radiodurans, 43
democracy, spaceflight, 254
Descent Imager/Spectral Radiometer, *52*
Didymos, 124
differentiated planets, 192
dimethyl sulfide, 45
Dimorphos, 124
direct imaging, 56
discarded relics, 41
distorted space-time, 285
DNA, 8; discovery of, 325n11; spaceflight damage to, 218; water and structural stability of, 15
docking, 296–97
Doherty, Peter, 115
Doomsday Clock, 74–75; graph, *75*; reductions in, 114
Doppler, Christian, 327n24
Doppler effect, 327n24
Double Asteroid Deflection Test (DART), 124–25
Drake, Frank, 22–24; SETI and, 31
Drake equation: intentions of, 24; Sagan on, 22–23; worked examples of, *23*
Drexler, Eric, 81–82
DSCOVR. *See* Deep Space Climate Observatory
Dyson, Freeman, 40, 112, 290
Dyson spheres, 28; extraterrestrial contact and, 39–40; first modern imagining of, 40; representation of, *40*

Earth: biosphere of, 10; confined environments on, 188–89; geological age of, 328n2; interstellar probes from, 41–43; Mars compared with, 191–93; Snowball, 13, 86, 324n2; terraforming, 120–21; Titan compared with, 52–53
Earth-Moon L1 point, 168–70
Earth-Moon L2 point, 168–70
ecological footprint analysis, 82–83
ecophagy, 81–82; likelihood of, 116–17
electromagnetic emissions: extraterrestrial life and, 28; SETI and, 31
El Niño–Southern Oscillation, 86
embryo ships, 268; challenges for, 266–67
Enceladus, 53–54
Endeavor, 236–38
Endurance (Kelly), 232
engineered negligible senescence, 227
engineered structures: energy requirements of, 39; extraterrestrials and, 28, 34–37; hypothetical, 39–40; observations of, 34; outside Solar System, 37
Engines of Creation (Drexler), 81–82
environment: of Enceladus, 53–54; on Moon, 47; of Venus, 51. *See also* confined environments
enzyme actions, 14
epigenetic changes, 227
Epsilon Eridani, 31
Eros, 46
Eta Carina, 100
eukaryotic cells, 12, 17; overview, 326n19
Europa, 53
event horizon, 334n2
exomoons, 62
Exoplanet Exploration Program, 212–13
exoplanets, 15; characteristics of, 57; chemical properties of, 61–62; classes of, *212*; closest, 209–13; common types of, 62; confirmed, 54; detection techniques, *54*, 55–57;

Index

discovery of, 203–4; distribution plots, *58*; extraterrestrial life on, 54–59, 62–63; first confirmation of, 54; in Goldilocks Zone, 59; light curve schematic, *61*; metallicity and, 205; nearest, 55; physical properties of, 61–62; potentially habitable, 212–13; properties of, 204–6; red dwarfs and, 211–12; reflectivity, 59; requirements for, 206–7; scouting expeditions, 208, 258; selecting targets for, 206–9, 257–58; space craft missions to, 54–59; statistics on, 204–6; summary of, 58–59; surface temperature of, 15–16, 59, *60*; in tidal locking zone, 59; 12 light years away, *210*; water on, 15–16

exovivarium, 253; characteristics of, 254

extinction, 13–14, 318–19; albedo change scenarios, 104; anthropogenic, 69, 72–95; artificial intelligence and, 80–81; background rate of, 70; chances of, 112–14; climate change and, 85–94, 118–21; comet impact, 7, 95–100, 122–25; Darwin on, 68; evolution through, 70, 97–98; extraterrestrial, 27; galaxy collision scenario, 107–9; gamma ray burst, 100–101, 121–22; genocide scenario, 78–80, 116; global disease pandemics, 75–78, 115–16; Hangenberg, 101; *Homo*, 21; *Homo neanderthalis*, 68; ice age, 70–71; Milankovitch cycle scenarios, 104–5; natural scenarios, 69, 72, 95–110; nuclear holocaust and, 74–75, 114–15; Ordovician-Silurian, 101; Permian, 2, 71–72; plate tectonic scenarios, 105–6; pollution scenario, 82–85, 117–18; population scenario, 82–85, 117–18; self replicating nanorobots and, 81–82, 116–17; Sun engulfment scenario, 109–10; sun output, 106–7; supernova scenario, 100–101, 121–22; timeline, *71*; Triassic, 71–72; volcano scenarios, 101–4, 121

extraterrestrial life, 15–26; alternate calculations for, 24–26; biosignatures of, 44–45; Bracewell probe and, 37–38; Clarke on, 7; common features with, 9; complex, 28–43; cross contamination and, 63–64; Davies on, 5; direct observation of, 27; Dyson spheres and, 39–40; electromagnetic emissions and, 28; Enceladus and, 53–54; engineered structures and, 28, 34–37; Europa and, 53; evidence of, 6, 27–28, 65; on exoplanets, 54–59, 62–63; extinction of, 27; genocide and, 79–80; interstellar probes and, 41–43; lunar rocks and, 46; Moon and, 46–47; von Neumann probe and, 37–38; nonbiological, 9; non-lunar bodies, 47–54; Oumuamua and, 40–41; outside Solar System, 37–43; panspermia and, 63–64; in popular culture, 9; resource exhaustion and, 27; seeded, 64; simple forms of, 43; space craft missions and, 46; theories about absence of, 24; Titan and, 51–53; Venus and, 51

extremophiles, 43

eyesight, 13

FAI. *See* Fédération aéronautique internationale

fake news, 29

faster-than-light travel (FTL), 284–86; distorted, 285

Fédération aéronautique internationale (FAI), 134
Fermi, Enrico, 26–27, 43
Fermi paradox, 26–27
51 Pegasi, 54
fine-tuning hypothesis, 61
first generation star death, 18
flare star, 333n7
Friedman, Louis, 333n4
FTL. *See* faster-than-light travel
Fukagima, 74
The Future of Humanity (Kaku), 1
future superhuman, 314

Gagarin, Yuri, 135
Gaia Observatory, 167
galaxy collisions extinction scenarios, 107–9
Galileo Galilei, 5
gamma ray burst (GRB): extinctions from, 100–101, 121–22; overview, 328n4; radiation from, 100, 328n4
generation starships, 252–54, 261–63; challenges for, 259–60
genetic engineering, 69, 233
genocide: defined, 78–79; definitional exceptions to, 328n7; examples of, 79; extinction from, 78–80, 116; extraterrestrials and, 79–80; justifications for, 78–79; likelihood of, 116
GHGs. *See* greenhouse gases
giant-impact hypothesis, 326n22
global winters, 87
Goddard, Robert, 221, 259, 275
Goldilocks zone, 2, 15–16, *16*; exoplanets in, 59; Venus and, 51
gravitational microlensing, 55–56
gravity assist propulsion, 276–77, 326n16
gravity wells, 286
GRB. *See* gamma ray burst

Great Dying, 71–72
Great Filter, 27, 74, 289
Great Plague of London, 77
Great Silence, 24, 34, 65
greenhouse gases (GHGs), 85–86; absorption and, 325n6; Venus and, 325n6
grey goo, 81–82, 116–17
Grissom, Gus, 251

Habitable Worlds Catalog, 209
Hamlin Pool, Western Australia, 11
Hangenberg extinction, 101
Hawking, Stephen, 32–33, 333n3; Reith Lectures by, 1
HDI. *See* Human Development Index
heliopause, 326n15
heliosphere, 326n14
helium, 14
Herschel, William, 57
Herschel Infrared Telescope, 167
Hertzsprung-Russell diagram, 330n20
hibernation, 264–65; spaceflight, 218
Hinton, Geoffrey, 306; on artificial intelligence, 80
Hiroshima, 74
Hiten, 170
Hohmann transfer orbit (HTO), 331n3
Homo: emergence of, 17, 67; extinctions of, 21; subspecies of, 67; technical competence of, 21
Homo neanderthalis, 68
Homo sapiens: emergence of, 17, 68; intelligent life defined by, 113
Homunculus Nebula, 100
HTO. *See* Hohmann transfer orbit
Hubble Space Telescope repairs, 173–74
Human Development Index (HDI), *83*; criticism of, 84
human spaceflight records, 140–42
Huygens, 52

359

hydrogen, 14; fusion, 17–18
hydrostatic equilibrium, 19

IBMP. *See* Russian Academy of Sciences Institute of Biomedical Problems
ice age, 68; extinctions during, 70–71; sea level and, 86
ichnofossil, 325n7
India Moon missions, 153
Industrial Revolution, 292; climate change and, 89–90
innovation, 129
inorganic transplants, 9–10
Intergovernmental Panel on Climate Change (IPCC), 119; construction of, 148–49
International Space Station (ISS), 135, 151; decommission plans, 174; funding for, 149; modules, 148; occupancy requirements of, 149–50; orbit of, 148; ownership of, 147; schematic, *149*; time dilation on, 228–29; visitors to, 141
interstellar arks, 268–69
interstellar spacecraft, 28; development of, 217; self replication and, 38
ion drive propulsion, 281—282
IPCC. *See* Intergovernmental Panel on Climate Change
isotope patterns, 44
ISS. *See* International Space Station
Itokawa, 46

James Webb Space Telescope, 167
Japan Moon missions, 153
Jeffryes, Sidney, 246–47
Joy, Bill, 82
Jupiter: impacts on, 96–97; orbit map, *169*
Justinian plague, 77

Kaku, Michio, 1
Kardashev, Nikolai, 325n12
Kardashev scale, 39
Kármán line, 134
Kelly, Mark, 239
Kelly, Scott, 232, 239
Kennedy, John F., 134; on space race, 151–52
Kepler, Johannes, 5, 279–80
Kepler-1606b, 213
Khiebtzevich, U., 151
Kordylewski clouds, 170
Krasnikov tubes, 285
Krueger, David, 80
Kuiper Belt, 126

Lagrange points: overview of, 332n2; schematic representation of, *166*, *170*
Lagrange point settlement, 165–70; disadvantages of, 172–73; Earth-Moon implications, 171–72; Sun-Earth implications, 171
landings, 297–98
last universal common ancestor (LUCA), 8
Late Heavy Bombardment, 98; overview, 329n15
launches, 297
Leibniz, Gottfried, 306
Lemkin, Raphael, 78–79
LEO. *See* low Earth orbit
Leonids, 95
life: biological platform for, 9–10; Bracewell probe classification of, 39; complex, 17–21; components of, 7; defining, 7–10; earliest evidence of, 10–11; event sequences, 60–61; expectancy, 333n1; higher morphological

complexity organization of, 13; *Homo sapiens* defining intelligent, 113; intelligence of, 17, 326n20; multicellular, 14; NASA on, 7; von Neumann probe as alternative form of, 38; originating in space, 64; requirements for, 9, 17–21; timeline, *11*; water and, 14–15. *See also* extraterrestrial life

life extension, 226–27; sleeper ships and, 263–66

life seeds, 64

life support: in confined environments, 186–90; Mars settlements, 194; spaceflight, 298–99

LightSail, 41, 209, 279–81

LightSail 2, 209, 280–81

light years, 325n10

lipids, 7–8

Lippershey, Hans, 5

liquid metallic core convection, 19

lithosphere, 324n5

Local Cluster, 329n17

LOFAR. *See* Low Frequency Array

Lovell, Jim, 143

Lovell Telescope, 32

low Earth orbit (LEO), 331n1

Lowell, Percival, 35

Lowell Observatory, 35

Low Frequency Array (LOFAR), 32

LUCA. *See* last universal common ancestor

Luna 1, 45, 151

Lunar Gateway, 175

lunar rocks: extraterrestrial life and, 46; Moon samples, 46, 200

Lunar Roving Vehicle, 46

machine learning, 10

Madigan, Cecil, 246

magnetite, 45

magnetosphere, 326n14; Mars, 183–84; terraforming, 183–84

magnetotail, 332n3

mammaliaformes, 98

Marcel, Jesse, 29

Mariner 2, 177

Mariner 4, 35

Markov chains, 80

Mars: atmosphere, 184, 192; biomarkers, 191; biosignatures, 191; destination time to, 224; Earth compared with, 191–93; magnetosphere, 183–84; methane on, 49; *Perseverance* landing on, 50–51; polar ice caps, 48; probing, 6; simulation chamber, 247, 261–62; space craft missions to, 46–54; stepping stone strategy, 226; terraforming, 183–84; toxicity of, 194; transport requirements to, 200–201; *Viking 1* landing on, 49; *Viking 2* landing on, 49; volcanoes on, 104; water on, 190–91

Mars-500, 261–62

Mars canals, 6, 28, 34, 47–48; comparison drawings, *36*; Lowell and, 35; *Mariner 4* and, 35; objections to, 35

Mars Dune Alpha, 262–63

Mars Face, 37; landform examination of, *37*; *Viking 1* capturing, 36

Mars Global Surveyor, 37

Mars Pathfinder/Sojourner, 49

Mars Reconnaissance Orbiter, 50

Mars settlements, 190–93, 195; advocates for, 189; challenges to, 194; conditions of, 182; life support, 194

Mawson, Douglas, 245–47, 249

Maxwell, James Clerk, 280

MDR-TB. *See* multidrug-resistant tuberculosis

mediocrity principle, 59–61; overview, 327n27
membrane lipids, 45
Mercury settlements, 195
Mercury terraforming, 195
Mertz, Xavier, 246
META, 31
metallicity, 205
metal shielding, 334n2
methane, 45; on Mars, 49
Microwave Observing Program (MOP), 32
Milankovitch, Milutin, 104–5
Milankovitch cycles, 94; extinction scenarios, 104–5
Miller-Urey experiment, 8
Milner, Yuri, 32–33, 333n3
mini-Neptunes, 62
Mir, 148
mitochondria, 12
mitochondrial dysfunction, 227
molecular fossil, 44–45
Moon: atmosphere of, 161; environment on, 47; extraterrestrial life and, 46–47; formation of, 20, 47; human time spent on, 153–59; India missions to, 153; Japan missions to, 153; landing sites, 153–57, *154*; lunar rock samples from, 46, 200; People's Republic of China missions to, 153; polar regions of, 156–57, *158*; as practice, 320; probing surface of, 6; Soviet Union missions to, 152; space craft missions beyond, 177–78; space craft missions to, 46–47; spin axis of, 162; territory claims, 163–65; United States missions to, 152–53; volcanos on, 47; water from, 46, 160
Moon settlements, 151–57; advantages of, 159–61; advocates for, 189; challenges of, 158–59, 161–63

Moore, William, 275
MOP. *See* Microwave Observing Program
morphological complexity, 13
Moscow Institute of Biomedical Problems, 247
Mount Pinatubo, 102
Mount Tambora, 102
multidrug-resistant tuberculosis (MDR-TB), 78
Murchison Widefield Array (MWA), 32
Murray, Bruce, 333n4
Musk, Elon, 112, 139, 271
MWA. *See* Murchison Widefield Array

Nagasaki, 74
NASA: life defined by, 7; Microwave Observing Program, 32
National Oceanic Atmospheric Administration (NOAA), 167–68
National Radio Astronomy Observatory, 32
National Radio Silence Day, 31
near-earth asteroids (NEA), 122; discoveries of, *123*
near-earth object (NEO), 122; estimations of, 123–24
near-Earth settlements: Lagrange point, 165–73; location options for, *146*; Moon, 151–63; space station, 147–51
NEO. *See* near-earth object
von Neumann, John, 38
von Neumann probe, 208, 325n11; as alternative life form, 38; Bracewell probe compatibility with, 39; extraterrestrial contact and, 37–38; self replication and, 38, 81
New Horizons, 39, 177; launch of, 42; location of, 42; speed of, 132; trajectory of, *42*
newton, 335n5
Night, Tom B., 291

Ninnis, Belgrave, 246
NOAA. *See* National Oceanic Atmospheric Administration
Northrop Grumman, 139
nova, 18; process of, 324n4
NTR. *See* nuclear thermal rocket
nuclear holocaust: accidental, 114–15; deliberate, 114–15; extinction from, 74–75, 114–15; impacts from, 74; likelihood of, 114–15
Nuclear Non-Proliferation and Disarmament Treaties, 114
nuclear powered propulsion, 280–81; photonic rocket, 279; pulsed, 277–78; thermal rocket, 278–79
nuclear thermal rocket (NTR), 278–79
nuclear winter, 74; overview, 328n5
nucleic acids, 8
nutrient sensing dysregulation, 227
nutrition, 218, 248–49

Oberth, Hermann, 275
Oberth effect, 331n3
object 21/Borisov, 41
OIB. *See* orbital insertion burn
On the Origin of Species (Darwin), 8, 68
Oort Cloud, 108; overview, 329n19
Opportunity, 49–50
orbital insertion burn (OIB), 276
Orbital Sciences Corporation, 139
Ordovician-Silurian boundary extinction, 101
organelles, 12
Orion, 159
Orion propulsion system, 278
Ort Cloud, 126
Oumuamua, 43; extraterrestrial contact and, 40–41; properties of, 41
outside Solar System settlements: challenges of, 217–20; destination times to, 224–26; scouting, 224–25

oxygen, 14; Cambrian period and availability of, 13; seafloor deficiency of, 13

Padalka, Gennady, 141, 187
pandemics: Doherty on, 115; examples of, *76*; extinction from, 75–78, 115–16; future candidates for, 77–78; historical, 75–76; likelihood of, 115–16
panspermia, 63–64
Pan-STARRS1 telescope, 40–41
paraterraforming: advantages of, 185; disadvantages of, 186
Parker Solar Probe, 38; speed of, 223, 335n3
Parkes radio telescope, 32
Partial Test Ban Treaty, 278
Pedalka, Gennady, 144
Penzias, Arno, 332n4
perchlorates, 194
Permian extinction, 2, 71–72
Perseverance: Mars landing of, 50–51; objectives of, 51
phosphene, 45
photosynthesis, 11
Piazzi, Giuseppe, 199
Pichai, Sundar, 307
Pioneer, 39
Pioneer 10: launch of, 41; trajectory of, *42*
Pioneer 11: launch of, 41; trajectory of, *42*
Planck Observatory, 167
planetary nebulae, 100
planetary position, 61
Planetary Society, 209, 280–81; founding of, 333n4
planets: chemical composition of, 19; rotation period, 20; surface distance of, 18–19. *See also* exoplanets; *specific planets*
plasma drive propulsion, 282

plate tectonics: Cambrian period and, 19; climate change and, 94; extinction scenarios, 105–6; liquid metallic core convection and, 19; overview, 324n5
Point Nemo, 174
Polk, J. D., 232, 255
pollution, 82–85, 117–18
Polyakov, Valeri, 140–41, 144, 187, 224
population growth, 318; challenges, 117–18; climate change and, 88–89, 95; extinction through, 82–85, 117–18; projected, *85*
posthumanism, 312–15
primary eclipse, 61
primates, 14
Project Phoenix, 32
prokaryotes, 17
propulsion: antigravity, 283–84; antimatter, 284; Bussard ramjet, 282–83; chemical rocket, 275–76; current, 274; domains, 272; faster-than-light, 284–86; futuristic methods of, 282–86; gravity assist, 276–77, 326n16; ion drive, 281—282; LightSail, 279–81; limitations with, 286–89; nuclear, 277–79; Orion, 278; plasma drive, 282; Q-thruster, 283; status of, *273*; types, *273*; warp drive, 284–86
protein: folding, 14; function dysregulation, 227; metabolism, 8; substrate binding, 14; transport, 14
Proxima b: journey to, 132; orbit of, 211; properties of, 211–12
Proxima Centauri, 34, 55, 132; destination time to, 225
PSR B1257+12, 54
Ptolemy, Claudius, 5

Q-thruster propulsion, 283

radial velocity, 55
radiation: atmosphere absorbing, 19–20, 85–86; degenerative tissue effects from, 243; from gamma ray burst, 100; spaceflight, 235–36, 242–44; from supernova, 100
radio emissions, 325n9
Rare Earth hypothesis, 332n1
Reade, Winwood, 300
red dwarf, 18; exoplanets and, 211–12; overview of, 333n6
red giant, 327n1
reflection and emission modulations, 56
resource exhaustion, 118; extraterrestrial life and, 27
retrograde, 329n14
RNA, 8; water and structural stability of, 15
robotics, 10, 302–4, 322; limitations of, 303–4
The Rock from Mars (Sawyer), 191
Roddenberry, Gene, 217
Roswell incident, 29–30
Russian Academy of Sciences Institute of Biomedical Problems (IBMP), 261
Ryugu, 46

Sagan, Carl, 31, 333n4; on Drake equation, 22–23; Venus terraforming and, 196
Sawyer, Kathy, 191
Schiaparelli, Giovanni, 34–35, 47
Schirra, Wally, 251
Schmidt, Harrison, 141
seafloor spreading, 105
Search for Extraterrestrial Intelligence (SETI), 28; Drake and, 31; electromagnetic emissions and, 31; expansion of, 31; outcomes to date, 31–34; Suitcase, 31; telescopes used for, 32

seasonal temperatures, 20, 86; on Titan, 52
secondary eclipse, 61
second generation star, 17–18
seed ships, 266–68
self-awareness, 302
self replication: extinction and, 81–82, 116–17; interstellar spacecraft and, 38; of nanorobots, 81–82; von Neumann probe and, 38, 81; objections to, 305; overview, 325n11
senescent cell accumulation, 227
SETI. *See* Search for Extraterrestrial Intelligence
SETI Institute of Mountain View, California, 32
sharks, 21
Shoemaker-Levy 9, 96, 98; impact of, 124
Siberian Traps, 71–72, 104
simulation, 315–16
single celled life, 43
singularity, 306–15
sleep deprivation, 159
sleeper ships, 263, 266; challenges for, 264–65
Snowball Earth, 13, 86; support for, 324n2
social issues, 219, 244–48
Solar and Heliospheric Observatory (SOHO), 167
solar sailing, 209
Solar System: engineered structures outside, 37; extraterrestrial life outside, 37–43; larger planets to clear, 20; space craft missions in, 45–66; space craft missions outside, 54–63
Solar System settlements, 178–202; characteristics, *179–81*; Mars, 182; opportunities for, 178–82; options, *179–81*
Solar Terrestrial Relations Observatory (STEREO), 170–71

Soviet Union: accomplishments of, 135–36; Moon missions, 152; spaceflight and, 135–36
space craft missions: to Bennu, 46; to Churyumov-Gerasimenko comet, 46; dual objectives of, 46; to Enceladus, 53–54; to Eros, 46; to Europa, 53; exoplanet, 54–59; extraterrestrial life and, 46; to Itokawa, 46; to Mars, 46–54, 191; beyond Moon, 177–78; to Moon, 46–47; outside Solar System, 54–63; to Tempel comet, 46; to Titan, 46, 51–53; total, 124; to Venus, 45, 51
Space Exploration Technologies Corporation (SpaceX), 139
spaceflight: behavioral health and, 244–48; central nervous system risks, 243; communication delays, 220; deaths, 297; defining, 134–35; democracy, 254; demographics of, 135; destination times, *222*; discontent and, 251–52; DNA damage from, 218; docking, 296–97; general challenges of, 235–40; governance and, 249–54; hibernation, 218; hollowed-out natural structures for, 268–69; human presence records, 140–42; interstellar arks, 268–69; landings, 297–98; launches, 297; life support, 298–99; long-term effects of, 236; multiples generations of, 252–54; non-government, 139–40; nutrition and, 218, 248–49; physical challenges, 240–44; physical impacts on, *238*; pioneer, 226; power supplies, 218–19; psychological impacts of, *238*; radiation during, 235–36, 242–44; resources for, 219; safety and, 235–40; small group governance during, 250–52;

social issues and, 219, *238*, 244–48; Soviet Union and, 135–36; speed, 223; stepping stone, 226, 269–70; structures, 298; territory claims and, 249–54; United States and, 136–39; vision impairment in, 241; weightlessness and, 240–42
Space Launch Program, 159
Space Nuclear Power Corporation, 279
space station settlements, 147; critical perspectives on, 150–51; requirements of, 149–50; social issues, 150–51
space tourism, 174
SpaceX, 139, 159
Spanish Flu, 77
special relativity theory, 227–29
spectral analysis, 62
Spirit, 49–50
sponges, 12
Sputnik 1, 15
Stapledon, Olaf, 40
star: energy stability of, 18; luminosity of, 18; mass of, 18; multiple systems, 18. *See also specific stars*
Star Maker (Stapledon), 40
Star Trek, 9
Star Wars, 9
stellar spectral classification types, 333n2
stem dell depletion, 227
STEREO. *See* Solar Terrestrial Relations Observatory
Stevenson, Adlai, 4
stony asteroids, 330n3
stromatolites, 11
Suez Canal, 35
sugars, 8
Suitcase SETI, 31
Sun-Earth L1 point, 167
Sun-Earth L2 point, 167
Sun-Earth L3 point, 167–68

Sun engulfment extinction scenario, 109–10
Sun output extinction scenario, 106–7
superbugs, 78
super-Earths, 62, 213
supernova, 18; extinction from, 100–101, 121–22; heavy elements and, 330n1; process of, 324n4; radiation from, 100
surface temperature: exoplanet, 15–16, 59, *60*; historic, *89*; planet distance and, 18–19
Surveyor, 49
survival paths, 301–4
suspended animation ships, 263–66
SWEEPS, 55
Swigert, John L., 134
swing-by maneuver. *See* gravity assist propulsion
synthetic biology, 233

Tau Ceti, 31
technical competence, 21
technological augmentation, 69
technological junk, 28
Technology Level of Readiness (TLR), 272
telescope, 5
telomere shortening, 227
Tempel comet, 46
tephra, 329n16
Terminator, 9
terraforming, 182; atmosphere, 184; candidates for, 190; driving force behind, 183; Earth, 120–21; initiatives for, 183–84; magnetosphere, 184; Mars, 183–84; Mercury, 195; time required for, 183; Venus, 196
territory claims: on Moon, 163–65; spaceflight and, 249–54

Theia, 47, 98; hypothetical background of, 326n22
Thermoanaerobacter, 45
tidal locking zone: exoplanets in, 59; overview, 327n25
time dilation, 227–29; on International Space Station, 228–29; problems with, 229
Titan: atmosphere of, 51, 197; Earth compared with, 52–53; extraterrestrial life and, 51–53; orbit of, 197; probing, 6; seasonal temperatures on, 52; space craft missions to, 46, 51–53
Titan settlements: advantages of, 196–97; disadvantages of, 198
TLR. *See* Technology Level of Readiness
Toba eruption, 102–3
trace fossils, 28, 325n7
tragedy of the commons, 118
transform boundary, 105
transformer neural network, 306
Transformers, 9
transit photometry, 55, 205
transit timing, 56
Triassic extinction, 71–72
Trinitrotoluene, 329n13
Tsiolkovsky, Konstantin, 131, 217, 275
twins study, 239
2001: A Space Odyssey, 9, 38, 39

UAPs. *See* unexplained aerial phenomena
UFOs. *See* unidentified flying objects
The Ultimate Migration (Goddard), 259
Ultra Safe Nuclear Corporation, 279
unexplained aerial phenomena (UAPs), 6
unidentified flying objects (UFOs), 6; crop circles and, 30; evidence of, 28–30; number of sightings, 28–29; Roswell incident and, 29–30; "War of the Worlds" broadcast and, 28–29

United Nations Convention on the Prevention and Punishment of the Crime of Genocide, 78–79
United States: accomplishments of, 136–39; Moon missions, 152–53; spaceflight and, 136–39
United States Communications Satellite Act of 1962, 139
universal common descent, 8

Venera 1, 177
Venus: climate change and, 91–92; environment of, 51; extraterrestrial life and, 51; Goldilocks zone and, 51; greenhouse gases and, 325n6; probing, 6; Sagan on, 196; space craft missions to, 45, 51; terraforming, 196
Viking 1: Mars Face captured by, 36; Mars landing of, 49
Viking 2, 49
Viking Lander 2, 48
Viking Orbiter 1, 48
viral hemorrhagic fever, 77–78
Virgin Galactic, 139, 174
virions, 324n3
viruses, 14; overview, 324n3; virions differentiated from, 324n3
vision impairment, 241
volcanoes, 13; extinction scenarios, 101–4, 121; on Mars, 104; on Moon, 47; super, 103–4, 121
Vostok 1, 135
Voyager, 39
Voyager 1, 177; launch of, 41; location of, 41–42; mission completion of, 178; time capsule, 178, 332n1; trajectory of, *42*
Voyager 2: launch of, 41; location of, 41; time capsule, 178, 332n1; trajectory of, *42*

The War of the Worlds (Wells), 29
warp drive propulsion, 284–86
water, 7; alternative chemistry to, 15; atmosphere preserving, 19–20; DNA structural stability and, 15; enzyme actions and, 14; on Europa, 53; on exoplanets, 15–16; life and, 14–15; on Mars, 48, 190–91; from Moon, 46; physiochemical properties of, 14–15; protein folding and, 14; protein substrate binding and, 14; protein transport and, 14; RNA structural stability and, 15; as universal solvent for polar molecules, 14
weightlessness exposure, 162, 218, 240–42; artificial gravity and, 241
Welles, Orson, 29
Wells, H. G., 29, 318

"What Is to Be Done" (Clarke), 145
Wilson, Robert, 332n4
worldhouses: advantages of, 185; disadvantages of, 186
worldships, 252–53, 259–63; characteristics of, 254
wormholes, 229–30
Wow! signal, 32; original printout of, *33*; plot of, *33*
Wright, Orville, 141–42, *142*
Wright Flier, 141–42

Yellowstone National Park, 121
Young, John, 224

Zuckerberg, Mark, 333n3
Zvezda, 148

About the Author

Dr. **Rod Hill** is a retired Chief Research Scientist and Group Executive at the Commonwealth Scientific and Industrial Research Organization (CSIRO Australia) and a former Pro Vice Chancellor for Industry Engagement and Commercialization at Monash University. He has PhD and DSc degrees from the University of Adelaide, South Australia, in crystallography, crystal chemistry, and mineralogy. He has published more than 110 research papers in the international literature in the primary areas of structural systematics and bonding in silicate minerals, and more recently on X-ray and neutron diffraction data collection and structure refinement optimization using the full-profile Rietveld method, including the *ab initio* determination of phase abundance in complex powder mixtures. He is a Fellow of the Australian Academy and Technological Sciences and Engineering, the Royal Australian Chemical Institute, and the Mineralogical Society of America. The mineral "Hillite" was named after him in 2003.

Dr. Hill has had a lifelong passion for astronomy and has constructed two 6-inch Newtonian telescopes. He has written several articles in popular astronomy journals, is a frequent speaker at meetings of the South Australian Astronomical Society, and has published two books on total solar eclipses. His most recent research publication was on gravity clearing of natural satellite orbits. He and his wife Monica are bona fide "eclipse chasers," having traveled to the far corners of the Earth to see 13 of these awe-inspiring natural wonders.